Annals of Mathematics Studies

Number 34

AUTOMATA STUDIES

W. R. ASHBY
J. MC CARTHY
J. T. CULBERTSON
M. L. MINSKY
M. D. DAVIS
E. F. MOORE
S. C. KLEENE
C. E. SHANNON
K. DE LEEUW
N. SHAPIRO
D. M. MAC KAY
A. M. UTTLEY
J. VON NEUMANN

Edited by

C. E. Shannon and J. McCarthy

Princeton, New Jersey
Princeton University Press

L.C. Card 56-7637

ISBN 0-691-07916-1

Second Printing, 1957

Third Printing, 1960

Fourth Printing, 1965

Fifth Printing, 1972

This research was supported in part by the Office of Naval Research.

Printed in the United States of America

PREFACE

Among the most challenging scientific questions of our time are the corresponding analytic and synthetic problems: How does the brain function? Can we design a machine which will simulate a brain? Speculation on these problems, which can be traced back many centuries, usually reflects in any period the characteristics of machines then in use. Descartes, in DeHomine, sees the lower animals and, in many of his functions, man as automata. Using analogies drawn from water-clocks, fountains and mechanical devices common to the seventeenth century, he imagined that the nerves transmitted signals by tiny mechanical motions. Early in the present century, when the automatic telephone system was introduced, the nervous system was often likened to a vast telephone exchange with automatic switching equipment directing the flow of sensory and motor data. Currently it is fashionable to compare the brain with large scale electronic computing machines.

Recent progress in various related fields leads to an optimistic view toward the eventual and not too remote solution of the analytic and synthetic problems. The fields of neuro-anatomy and neuro-physiology, using powerful tools from electronics in encephalographic studies, have illuminated many features of the brain's operation.

The development of large scale computers has led to a clearer understanding of the theory and design of information processing devices. Programming theory, still in its infancy, is already suggesting the tremendous potential versatility of computers. Electronic computers are also valuable tools in the study of various proposed models of nervous systems. Often such models are much too complex for analytic appraisal, and the only available method of study is by observation of the model in operation.

On the mathematical side, developments in symbolic logic, recursive function theory and Turing machine theory have led to deeper understanding of the nature of what is computable by machines. Neighboring fields of game theory and information theory have developed concepts relating to the nature and coding of information that may prove useful in automata research.

The problem of giving a precise definition to the concept of "thinking" and of deciding whether or not a given machine is capable of thinking has aroused a great deal of heated discussion. One interesting definition has been proposed by A. M. Turing: a machine is termed capable of thinking if it can, under certain prescribed conditions, imitate a human being by answering questions sufficiently well to deceive a human questioner for a reasonable period of time. A definition of this type has the advantages of being operational or, in the psychologists' term, behavioristic. No metaphysical notions of consciousness, ego and the like are involved. While certainly no machines at the present time can even make a start at satisfying this rather strong criterion, Turing has speculated that within a few decades it will be possible to program general purpose computers in such a way as to satisfy this test.

PREFACE

A disadvantage of the Turing definition of thinking is that it is possible, in principle, to design a machine with a complete set of arbitrarily chosen responses to all possible input stimuli (see, in this volume, the Culbertson and the Kleene papers). Such a machine, in a sense, for any given input situation (including past history) merely looks up in a "dictionary" the appropriate response. With a suitable dictionary such a machine would surely satisfy Turing's definition but does not reflect our usual intuitive concept of thinking. This suggests that a more fundamental definition must involve something relating to the manner in which the machine arrives at its responses--something which corresponds to differentiating between a person who solves a problem by thinking it out and one who has previously memorized the answer.

The present volume is a collection of papers which deal with various aspects of automata theory. This theory is of interest to scientists in many different fields and, correspondingly, among the authors are workers who are primarily logicians, mathematicians, physicists, engineers, neurologists and psychologists. The papers include some which are close to pure mathematics; others are essentially directed to the synthesis problem and some relate largely to philosophic questions. There is also a certain amount of overlap, the same problem being handled from somewhat different points of view by different authors.

The papers have been divided into three groups. The first group consists of papers dealing with automata having a finite number of possible internal states. In the usual quantized model of this type, the automaton has a finite number of inputs and outputs and operates in a quantized time scale. Thus, such a device is characterized by two functions of the current state and input, one function giving the next state and the other the next output. Although seemingly trivial at this level of description, many interesting problems arise in the detailed analysis of such machines. Indeed, it should be remembered that essentially all actual physical machines and even the brain itself are, or can be reasonably idealized to be, of this form.

Neurophysiologists have proposed a number of models for the neuron and Kleene, in his paper, investigates the capabilities and limitations of automata constructed from these idealized components. von Neumann, using similar components, allows the possibility of statistically unreliable operation and shows that under certain conditions it is possible, with unreliable components, to construct large and complex automata with as high a reliability as desired. In Culbertson's paper a simple construction is given for an idealized neural network which will react in an arbitrary prescribed manner for an arbitrary lifetime. In all of these papers the notion of universal components plays a significant role. These are components which, roughly speaking, are sufficiently flexible to form devices capable of acting like any machine. Minsky considers the problem of universality of components and finds conditions which ensure this property. In a paper of a somewhat different type, Moore studies what can be learned about finite state automata by experiments performed on the inputs and outputs of the machine (without direct observation of its interior).

The second group of papers deals with the theory of Turing machines and related questions, that is to say, with automata having an unlimited number of possible states. The original Turing machine (since then, recast in many different forms) may be described as follows. Let there be given a tape of infinite length which is divided into squares and a finite list of symbols which may be written on these squares. There is an additional mechanism, the head, which may read the symbol on a square, replace it by another or the same symbol and move to the adjoining square to the left or right. This is accomplished as follows: At any given time the head is in one of a finite number of internal states. When it reads a square it prints a new symbol, goes into a new internal state and moves to the right or left depending on the original internal state and the symbol read. Thus a Turing machine is described by a finite list of quintuplets such as 3, 4, 3, 6, R which means: If the machine is in the third internal state and reads the fourth symbol it prints the third symbol, goes into the sixth internal state and moves to the right on the tape. There is a fixed initial internal state and the machine is supposed to start on a blank tape. One of the symbols represents a blank square and there may be given a state in which the machine stops.

Turing gave a convincing argument to the effect that any precisely defined computation procedure could be carried out by a machine of the type described above. He also showed that the Turing machines can be enumerated and that a universal machine could be made which, when it read the number of any Turing machine, would carry out the computation that that machine would have carried out were it put on a blank tape. His final result was to show that there did not exist a Turing machine which when confronted with the number of another machine would decide whether that machine would ever stop.

Any of the present automatic electronic computers is equivalent to a universal Turing machine if it is given, for example, a means of asking for more punched cards and for the return of cards it has already punched. In Shannon's paper it is shown that a universal Turing machine can be constructed with only two internal states, or alternatively, with only two tape symbols. Davis gives a general definition of a universal Turing machine and establishes some results to make this definition appear reasonable. McCarthy discusses the problem of calculating the inverse of the function generated by a Turing machine, after some argument to the effect that many intellectual problems can be formulated in this way. Finally, De Leeuw, Moore, Shannon and Shapiro investigate whether machines with random elements can compute anything uncomputable by ordinary Turing machines.

PREFACE

The third section of the book contains papers relating more directly to the synthesis of automata which will simulate in some sense the operation of a living organism. Ashby discusses the problem of designing an intelligence amplifier, a device which can solve problems beyond the capacities of its designer. MacKay, dealing with the same general problem, suggests means for an automaton to symbolize new concepts and generate new hypotheses. Uttley studies from a still different point of view the problem of the abstraction of temporal and spatial patterns by a machine, that is, the general problem of concept formation.

It gives us pleasure to express our gratitude to all those who have contributed to the preparation of this volume. The work was supported in part by the Princeton Logistics Project sponsored by the Office of Naval Research. Professor A. W. Tucker, directing this project, has been most helpful. H. S. Bailey, Jr. and the staff of the Princeton University Press, particularly Mrs. Dorothy Stine and Mrs. Jean Muiznieks have been efficient and cooperative. Thanks are also due Dr. Julia Robinson for help with the reviewing and Mrs. E. Powanda for secretarial services.

John McCarthy
Claude Shannon

CONTENTS

FINITE AUTOMATA

REPRESENTATION OF EVENTS IN NERVE NETS AND FINITE AUTOMATA[1]

S. C. Kleene

INTRODUCTION

1. Stimuli and Response

An organism or an automaton receives stimuli via its sensory receptor organs, and performs actions via its effector organs. To say that certain actions are a response to certain stimuli means, in the simplest case, that the actions are performed when and only when those stimuli occur.

In the general case both the stimuli and the actions may be very complicated.

In order to simplify the analysis, we may begin by leaving out of account the complexities of the response. We reason that any sort of stimulation, or briefly any event, which affects action in the sense that different actions ensue according as the event occurs or not, under some set of other circumstances held fixed, must have a representation in the state of the organism or automaton, after the event has occurred and prior to the ensuing action.

So we ask what kind of events are capable of being represented in the state of an automaton.

For explaining actions as responses to stimuli it would remain to study the manner in which the representations of events (a kind of internal response) lead to the overt responses.

Our principal result will be to show (in Sections 7 and 9) that all and only the events of a certain class called "regular events" are representable.

---

[1] The material in this article is drawn from Project RAND Research Memorandum RM-704 (15 December 1951, 101 pages) under the same title and by the author. It is used now by permission of the RAND Corporation. The author's original work on the problem was supported by the RAND Corporation during the summer of 1951.

2. Nerve Nets and Behavior

McCulloch and Pitts [1943] in their fundamental paper on the logical analysis of nervous activity formulated certain assumptions which we shall recapitulate in Section 3.

In showing that each regular event is representable in the state of a finite automaton, the automaton we use is a McCulloch-Pitts nerve net. Thus their neurons are one example of a kind of "universal elements" for finite automata.

The McCulloch-Pitts assumptions were put forward as an abstraction from neuro-physiological data. We shall not be concerned with the question of how exactly the assumptions fit. They seem to fit roughly up to a point, though one of McCulloch's and Pitts' results is that certain alternative assumptions can explain the same kind of behavior. With increasing refinement in the neuro-physiological data the emphasis is no doubt on respects in which the assumptions do not fit.

Our theoretical objective is not dependent on the assumptions fitting exactly. It is a familiar strategem of science, when faced with a body of data too complex to be mastered as a whole, to select some limited domain of experiences, some simple situations, and to undertake to construct a model to fit these at least approximately.

Having set up such a model, the next step is to seek a thorough understanding of the model itself. It is not to be expected that all features of the model will be equally pertinent to the reality from which the model was abstracted. But after understanding the model, one is in a better position to see how to modify or adapt it to fit the limited data better or to fit a wider body of data and when to seek some fundamentally different kind of explanation.

McCulloch and Pitts in their original paper give a theory for nerve nets without circles [Part II of their paper] and a theory for arbitrary nerve nets [Part III]. The present article is partly an exposition of their results; but we found the part of their paper dealing with arbitrary nerve nets obscure, so we have proceeded independently there.

Although we are concerned with the model itself rather than its application, a few remarks on the latter may prevent misunderstanding.

To take one example, as consideration of the model shows, memory can be explained on the basis of reverberating cycles of nerve impulses. This seems a plausible explanation for short-term memories. For long-term memories, it is implausible on the ground of fatigue, also on the ground that calculations on the amount of material stored in the memory would call for too many neurons [McCulloch, 1949], and also on the basis of direct experimental evidence that temporary suppression of nervous activity does not cut off memory [Gerard, 1953].

The McCulloch-Pitts assumptions give a nerve net the character of a digital automaton, as contrasted to an analog mechanism in the sense familiar in connection with computing machines. Some physiological processes of control seem to be analog. Just as in mathematics continuous processes can be approximated by discrete ones, analog mechanisms can be approximated in their effect by digital ones. Nevertheless, the analog or partly analog controls may for some purposes be the simplest and most economical.

An assumption of the present mathematical theory is that there are no errors in the functioning of neurons. Of course this is unrealistic both for living neurons and for the corresponding units of a mechanical automaton. It is the natural procedure, however, to begin with a theory of what happens assuming no malfunctioning. Indeed in our theory we may represent the occurrence of an event by the firing of a single neuron. Biologically it is implausible that important information should be represented in an organism in this way. But by suitable duplication and interlacing of circuits, one could then expect to secure the same results with small probability of failure in nets constructed of fallible neurons.

Finally, we repeat that we are investigating McCulloch-Pitts nerve nets only partly for their own sake as providing a simplified model of nervous activity, but also as an illustration of the general theory of automata, including robots, computing machines and the like. What a finite automaton can and cannot do is thought to be of some mathematical interest intrinsically, and may also contribute to better understanding of problems which arise on the practical level.

PART I: NERVE NETS

3. McCulloch-Pitts Nerve Nets

Under the assumptions of McCulloch and Pitts [1943], a nerve cell or _neuron_ consists of a body or _soma_, whence nerve fibers (_axons_) lead to one or more _endbulbs_.

A _nerve net_ is an arrangement of a finite number of neurons in which each endbulb of any neuron is adjacent to (_impinges_ on) the soma of not more than one neuron (the same or another); the separating gap is a _synapse_. Each endbulb is either _excitatory_ or _inhibitory_ (not both).

We call the neurons (zero or more) on which no endbulbs impinge _input neurons_; the others, _inner neurons_.

At equally separated moments of time (which we take as the integers on a time scale, the same for all neurons in a given net), each neuron of the net is either _firing_ or _not firing_ (being _quiet_). For an input neuron, the firing or non-firing at any moment t is determined by conditions outside the net. One can suppose each is impinged on by a sensory receptor organ, which under suitable conditions in the environment causes the neuron to fire at time t. For an inner neuron, the condition for firing at time t

is that at least a certain number h (the threshold of that neuron) of the excitatory endbulbs and none of the inhibitory endbulbs impinging on it belong to neurons which fired at time $t - 1$.

For illustration, consider the nerve net shown in Figure 1, with input neurons $\mathcal{J}$, $\mathcal{K}$, $\mathcal{L}$, $\mathcal{M}$ and $\mathcal{N}$, and inner neuron $\mathcal{P}$. Excitatory endbulbs are shown as dots, and inhibitory as circles. The threshold of $\mathcal{P}$ is 3, as shown by the number in the triangle representing its soma. The formula written below the net expresses in logical symbolism that neuron $\mathcal{P}$ fires at time t, if and only if all of $\mathcal{J}$, $\mathcal{K}$ and $\mathcal{L}$ and none of $\mathcal{M}$ and $\mathcal{N}$ fired at time $t - 1$. We are writing "P(t)" to say that neuron $\mathcal{P}$ fires at time t, "J(t - 1)" to say that $\mathcal{J}$ fired at $t - 1$, etc. The symbol "≡" means if and only if (or is equivalent to), "&" means and, "∨" means or (in the non-exclusive sense), and "¯" means not.

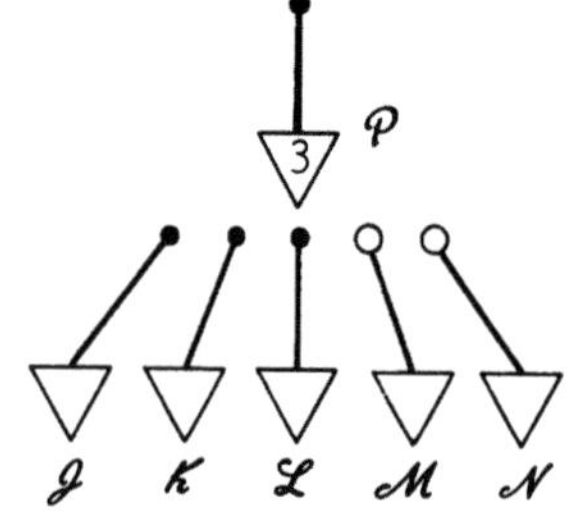

$$P(t) \equiv J(t-1) \,\&\, K(t-1) \,\&\, L(t-1) \,\&\, \overline{M(t-1)} \,\&\, \overline{N(t-1)}.$$

FIGURE 1: Conjunctive Net

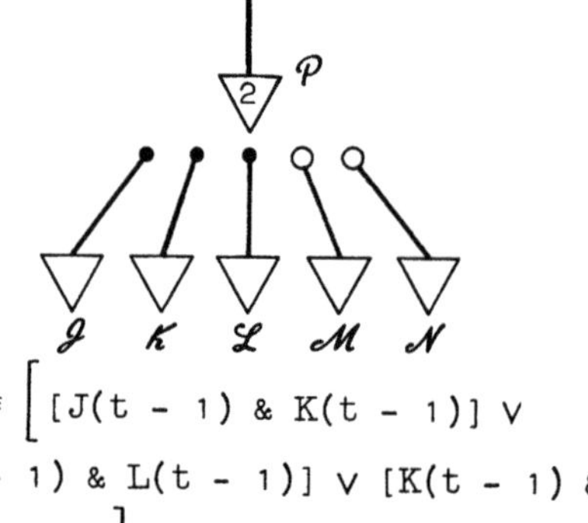

$$P(t) \equiv \Big[[J(t-1) \,\&\, K(t-1)] \vee [J(t-1) \,\&\, L(t-1)] \vee [K(t-1) \,\&\, L(t-1)]\Big] \,\&\, \overline{M(t-1)} \,\&\, \overline{N(t-1)}.$$

FIGURE 2

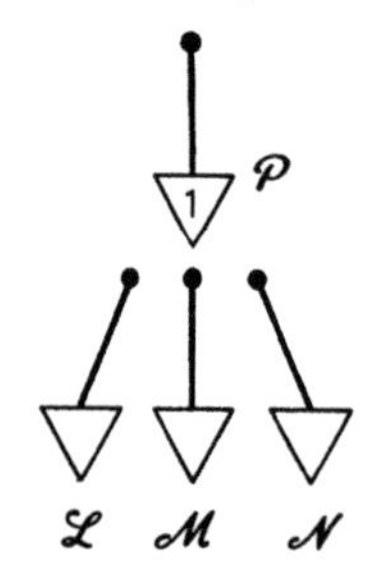

$$P(t) \equiv L(t-1) \vee M(t-1) \vee N(t-1).$$

FIGURE 3: Disjunctive Net

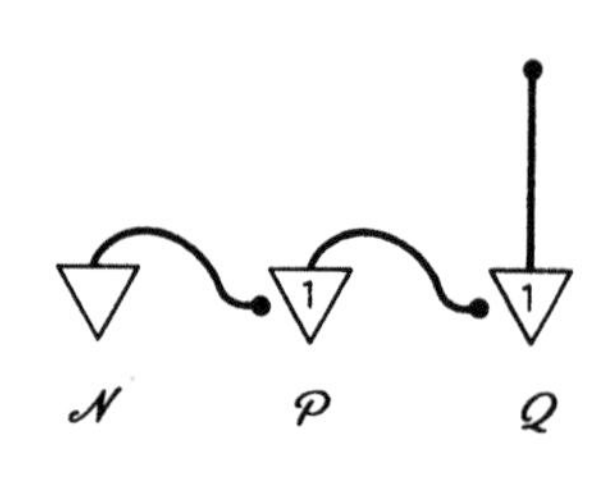

$$P(t) \equiv N(t-1),$$
$$Q(t) \equiv N(t-2).$$

FIGURE 4: Delay Net

4. The Input to a Nerve Net

Consider a nerve net with k input neurons $\mathcal{N}_1, \ldots, \mathcal{N}_k$. We assume $k \geq 1$, until 6.3. The input (or experience) over all past time up to the present moment inclusive can be described by a table with k columns corresponding to the input neurons, and with rows corresponding to the moments counting backward from the present moment p. The positions are filled with 0's and 1's, where 0 is to stand for quiescence, and 1 for firing, of the neuron in question at the moment in question.

For example with $k = 2$ the table might be like Figure 5. The 1 in the first row and first column means that $\mathcal{N}_1$ fires at time p, the 0 in the third row and first column that $\mathcal{N}_1$ did not fire at time $p - 2$, etc. If this table is extended down infinitely, we have a representation of the input, thought of as extending over all past time, which for the time being we treat as infinite. (In Section 6 we shall reconsider the matter.)

t	$\mathcal{N}_1$	$\mathcal{N}_2$
p	1	0
p - 1	1	1
p - 2	0	1
.	.	.

FIGURE 5

By an <u>event</u> we shall mean any property of the input. In other words, any subclass of the class of all the possible tables describing the input over all past time (and ending with the present p inclusive, except when otherwise stated) constitutes an event, which <u>occurs</u> when the table describing the actual input belongs to this subclass.

Examples of events with two input neurons $\mathcal{N}_1$ and $\mathcal{N}_2$ are:

(1) $\mathcal{N}_1$ fires at time p.

(2) $\mathcal{N}_2$ does not fire at time p, and $\mathcal{N}_1$ fired at time $p - 1$.

(3) One of $\mathcal{N}_1$ and $\mathcal{N}_2$ fires at time p.

(4) $\mathcal{N}_1$ and $\mathcal{N}_2$ both fire at time p.

(5) $\mathcal{N}_2$ fired at some time.

(6) $\mathcal{N}_2$ fired at every time except p.

Of these, the input described by the table of Figure 5 constitutes an occurrence of events (1), (2), (3) and (5), but not of (4), while we need to know the rest of the table to know whether it constitutes an occurrence of (6).

5. Definite Events

5.1 "DEFINITE EVENTS" DEFINED. We shall discuss first events which refer to a fixed period of time, consisting of some ℓ (≥ 1) consecutive moments $p - \ell + 1, \ldots, p$ ending with the present. We call such events <u>definite of length</u> (or <u>duration</u>) ℓ. Of the preceding examples, (1) — (4) are definite, but not (5) and (6).

Then in a table such as Figure 5 we need consider only the uppermost ℓ rows; e.g., that table for $\ell = 3$ then describes an event also described by the formula $N_1(p) \;\&\; \overline{N_2(p)} \;\&\; N_1(p-1) \;\&\; N_2(p-1) \;\&\; \overline{N_1(p-2)} \;\&$ $\&\; N_2(p-2)$.

There are exactly $k\ell$ entries in a table describing the input on k neurons for the ℓ moments $p - \ell + 1, \ldots, p$. Therefore there are exactly $2^{k\ell}$ possible such tables. Therefore there are exactly $2^{2^{k\ell}}$ definite events on k input neurons of length ℓ, since any particular one is determined by saying which of the inputs described by the $2^{k\ell}$ $k \times \ell$ tables would constitute an occurrence of the event.

We call a definite event <u>positive</u>, if it occurs only when at least one input neuron fires during the period to which the event refers. There are exactly $2^{2^{k\ell}-1}$ positive definite events on k input neurons of length ℓ, since that input described by the table of all 0's is excluded as an occurrence.

5.2 REPRESENTABILITY OF DEFINITE EVENTS: AN ILLUSTRATION. Consider the definite event which occurs when the pattern of firings fits either the table of Figure 5 (stopped at three rows) or that of Figure 6.

t	$\mathcal{N}_1$	$\mathcal{N}_2$
p	1	0
p - 1	1	0
p - 2	1	0

$$N_1(p) \mathbin{\&} \overline{N_2(p)}$$
$$\mathbin{\&} N_1(p-1) \mathbin{\&} \overline{N_2(p-1)}$$
$$\mathbin{\&} N_1(p-2) \mathbin{\&} \overline{N_2(p-2)}.$$

FIGURE 6

That is, exactly these two (out of the $2^{2\cdot 3} = 64$) 2×3 tables are to constitute an occurrence of the event. The event is described by the right member of the equivalence in Figure 7, which is obtained by combining disjunctively the conjunctions describing the respective tables separately. In the nerve net of Figure 7, the neuron $\mathcal{P}$ fires at time $p + 2$, if and only if the event occurs ending at time p; or briefly, the net represents the event by the firing of $\mathcal{P}$ with <u>lag</u> 2. The neurons $\mathcal{N}_1$, $\mathcal{N}_1'$, $\mathcal{N}_1''$ with just the axons connecting them are a "delay net" (cf. Figure 4). The synapse at $\mathcal{M}_1$ with the seven neurons involved is a "conjunctive net" (cf. Figure 1). That at $\mathcal{P}$ is a "disjunctive net" (cf. Figure 3).

The method of this illustration applies to every positive definite event which occurs for some one or more tables.

There remains the case of the event which never occurs. This is represented by the firing of $\mathcal{P}$ say with lag 2 in the net of Figure 8. The neuron $\mathcal{M}_2$ is inserted to show that we can have the net <u>connected</u> (in the obvious sense); otherwise $\mathcal{M}_2$ could be omitted. $\mathcal{M}_1$ could be the $\mathcal{P}$.

We have thus already proved that any positive definite event is representable by firing a neuron with lag 2. However, we shall give a more flexible treatment, listing this result as part of Theorem 1 Corollary 1.

5.3 REPRESENTABILITY OF DEFINITE EVENTS: GENERAL THEORY. We consider logical expressions constructed using & and $\vee$ from the expressions symbolizing

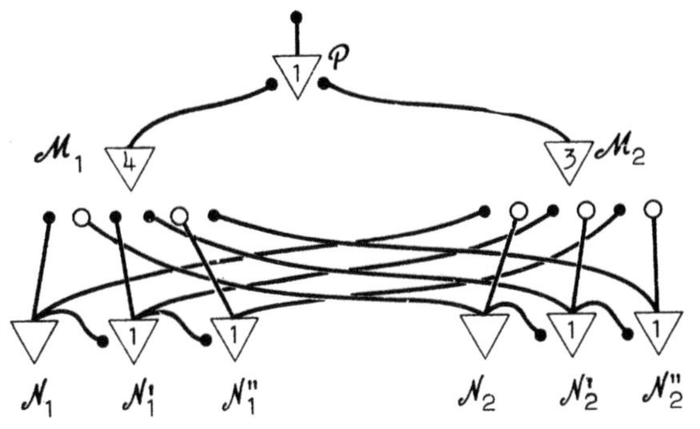

$P(p+2) \equiv [N_1(p) \,\&\, \overline{N_2(p)} \,\&\, N_1(p-1) \,\&\, N_2(p-1) \,\&\, \overline{N_1(p-2)} \,\&\, N_2(p-2)] \vee [N_1(p) \,\&\, \overline{N_2(p)} \,\&\, N_1(p-1) \,\&\, \overline{N_2(p-1)} \,\&\, N_1(p-2) \,\&\, \overline{N_2(p-2)}]$.

FIGURE 7

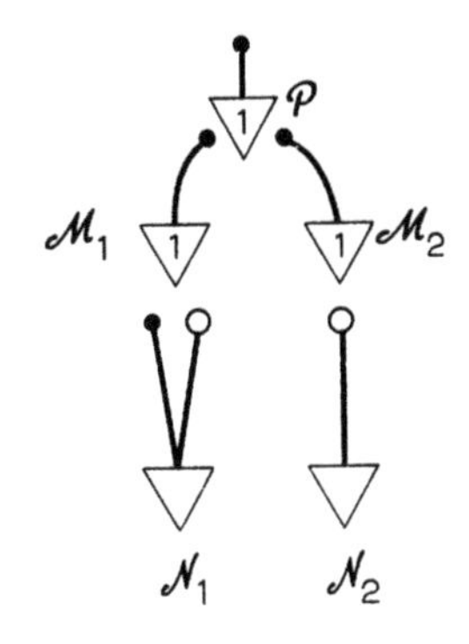

$\mathcal{P}$ does not fire at time $p + 2$, or in symbols $P(p+2) \equiv N_1(p) \,\&\, \overline{N_1(p)}$.

FIGURE 8

the firing or non-firing of one of the k input neurons $\mathcal{N}_1, \ldots, \mathcal{N}_k$ at one of the ℓ moments $p - \ell + 1, \ldots, p$. Such an expression we call a $k\ell$ - formula, of depth equal to the greatest number of successive times $\&$ or $\vee$ is used in its construction. Here we allow any bracketing in conjunctions and disjunctions of more than two members; so e.g., $[[N_1(p) \,\&\, \overline{N_2(p)}] \,\&\, N_2(p-1]] \vee \overline{N_1(p)}$ as written is of depth 3, but it can be rewritten as $[N_1(p) \,\&\, \overline{N_2(p)} \,\&\, N_2(p-1)] \vee \overline{N_1(p)}$ with depth 2.

These definitions can be given by mathematical induction on s, thus:

(1) For each i and j $(i = 1, \ldots, k;\ j = 1, \ldots, \ell)$, $N_i(p - j + 1)$ and $\overline{N_i(p - j + 1)}$ are $k\ell$ - formulas of depth 0.

(2) For $s > 0$, if $G_1, \ldots, G_n$ $(n \geq 2)$ are $k\ell$ - formulas of depths $< s$, at least one of them being of depth $s - 1$, then $G_1 \,\&\, \ldots \,\&\, G_n$ and $G_1 \vee \ldots \vee G_n$ are $k\ell$ - formulas of depth s. (Here each G_e not of depth 0 is to be enclosed in brackets when written out.)

Since the truth or falsity of a $k\ell$ - formula F is determined logically from only the truth or falsity of the $N_i(p - j + 1)$ which enter into it as prime components, each F expresses a definite event E on k input neurons of length ℓ.

We lose nothing essential by applying the negation symbol $\overline{}$ only directly to the prime components. For by repeated use of the logical identities $\overline{\overline{G}} \equiv G$, $\overline{G_1 \,\&\, \ldots \,\&\, G_n} \equiv \overline{G_1} \vee \ldots \vee \overline{G_n}$ and $\overline{G_1 \vee \ldots \vee G_n} \equiv \overline{G_1} \,\&\, \ldots \,\&\, \overline{G_n}$, negation symbols used otherwise could be moved inward to the prime components without changing the depth. The only other operations commonly employed in the two-valued propositional calculus, namely $\rightarrow$ (implies) and $\equiv$, can be expressed in terms of $\overline{}$, & and $\vee$ thus: $G \rightarrow H \equiv \bar{G} \vee H$, $(G \equiv H) \equiv (G \rightarrow H) \;\&\; (H \rightarrow G)$.

A circle (of length c) in a nerve net is a set of distinct neurons $\mathcal{N}_1, \ldots, \mathcal{N}_c$ $(c \geq 1)$ such that $\mathcal{N}_i$ has an endbulb on $\mathcal{N}_{i+1}$ for each i $(i = 1, \ldots, c - 1)$ and $\mathcal{N}_c$ has an endbulb on $\mathcal{N}_1$. The nets so far considered are without circles, including the conjunctive, disjunctive and delay nets (Figures 1, 3 and 4), and certain nets composed thence.

THEOREM 1. Let F be any $k\ell$ - formula of depth s, and let E be the definite event on k input neurons of length ℓ which F expresses. There is a nerve net of structure corresponding to F (and therefore without circles) which represents E by firing or by not firing, according as E is positive or non-positive, a certain neuron $\mathcal{P}$ (inner if $s > 0$) at time $p + s$.

By saying that the net is of "structure corresponding to F", we mean that it is composed out of conjunctive and disjunctive nets (together with delay nets) corresponding to the operations used in constructing F, as will be indicated in the proof.

PROOF, by induction on s. Under our definition of $k\ell$ - formula not all of the symbols N_i $(i = 1, \ldots, k)$ need occur in F. In showing by induction how to construct the net to correspond to the logical structure of F, we incorporate only the neurons $\mathcal{N}_i$ for which N_i occurs in F. The others can be considered as floating around, unless one wishes them connected to the rest of the net, in which case if $s > 1$ they can be, e.g., as illustrated for $\mathcal{N}_2$ in Figure 8.

BASIS: $s = 0$. Then F is $N_i(p - j + 1)$ or $\overline{N_i(p - j + 1)}$ for some i and j. Then $\mathcal{N}_i$ is also the $\mathcal{P}$, if $j = 1$; and otherwise the $\mathcal{P}$ is a neuron coming from $\mathcal{N}_i$ by a suitable delay net (Figure 4).

INDUCTION STEP: $s > 0$. Then F is $G_1 \;\&\; \ldots \;\&\; G_n$ or $G_1 \vee \ldots \vee G_n$. For $e = 1, \ldots, n$, let M_e be G_e or $\overline{G_e}$ according as (the event described by) G_e is positive or non-positive, as will be known from the case which applied to G_e. Then G_e (of depth $s_e < s$) is equivalent to M_e or $\overline{M}_e$, respectively, and by the hypothesis of the induction, there is a nerve net with a neuron $\mathcal{G}_e$ the firing or non-firing, respectively, of which at time $p + s_e$ represents G_e. Thence we obtain a neuron $\mathcal{M}_e$ the firing of which at time $p + s - 1$ represents M_e; this $\mathcal{M}_e$ is $\mathcal{G}_e$ itself if $s_e = s - 1$, and otherwise a neuron coming from $\mathcal{G}_e$ by a suitable delay net. Now we have four cases, according to how F is composed out of $M_1, \ldots, M_n$, via the construction of F from $G_1, \ldots, G_n$ and the equivalence of each G_e to one of M_e and $\overline{M}_e$.

CASE 1: A conjunction containing at least one unnegated factor, e.g., $M_1 \& M_2 \& M_3 \& \overline{M}_4 \& \overline{M}_5$. The event is then positive; so we wish to represent it by the firing of a neuron $\mathcal{P}$ at time $p + s$. A conjunctive net (cf. Figure 1) gives us this neuron.

CASE 2: A conjunction containing only negated factors, e.g., $\overline{M}_1 \& \overline{M}_2 \& \overline{M}_3$. The event is then non-positive. But its negation $\overline{\overline{M}_1 \& \overline{M}_2 \& \overline{M}_3}$ is positive. The latter is equivalent to $M_1 \vee M_2 \vee M_3$. A disjunctive net (cf. Figure 3) represents the latter by firing a neuron $\mathcal{P}$ at $p + s$; the net then represents the original event by the non-firing of $\mathcal{P}$ at $p + s$, which is how we wished it to be represented.

CASE 3: A disjunction containing at least one negated term, e.g., $\overline{M}_1 \vee \overline{M}_2 \vee \overline{M}_3 \vee M_4 \vee M_5$. The event is non-positive. But its negation is positive and equivalent to $M_1 \& M_2 \& M_3 \& \overline{M}_4 \& \overline{M}_5$. A conjunctive net represents the latter by firing a neuron $\mathcal{P}$ at $p + s$; then the original event is represented by the non-firing of $\mathcal{P}$ at $p + s$.

CASE 4: A disjunction containing only unnegated terms, e.g., $M_1 \vee M_2 \vee M_3$. The event is positive. A disjunctive net represents it as desired by firing a neuron $\mathcal{P}$ at $p + s$.

EXAMPLES. The net of Figure 7 is what the present method gives for the formula there. Another illustration is in Figure 9. Treating the three formulas $N_1(p) \vee N_2(p) \vee N_3(p)$, $\mathcal{N}_1(p) \vee \overline{\mathcal{N}_2(p)} \vee \overline{\mathcal{N}_3(p)}$ and $\overline{\mathcal{N}_1(p)} \vee \mathcal{N}_2(p) \vee \mathcal{N}_3(p)$ gives us respective neurons $\mathcal{M}_1, \mathcal{M}_2$ and $\mathcal{M}_3$ which represent the events expressed, the first which is positive by firing (Case 4), the second and third which are non-positive by non-firing (Case 3), at time $p + 1$. Then $\mathcal{P}$ is obtained to represent the entire event which is positive by firing at time $p + 2$ (Case 1).

COROLLARY 1. To each positive (non-positive) definite event, there is a nerve net without circles which represents the event by firing (not firing) a certain inner neuron at time $p + 2$.

The result was stated (for positive events and without the remark on the lag) by McCulloch and Pitts [1943].

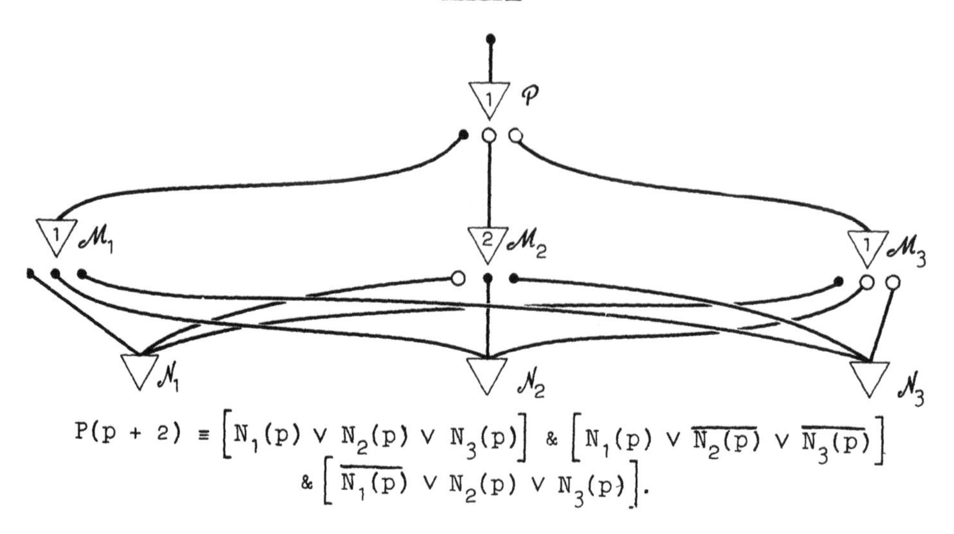

FIGURE 9

PROOF. To infer this from the theorem, we need merely observe that the method of 5.2 gives a $k\ell$ - formula of depth ≤ 2 to every definite event on k input neurons of length ℓ. (If the depth is < 2, a delay net may be used to increase the lag to 2.)

DISCUSSION. Readers familiar with symbolic logic will recognize the $k\ell$ - formula so obtained as a <u>principal disjunctive normal form</u> of Hilbert and Ackermann [1928]; it is "principal" because in each of its terms every one of $N_1(p), \ldots, N_k(p - \ell + 1)$ occurs negated or unnegated (with an exception in the case of Figure 8).

The formula of Figure 9 is a <u>principal conjunctive normal form</u>. If the *p.d.n.f.* has $n < 2^{k\ell}$ terms, the *p.c.n.f.* has $2^{k\ell} - n$ factors. (The *p.c.n.f.* is obtained by evaluating the negation of the *p.d.n.f.* of the negation of the event.)

The use of the *p.d.n.f.* simplifies the proof of representability (cf. 5.2), and gives the fact that the lag can be held to 2, but the net constructed may be unnecessarily complicated. The event may admit of being described more simply by a disjunctive or conjunctive normal form not principal (which still enables the lag to be held to 2). For example (with $k = 2$, $\ell = 3$), $[\overline{N_2(p)} \,\&\, N_2(p - 2)] \vee N_2(p - 1)$ is a *d.n.f.*, the *p.d.n.f.* for which would have 40 terms (the *p.c.n.f.* 24 factors). There may be simpler equivalents not disjunctive or conjunctive normal forms.

Since the theorem gives a representing net of corresponding structure to the formula, the problem of finding as simple nets as possible to represent definite events is correlated to the problem of finding simplest equivalents of an expression in the propositional calculus, which has recently been treated by Quine [1952].

In special cases the net may be constructed more simply than corresponding to the formula; e.g., in Figure 2 taking $p = t - 1$ the net represents the event with lag 1, although the formula is of depth 3, and no equivalent formula of depth 1 exists.

Reduction of the lag below 2 is not possible in general. For example (with $k = 3$, $\ell = 1$) the event $N_1(p)$ & $\overline{(\overline{N_2(p)} \vee \overline{N_3(p)})}$ is not representable with lag 1. For it is easily seen that no net consisting of a neuron $\mathcal{P}$ impinged on only by endbulbs belonging directly to $\mathcal{N}_1$, $\mathcal{N}_2$ and $\mathcal{N}_3$ can represent this event.

To hold the lag to 2, we may be obliged to have very large numbers of endbulbs originating from or impinging on a given soma.

COROLLARY 2. To each positive (non-positive) definite event, there is a number s and a nerve net without circles, composed of neurons each having, and impinged upon by, at most two endbulbs and of threshold at most 2, which represents the event by the firing (non-firing) of a certain neuron at time $p + s$.

PROOF. By using & and $\vee$ only as binary operations in the $k\ell$ - formula, no neuron in the net construction for the theorem will be impinged upon by more than two endbulbs. Each inner neuron outside of the delay nets has only one endbulb. Each input neuron and delay net can if necessary be replaced (increasing the lag) by a tree of neurons each with at most two endbulbs.

We have been considering representation of an event ending with time p by the firing or by the non-firing of a certain neuron at a certain time $p + s$ $(s \geq 0)$. More generally we can consider representation by a property of the state of the net (i.e., the firing or non-firing of each of its neurons) at $p + s$; i.e., the state of the net is to have or not have this property at time $p + s$, according as the event did or did not occur ending with time p. In the following lemma and corollary, it is not being assumed that the event is definite or the net without circles.

LEMMA 1. An event which is representable in a nerve net by a property of the state at time $p + s$ for a given $s > 0$ is representable by a property of the state of the same net at time p.

PROOF. What happens at times $\leq p$ can only affect the state of the net at time $p + s$ via the state of the entire net, including both the k input neurons and, say, m inner neurons, at time p.

COROLLARY 3. An event which is representable in a nerve net by a property of the state at time $p + s$ for a given $s \geq 0$ is representable by the firing or the non-firing (according to the nature of the property) of a certain inner neuron in a suitable net at time $p + 2$.

PROOF. We can treat the $k + m$ neurons as though all of them were input neurons, for the purpose of applying Corollary 1. By Lemma 1, the property in question is equivalent to a property of the $k + m$ neurons at time p. The latter constitutes a definite event of length 1 on the $k + m$ neurons.

5.4 NERVE NETS WITHOUT CIRCLES.

THEOREM 2. Given any nerve net without circles and given any inner neuron $\mathcal{N}$ in that net, the firing (non-firing) of that neuron at time $p + 1$ is equivalent to the occurrence of a positive (non-positive) definite event.

This theorem was stated (for positive events) by McCulloch and Pitts [1943].

PROOF. Whether or not $\mathcal{N}$ fires at $p + 1$ is completely determined by the state (firing or non-firing) at p of those neurons $\mathcal{N}'_1, \ldots, \mathcal{N}'_r$ having endbulbs impinging on $\mathcal{N}$. Consider those of $\mathcal{N}'_1, \ldots, \mathcal{N}'_r$ which are inner neurons, and repeat the argument. Since there are no circles, any chain of neurons beginning with $\mathcal{N}$ and each impinged on by an endbulb of the next must terminate (with an input neuron). Let $\ell + 1$ be the greatest of the lengths of these chains; since $\mathcal{N}$ is inner, $\ell \geq 1$. After ℓ steps, no inner neurons remain to be considered. Thus whether or not $\mathcal{N}$ fires at time $p + 1$ is completely determined by the state of certain input neurons at certain times between $p - \ell + 1$ and p inclusive; i.e., $\mathcal{N}$'s firing at $p + 1$ is equivalent to a definite event of length ℓ. This event is positive, as firing can only be propagated but not originated under the law for an inner neuron's firing.

REMARK. Any definite event is expressible by a logical formula, e.g., by a principal disjunctive normal form as in 5.2. So a priori there is a formula to express the event of the theorem. By utilizing the condition for firing at each synapse, which can be formulated in logical symbols depending on the threshold and the numbers and kinds of the endbulbs (cf. Figures 1 — 3 for several examples), one can build up a formula directly in ℓ steps, as McCulloch and Pitts indicate.

COROLLARY 1. Any event which is representable in a nerve net without circles by the firing (non-firing) of a given inner neuron $\mathcal{N}$ at time $p + s$ for a given $s \geq 1$ is positive (non-positive) definite.

PROOF. By the theorem, the firing of $\mathcal{N}$ at time $p + s$ is equivalent to the occurrence of a positive definite event ending with time $p + s - 1$. But by the hypothesis that $\mathcal{N}$'s firing at time $p + s$ represents an event, i.e., one ending with time p (cf. Section 4), the input over the moments $p + 1, \ldots, p + s - 1$ has no effect on whether $\mathcal{N}$ fires at time $p + s$.

COROLLARY 2. Any event which is representable in a nerve net without circles by a property of the state at time $p + s$ for a given $s \geq 0$ is definite.

PROOF. By Corollary 1, with Theorem 1 Corollary 3 (which does not introduce circles).

6. Indefinite Events: Preliminaries

6.1 EXAMPLES. Let "(Et)" mean there exists a t ... such that, "(t)" mean for all t, and "$\rightarrow$" mean implies. The net in Figure 10 has a circle of length 1. If at some time $t \leq p$ the input neuron $\mathcal{N}$ fires, then $\mathcal{M}$ will fire at every subsequent moment, in particular at $p + 1$ as the formula expresses. But the firing of $\mathcal{M}$ at time $p + 1$ does not represent the indefinite event $(Et)_{t \leq p}N(t)$ (i.e., we do not have $M(p + 1) \equiv (Et)_{t \leq p}N(t)$), if past time is infinite, because the firing of $\mathcal{M}$ at time $p + 1$ can also be explained by $\mathcal{M}$ having fired at every past moment, without $\mathcal{N}$ having ever fired. Similar examples are given in Figures 11 and 12.

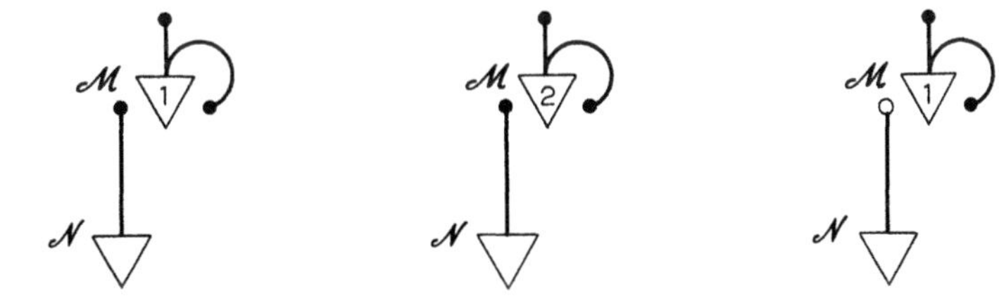

$(Et)_{t \leq p}N(t) \rightarrow M(p + 1)$. $M(p + 1) \rightarrow (t)_{t \leq p}N(t)$. $(Et)_{t \leq p}N(t) \rightarrow \overline{M(p + 1)}$.

FIGURE 10 FIGURE 11 FIGURE 12

This difficulty is not escapable by using other nets to represent the events, or in other examples of indefinite events, but constitutes the general rule, by Theorem 6 in Section 10 below with Lemma 1.

Of course any living organism or constructed robot has only a finite past. Theorem 6 shows that we must take this into account in the theory; otherwise we might be tempted to simplify the theory by the fiction of an infinite past, as we did in Section 5.

6.2 INITIATION. Accordingly we shall hereafter assume (except when we indicate otherwise) that the past for our nerve nets goes back from p (the present) a finite time only, the first moment of which shall be 1 on our time scale. The range of the time variable in our logical formulas shall be the integers from 1 forward.

Now if in the net of Figure 10 $\mathscr{M}$ is quiet at time 1, we do have $M(p + 1) \equiv (Et)_{t \leq p}N(t)$; in Figure 11 if $\mathscr{M}$ fires at time 1, $M(p + 1) \equiv (t)_{t \leq p}N(t)$; and in Figure 12 if $\mathscr{M}$ fires at time 1, $\overline{M(p + 1)} \equiv (Et)_{t \leq p}N(t)$. Thus the nets of Figures 10 and 12 are able to remember that $\mathscr{N}$ has fired since their beginning by changing $\mathscr{M}$ from the state it had initially; while the net of Figure 11 is able to recognize that $\mathscr{N}$ has never failed to fire by preserving $\mathscr{M}$ in the state it had originally, as Householder and Landahl [1945] have commented [p. 109].

The nets in question only represent the events in question, when the inner neuron $\mathscr{M}$ has the state mentioned at time 1.

This again is the general rule for indefinite events, by Theorem 7 with Lemma 1. (Lemma 1 holds for finite past.)

We illustrate this now by showing that to represent the event $(t)_{t \leq p}N(t)$ at least one inner neuron must fire at time 1. For let the proposed representation be by a property of the state at time p (Lemma 1), i.e., solely of this state and not also of the value of p. Say $\mathscr{N}$ is the only input neuron. Were all the inner neurons quiet at $t = 1$, then with the input shown in the table of Figure 13 all inner neurons would be quiet at $t = 2$, so the state at $t = 2$ would be indistinguishable from that with Figure 14 at $t = 1$. Hence with Figure 13 the net would have the same state for $p = 2, 3, 4, \ldots$ as with Figure 14 for $p = 1, 2, 3, \ldots$, respectively, though with the former $(t)_{t \leq p}N(t)$ is false, with the latter true.

Accordingly in studying the representability of events, we shall hereafter not only choose a net for the purpose but also choose the state (firing or non-firing) of each inner neuron at time 1.

As the example of $(t)_{t \leq p}N(t)$ shows, for some events it will not be sufficient to have all the inner neurons quiet initially.

We are developing the theory of McCulloch-Pitts nerve nets as an illustrative case of the theory of finite automata. From the standpoint of the latter theory, one initial state is as reasonable as another. The alternative of excluding such events as $(t)_{t \leq p}N(p)$ from the class of representable events would be more awkward.

t	1	2	3	4	...
$\mathcal{N}$	0	1	1	1	...

FIGURE 13

t	1	2	3	4	...
$\mathcal{N}$	1	1	1	1	...

FIGURE 14

To one who feels that the firing of inner neurons at time 1 requires explanation under the McCulloch-Pitts laws of neural behavior, we need merely say that we have isolated certain input neurons $\mathcal{N}_1, \ldots, \mathcal{N}_k$ and a certain portion $t = 1, 2, 3, \ldots$ of time for the input for the events to be represented. We can go outside those neurons and that part of time to bring about any assumed state of our inner neurons at time $t = 1$. This is most simply accomplished by adding an extra input neuron $\mathcal{N}_{k+1}$, which is to fire at $t = 0$ and then only, and which is to have on each of the inner neurons we wish fired at $t = 1$ a number of excitatory endbulbs equal to the threshold of that neuron, but no other endbulbs.

A more complicated device, which requires an extra input neuron $\mathcal{K}$ but no extra moment of time, is illustrated in Figure 15 for the event $(t)_{t \leq p} N(p)$. The event is represented by firing $\mathcal{P}$ at $t = p + 2$, if $\mathcal{K}$ is fired, but all inner neurons are quiet, at $t = 1$. We can imagine $\mathcal{K}$ exposed to continual environmental stimulation which guarantees its firing at $t = 1$; but its firing at later times does not interfere with the representation. If $\mathcal{K}$ did not fire at $t = 1$, but first at some later time $t = u$, $\mathcal{P}$'s firing at $p + 2$ would represent that $\mathcal{N}$ had fired at all moments from u to p inclusive.[2]

6.3 **DEFINITE EVENTS RECONSIDERED.** An event is a partition of the class of all the possible inputs over the past (including the present) into two subclasses, those inputs for which the event occurs and those inputs for which

[2] McCulloch and Pitts consider the problem of "solving" nets with their initial state unspecified. To "solve" for a given inner neuron $\mathcal{P}$, say at time $p + 1$, means then to find for which inputs over time $1, \ldots, p$ and initial states of the inner neurons $\mathcal{P}$ will fire at time $p + 1$. In the following net, the necessary and sufficient condition that $\mathcal{P}$ fire at $p + 1$ is that $\mathcal{N}$ fire at all times $\leq p$ and both $\mathcal{P}$ and $\mathcal{Q}$ fire

$$P(p + 1) \equiv (t)_{t \leq p} N(t) \ \& \ P(1) \ \& \ Q(1).$$

at time 1. This seems to be a counterexample to the formula next after (9) on p. 126 of McCulloch-Pitts [1943], the proof of which we did not follow; for if we understand the formula correctly, it implies that the condition for firing of $\mathcal{P}$ should only require the existence of one neuron that fires initially. (Their 0 seems to be our 1.) This apparent counterexample discouraged us from further attempts to decipher Part III of McCulloch-Pitts [1943].

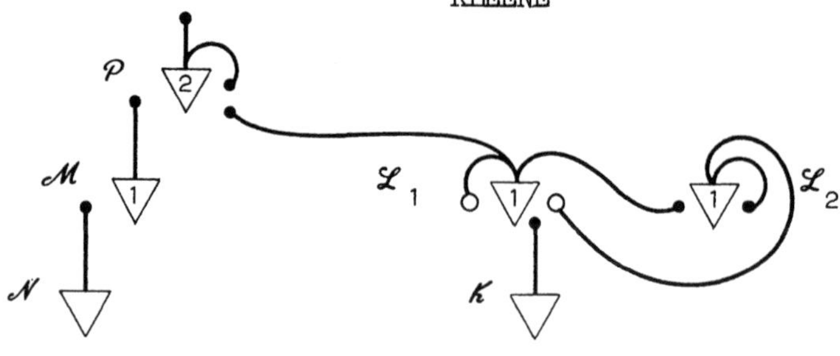

FIGURE 15

the event does not occur. The possible inputs on k neurons $\mathcal{N}_1, \ldots, \mathcal{N}_k$ are described by $k \times p$ tables of 0's and 1's with columns for $\mathcal{N}_1, \ldots, \mathcal{N}_k$ and rows for $t = p, \ldots, 1$. As p varies over all positive integers, these are all tables of 0's and 1's with k columns and any finite number of rows.

In Section 5 we used $k \times \ell$ tables to describe inputs over the last ℓ moments ending with the present.

Now that our time has an initial moment 1, we must be careful whenever we give a table with k columns and a finite number, say ℓ, of rows to describe an input, to make it clear whether we intend it to describe the input over the complete past (so $p = \ell$) or only over the last ℓ moments of the past (so $p \geq \ell$). In the one case we call the table <u>initial</u>, in the other <u>non-initial</u>. The table may be thought of as carrying a tag saying, respectively, $p = \ell$ or $p \geq \ell$. There was no necessity for this in Sections 4 and 5, as there tables referring to the complete past were infinite.

A definite event of length ℓ in Section 5 was one in which the partition of the inputs over the complete past is such that any two inputs which agree in the upper ℓ rows of their tables always fall into the same one of the subclasses. But now when $p < \ell$ there won't be ℓ rows in the table describing the input. For such a p, can the event occur? The convention we adopt is that the event shall not occur in this case. Thus the inputs of the first subclass for a definite event of length ℓ are those described by a set of non-initial $k \times \ell$ tables. If E_1 is the logical formula we used in Section 5 to describe a definite event, the event is now described by $E_1 \ \& \ p \geq \ell$. The negation of this is $\bar{E}_1 \vee p < \ell$, while the formula for the <u>complementary</u> definite event of length ℓ is $\bar{E}_1 \ \& \ p \geq \ell$, which is not equivalent, except for $\ell = 1$ when the "$\& \ p \geq \ell$" and "$\vee \ p < \ell$" are superfluous.

The <u>identical event</u> (written I_k or briefly I) which occurs no matter what the input (the second subclass of the partition being empty) is a definite event of length 1; the <u>improper event</u> (written $\bar{I}_k$ or $\bar{I}$) which never occurs (the first subclass being empty) can be considered as a definite event of length ℓ for every ℓ.

With the sole exception of the improper event, a given event can be definite of length ℓ for only one ℓ, and the set of the $k \times \ell$ tables (all of them non-initial) which describes it is unique. This was not the case in Section 5, as there a definite event of length ℓ was also definite of length m for each $m > \ell$; but now this would be absurd (except for the improper event) as the extra specification that $p \geq m$ would contradict that the event can occur for $p = \ell, \ldots, m - 1$.

The definite events we have just finished describing ($2^{2^{k\ell}}$ of them for a given k and ℓ) are those which arise most naturally from those considered in Section 5 by taking into account that now the past may not include ℓ moments.

We now find it advantageous to introduce also a new kind of definite event on k neurons of length ℓ, by changing the specification for all the tables that $p \geq \ell$ to $p = \ell$; we do not include the improper event among these. These definite events we call <u>initial</u>. For a given k and ℓ, there are $2^{2^{k\ell}} - 1$ of them. An event can be an initial definite event for only one ℓ, and the set of the $k \times \ell$ tables (all of them initial) which describes it is unique. If E_1 & $p \geq \ell$ is a given non-initial definite event not improper, E_1 & $p = \ell$ is the corresponding initial definite event.

In Section 5 p entered the formulas for events only relatively, but now the events can refer to the value of p. This may seem somewhat unnatural; but, reversing the standpoint from which we were led to this in 6.1 and 6.2, if we are to analyze nerve nets in general, starting from arbitrary initial states, we are forced to give p an absolute status. This is illustrated in Figures 16-21, where the formula gives for each net the "solution" for L_1, i.e., the condition for its firing. The "+" indicates initial firing of the indicated neuron; inner neurons not bearing a "+" are initially quiet.

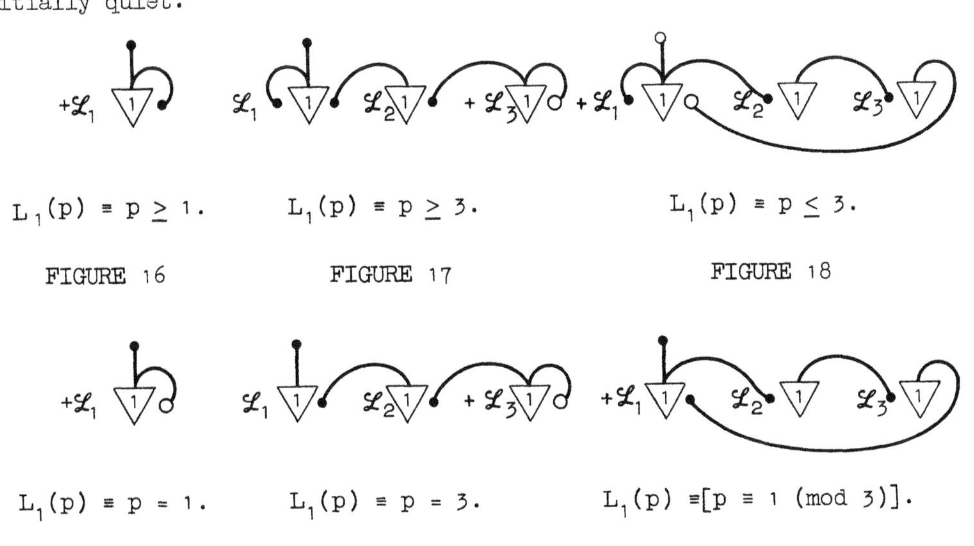

$L_1(p) \equiv p \geq 1$.

FIGURE 16

$L_1(p) \equiv p \geq 3$.

FIGURE 17

$L_1(p) \equiv p \leq 3$.

FIGURE 18

$L_1(p) \equiv p = 1$.

FIGURE 19

$L_1(p) \equiv p = 3$.

FIGURE 20

$L_1(p) \equiv [p \equiv 1 \pmod 3]$.

FIGURE 21

Our theory can now include the case $k = 0$. (In Sections 4 and 5 we were assuming $k \geq 1$, which of course is required for nets without circles; Lemma 1 and Theorem 1 Corollary 3 hold for $k = 0$.) For $k = 0$ there are exactly three definite events of a given length ℓ, namely $p \geq \ell$, $p = \ell$ and $p \neq p$; only $p \neq p$ is positive. The nets of Figures 16 – 21 can be considered as representing events for $k = 0$.

6.4 REPRESENTABILITY OF DEFINITE EVENTS. In Section 5 we showed how to construct nets which represent definite events on $k > 0$ input neurons of length ℓ under the assumption of an infinite past. The proof there that the nets represent the events is valid now for non-initial (initial) definite events for values of $p \geq \ell$ $(p = \ell)$, when the nets are started with all inner neurons quiet at time $t = 1$. To make the following discussion general, we can take the representation to be by a property of the state of the net at $t = p$ (cf. Lemma 1).

Using these nets now can sometimes give rise to a "hallucination" in the sense that the state of the net at $t = p$ has the property without the event having occurred. By reasoning similar to that in 6.2 in connection with Figures 13 and 14, this will happen for suitable inputs, when $p > \ell$ in the case of an initial definite event, and when $p < \ell$ (so $\ell > 1$) in the case of a definite event which can occur without the firing of an input neuron at its first moment $t = p - \ell + 1$.

Conversely these are the only cases in which it will happen. For consider any nerve net a property of which at $t = p$ represents a non-initial (initial) definite event correctly for $p \geq \ell$ $(p = \ell)$, when the net is started at $t = 1$ with all inner neurons quiet. For there to be a hallucination when $p = m < \ell$ (so $\ell > 1$) means that for some input $\mathcal{C}_1 \ldots \mathcal{C}_m$ over $t = 1, \ldots, m$ the net has at $t = m$ a state having the property which goes with occurrence of the event. Now let the input $\mathcal{C}_1 \ldots \mathcal{C}_m$ be assigned instead to $t = \ell - m + 1, \ldots, \ell$ and an input consisting of only non-firings $\mathcal{C}'_1 \ldots \mathcal{C}'_{\ell-m}$ to $t = 1, \ldots, \ell - m$. With the input $\mathcal{C}'_1 \ldots \mathcal{C}'_{\ell-m}$ the state of the inner neurons at $t = \ell - m + 1$ must consist of all non-firings, as it did before at $t = 1$. So with the input $\mathcal{C}'_1 \ldots \mathcal{C}'_{\ell-m} \mathcal{C}_1 \ldots \mathcal{C}_m$ the net will have at $t = \ell$ the state it had before at $t = m$, which shows that $\mathcal{C}'_1 \ldots \mathcal{C}'_{\ell-m} \mathcal{C}_1 \ldots \mathcal{C}_m$ constitutes an occurrence of the event.

Call a definite event of length ℓ prepositive, if the event is not initial, and either $\ell = 1$ or the event only occurs when some input neuron fires at $t = p - \ell + 1$. (For $k = 0$, then only $p \geq 1$ and $p \neq p$ are prepositive.) Prepositiveness is a necessary and sufficient condition for representability in a nerve net with all the inner neurons quiet initially.

This result suggests our first method for constructing nets to represent non-prepositive definite events. Say first the event is non-initial and $\ell > 1$. We supply the $\mathcal{L}_1$ of Figure 16, and treat it as though it were a $k+1$ - st input neuron, required to fire at $t = p - \ell + 1$ (but otherwise not taken into account) in reconsidering the event as on the $k+1$ neurons $\mathcal{N}_1, \ldots, \mathcal{N}_k, \mathcal{L}_1$. This event on $k+1$ neurons is prepositive, so our former methods of net construction (Section 5) apply.

A second method is to use the net of Figure 17 or 18; e.g., if the representation is by firing an inner neuron $\mathcal{P}$ (quiet at $t = 1$) at $t = p + s$, the inhibitory endbulb of $\mathcal{L}_1$ in Figure 18 shall impinge upon $\mathcal{P}$, and the figure is for $\ell + s = 5$. If the representation is by a property of the state at $t = p$, that property shall include that $\mathcal{L}_1$ of Figure 17 fire (for $\ell = 3$) or that $\mathcal{L}_1$ of Figure 18 not fire (for $\ell = 4$).

For initial definite events, the respective methods apply using Figures 19 and 20 instead of Figures 16 and 17, respectively.

The upshot is that only by reference to artifically produced firing of inner neurons at $t = 1$ could an organism recognize complete absence of stimulation of a given duration, not preceded by stimulation; otherwise it would not know whether the stimulation had been absent, or whether it had itself meanwhile come into existence.

As already remarked in 6.2, instead of an initially fired inner neuron as in Figures 16-20, we could use an additional input neuron $\mathcal{K}$ subject to continual environmental stimulation.

A hallucination of the sort considered would be unlikely to have a serious long-term or delayed effect on behavior; but when definite events are used in building indefinite ones, this cannot be ruled out without entering into the further problem of how the representation of events is translated into overt responses.

For organisms, the picture of the nervous system as coming into total activity at a fixed moment $t = 1$ is implausible in any case. But this only means that organisms (at least those which survive) do solve their problems for their processes of coming into activity. For artificial automata or machines generally it is familiar that starting phenomena must be taken into account.

Of course our analysis need not apply to the whole experience and the entire nerve net of an organism, but $t = 1$ can be the first moment of a limited part of its experience, and the nerve net considered a sub-net of its whole nerve net.

7. Regular Events

7.1 REGULAR SETS OF TABLES AND REGULAR EVENTS. In this section as in 6.3 we shall use $k \times \ell$ tables (for fixed k and various ℓ, with each table tagged as either non-initial or initial) to describe inputs on k neurons $\mathcal{N}_1, \ldots, \mathcal{N}_k$ over the time $t = p - \ell + 1, \ldots, p$ for which an event shall occur. But we shall not confine our attention to the case of 6.3 that the set of tables describing when the event occurs are all of them $k \times \ell$ tables for the same ℓ and either all non-initial or all initial.

First we define three operations on sets of tables. If E and F are sets of tables, $E \vee F$ (their sum or disjunction) shall be the set of tables to which a table belongs exactly if it belongs to E or belongs to F.

If E and F are sets of tables, EF (their product) shall be the set of tables to which a table belongs exactly if it is the result of writing any table of F next below any non-initial table of E; if the table of E has ℓ_1 rows and that of F has ℓ_2 rows, the resulting table has $\ell_1 + \ell_2$ rows, is initial exactly if the table of F is initial, and describes an occurrence of an event consisting in the event described by F having occurred ending with $t = p - \ell_1$, as evidenced by the input over $t = p - \ell_1 - \ell_2 + 1, \ldots, p - \ell_1$, followed by the event E having occurred ending with $t = p$, as evidenced by the input over $t = p - \ell_1 + 1, \ldots, p$. The notation EF is written so that we proceed backward into the past in reading from left to right.

Obviously $E \vee F$ and EF are associative operations. We may write E^0F for F, E^1 for E, E^2 for EE, E^3 for EEE, etc.

If E and F are sets of tables, $E*F$ (the iterate of E on F, or briefly E iterate F) shall be the infinite sum of the sets F, EF, EEF, ..., or in self-explanatory symbolism $F \vee EF \vee EEF \vee \ldots$ or $\sum_{n=0}^{\infty} E^nF$.

The regular sets (of tables) shall be the least class of sets of tables which includes the unit sets (i.e., the sets containing one table each) and the empty set and which is closed under the operations of passing from E and F to $E \vee F$, to EF and to $E*F$.

An event shall be regular, if there is a regular set of tables which describes it in the sense that the event occurs or not according as the input is described by one of the tables of the set or by none of them.

To include the case $k = 0$ under these definitions, we shall understand that for $k = 0$ and each $\ell \geq 1$ there are two $k \times \ell$ tables, one non-initial and one initial.[3]

Any finite set of tables is obviously regular, in particular the empty set, and the sets of $k \times \ell$ tables all with a given ℓ and either all non-initial or all initial; so every definite event is regular.

In writing expressions for regular sets or the events they describe we may omit parentheses under three associative laws ((3) – (5) in 7.2),

---

[3] McCulloch and Pitts [1943] use a term "prehensible", introduced quite differently, but in what seems to be a related role. Since we do not understand their definition, we use another term.

besides which we also omit parentheses under the conventions of algebra treating $E \vee F$, EF and $E*F$ as analogous to $e + f$, ef and $e^n f$. For example, $N \vee NI*I$ means $N \vee (N(I*I))$. We may use the same letter to designate a definite event or the set of tables for it or the table itself in the case of a unit set of tables.

We write $E = F$ to say that E and F are the same set of tables; $E \equiv F$ (E is equivalent to F) to say that they describe the same event. Obviously $E = F \rightarrow E \equiv F$. The converse is not true in general, as we illustrate now for regular sets of tables.

Thus with $k = 1$, if N is the non-initial 1×1 table consisting of 1 (which describes the definite event symbolized in Section 5 by $N(p)$), and I is the set of all 1×1 non-initial tables (cf. 6.3), then $N \vee NI*I$ is the set of all non-initial $1 \times \ell$ tables (for all ℓ) with 1 in the top row. Now $N \vee NI*I \equiv N$ but $N \vee NI*I \neq N$.

We can also give counterexamples involving the distinction between non-initial and initial tables. If E is a set of tables, by E^o we denote the set of tables resulting from the tables of E by redesignating as initial those which are not already initial. For any E, $E \equiv E \vee E^o$. But e.g., $I \not\equiv I \vee I^o$.

From $E = F$ we can infer $E \vee G = F \vee G$, $G \vee E = G \vee F$, $EG = FG$, $GE = GF$, $E*G = F*G$ and $G*E = G*F$ by the general replacement theory of equality, since $E \vee G$, EG and $E*G$ are defined as univalent operations on sets of tables. If $=$ is replaced by $\equiv$, the third and fifth of these inferences fail to be valid in general, because the lengths of the tables in E apart from the event described by them enter into the meaning of EG and $E*G$. Thus e.g., $N \vee NI*I \equiv N$ but $(N \vee NI*I)N \not\equiv NN$ and $(N \vee NI*I)*N \not\equiv N*N$.

We have now two systems of notation for describing events:

(A) The logical symbolism as used for definite events in Section 5 (supplemented in Section 6 by suffixing $p \geq \ell$ or $p = \ell$) and for some examples of indefinite events in Section 6,

(B) the symbolism, usually starting with capital letters to represent definite events, by which we describe regular events (via regular sets of tables) in this section.

The question of translatability between the two systems has not yet been thoroughly investigated. By Theorem 8 in Section 12, to any expression (B) there is a logical notation (A), provided sufficient mathematical symbolism is included. Of course we have given no exact delimitation of what symbolism is to be included under (A), so the problem of translatability is not precise. In any case, with very limited mathematical symbolism included under (A), a non-regular event can be expressed, as we shall see in Section 13. It is an open problem whether there is any simple characterization of regularity of events directly in terms of the symbolism (A).

Some examples of translation from (A) to (B) follow. In the examples not involving $\mathcal{N}$, k can have any fixed value ≥ 0; involving $\mathcal{N}$ only, ≥ 1; involving $\mathcal{K}$ also, ≥ 2. Sets of non-initial $k \times 1$ tables expressing that $\mathcal{N}$ fires at time p, that $\mathcal{K}$ fires at p, and that $\mathcal{K}$ and $\mathcal{N}$ both fire at p, are denoted by N, K, and L, respectively. Also I is to be all non-initial $k \times 1$ tables; and to any set E of $k \times 1$ non-initial tables, $\bar{E}$ is the complementary set of $k \times 1$ non-initial tables, in particular $\bar{I}$ is the empty set (cf. 6.3).

$(Et)_{t \leq p} N(t)$	$I*N$
$(t)_{t \leq p} N(t)$	$N*N^{o}$
$(Eu)_{u \leq p}\left[K(u) \;\&\; (t)_{u \leq t \leq p} N(t)\right]$	$N*L$
$(Eu)_{u \leq p}\left[K(u) \;\&\; (s)_{s<u} \overline{K(s)} \;\&\; (t)_{u \leq t \leq p} N(t)\right]$	$N*L^{o} \vee N*L\bar{K}*\bar{K}^{o}$
$N(t)$ for at least two values of $t \leq p$	$I*NI*N$
$N(t)$ for exactly one value of $t \leq p$	$\bar{N}*N^{o} \vee \bar{N}*N\bar{N}*\bar{N}^{o}$, call this M
$N(t)$ for an odd number of values of $t \leq p$	$(\bar{N}*N\bar{N}*N)*M$
$p \geq 3$	I^{3}
$p = 1$	I^{o}
$p \equiv 1 \pmod{3}$	$(I^{3})*I^{o}$
$p \leq 3$	$I^{o} \vee II^{o} \vee I^{2}I^{o}$

7.2 ALGEBRAIC TRANSFORMATIONS OF REGULAR EXPRESSIONS. We list some equalities for sets of tables. (We have scarcely begun the investigation of equivalences.)

(1) $E \vee E = E$. (2) $E \vee F = F \vee E$.

(3) $(E \vee F) \vee G = E \vee (F \vee G)$. (4) $(EF)G = E(FG)$. (5) $(E*F)G = E*(FG)$.

(6) $(E \vee F)G = EG \vee FG$. (7) $E(F \vee G) = EF \vee EG$. (8) $E*(F \vee G) = E*F \vee E*G$.

(9) $E*F = F \vee E*EF$. (10) $E*F = F \vee EE*F$.

(11) $E*F = E^{s}*(F \vee EF \vee E^{2}F \vee \ldots \vee E^{s-1}F)$ $(s \geq 1)$.

(12) $E \vee \bar{I} = \bar{I} \vee E = E$. (13) $E\bar{I} = \bar{I}E = \bar{I}$. (14) $E*\bar{I} = \bar{I}$. (15) $\bar{I}*E = E$.

(16) $E^{o} \vee F^{o} = (E \vee F)^{o}$. (17) $EF^{o} = (EF)^{o}$. (18) $E^{o}F = \bar{I}$.

(19) $E*F^{o} = (E*F)^{o}$. (20) $E^{o}*F = F$. (21) $(E \vee F^{o})*G = E*G$.

(22) $E^{oo} = E^{o}$. (23) $\bar{I}^{o} = \bar{I}$.

To prove (11), we have

$$E*F = \sum_{n=0}^{\infty} E^{n}F = \sum_{q=0}^{\infty} \sum_{r=0}^{s-1} E^{sq+r}F = \sum_{q=0}^{\infty} E^{sq} \sum_{r=0}^{s-1} E^{r}F.$$

In this subsection, we shall deal with particular ways of expressing a regular set of tables under the definition in 7.1. As we saw there, we can as well start with any sets of tables for definite events, instead of simply with the unit sets and the empty set. By a <u>regular expression</u>, we shall mean a particular way of expressing a regular set of tables starting with sets of tables for definite events and applying zero or more times the three operations (passing from E and F to $E \vee F$, EF or E*F); the occurrences of sets of tables for definite events with which the construction starts we call the <u>units</u>. A unit is of <u>length</u> ℓ, if the definite event described by it is of length ℓ; <u>initial</u>, if that is <u>initial</u>. (It comes to almost the same to let "regular expression" mean a notation for a regular set of tables obtained by starting with symbols for definite events and combining them by use of the notations "$E \vee F$", "EF" and "E*F", and most of what we say can be read either way. But when we say that $\bar{I}$ does not occur in a regular expression as a unit, in terms of notation we would mean that neither "$\bar{I}$" nor any other symbol for the empty set occurs as a unit. Also, in terms of notation we would have to identify the units whenever they are not all of them single letters.)

LEMMA 2. Each regular expression is reducible either to $\bar{I}$ or to a regular expression in which $\bar{I}$ does not occur as a unit.

PROOF. By repeated use of (12) – (15).

LEMMA 3. Each regular expression G is reducible to the form $G_1 \vee G_2^0$ where G_1 contains no initial units.

PROOF, by induction on the number n of units in G.

BASIS: $n = 1$. Then G is a unit. If G is not initial, let $G_1 = G$ and use (12) and (23). If G is initial, let $G_2^0 = G$ and use (12).

INDUCTION STEP: $n > 1$.

CASE 1: G is $E \vee F$. By the hypothesis of the induction, $E = E_1 \vee E_2^0$ and $F = F_1 \vee F_2^0$. Thence, using (2), (3) and (16), $G = (E_1 \vee F_1) \vee (E_2 \vee F_2)^0$.

CASE 2: G is EF. Using the hypothesis of the induction, (6) and (7), (18) and (12), and (17), $G = E_1F_1 \vee (E_1F_2)^0$.

CASE 3: G is E*F. By the hypothesis of the induction, (21), (8) and (19), $G = E_1{*}F_1 \vee (E_1{*}F_2)^0$.

We define recursively the "earliest units" of a regular expression, thus.

(1) A regular expression consisting of only one unit is its <u>earliest unit</u>.

(2) The <u>earliest units</u> of E and the <u>earliest units</u> of F are the <u>earliest units</u> of $E \vee F$.

(3) The <u>earliest units</u> of F are the earliest units of EF and of $E*F$.

LEMMA 4. Each regular expression G is reducible to $\bar{I}$ or to a regular expression in which $\bar{I}$ does not occur as a unit and only earliest units are initial.

PROOF, by induction on the number n of units in G.

BASIS: $n = 1$. Then G is $\bar{I}$, or not $\bar{I}$ but a unit and therefore earliest.

INDUCTION STEP: $n > 1$.

CASE 1: G is $E \vee F$. Use the hypothesis of the induction.

CASE 2: G is EF. By Lemma 3, $E = E_1 \vee E_2^0$. Thence, using (6), (18) and (12), $G = E_1F$. Now apply the hypothesis of the induction to F.

CASE 3: G is $E*F$. Using Lemma 3 and (21), $G = E_1*F$. Apply the hypothesis of the induction to F.

In transforming regular expressions we may reconstitute the units; e.g., when E_1 and E_2 are units of lengths ℓ_1 and ℓ_2, E_1 not being initial and neither being $\bar{I}$, then E_1E_2 can be taken as a new unit, which is of length $\ell_1 + \ell_2$ and initial or not according as E_2 is initial or not.

LEMMA 5. For each $s \geq 1$: Each regular expression is reducible either to $\bar{I}$ or else to a regular expression not containing $\bar{I}$ as a unit, having initial units only as earliest units, and having the form of a disjunction of one or more terms of two kinds: a unit of length $< s$, or a regular expression composed of units all of length $\geq s$.

One can always take the number of terms of the second kind to be one, since a disjunction of terms of the second kind is a term of the second kind (cf. (2) and (3)).

PROOF. For $s = 1$, Lemma 5 coincides with Lemma 4. Now take a fixed $s \geq 2$, and suppose that after applying Lemma 4 we have a regular expression G of the second type there. Transformation of G by any of (1) – (11), which include all individual transformation steps used in what follows, preserves this type. This enables us below to reconstitute $E_1E_2 \ldots E_m$, where $E_1, \ldots, E_m$ are units, as a new unit. Now we show by induction on the number n of units in G, that G can be transformed into a disjunction of terms of the two kinds for Lemma 5.

BASIS: $n = 1$. Then G is of the first or second kind according as its length is $< s$ or $\geq s$.

INDUCTION STEP: $n > 1$.

CASE 1: G is $E \vee F$. Then E and F will each be of the second type of Lemma 4. So by the hypothesis of the induction, E and F are both expressible as disjunctions of terms of only the two kinds. Thence so is $E \vee F$, by combining the two disjunctions as one disjunction.

CASE 2: G is EF. Using the hypothesis of the induction, (6) and (7), EF is then equal to a disjunction of terms each of which is of one of the four types

$$E'F',\ E''F'',\ E''F',\ E'F''$$

where ' identifies a factor (originally a term of the disjunction for E or for F) of the first kind, " of the second kind. By the reasoning of Case 1, it will suffice to show that each of these four types of products is expressible as a disjunction of terms of the two kinds.

But $E'F'$ can be reconstituted as a unit, and according as this new unit is of length $< s$ or $\geq s$, $E'F'$ becomes of the first or of the second kind.

The product $E''F''$ is of the second kind.

Now consider $E''F'$. Using (4) – (6), the F' can be moved progressively inward until finally F' occurs only in parts of the form HF' where H is a unit of length $\geq s$. Each such part can be reconstituted as a unit of length $\geq s + 1$, so that $E''F'$ becomes of the second kind.

For $E'F''$ we proceed similarly, using (4) (from right to left), (7), and (10) followed by (7) and (4).

CASE 3: G is $E*F$. Applying to $E*F$ successively (11) and (9),

$$E*F = F \vee EF \vee E^2F \vee \ldots \vee E^{s-1}F \vee E^s*E^s(F \vee EF \vee E^2F \vee \ldots \vee E^{s-1}F).$$

Since $E^2F, \ldots, E^{s-1}F$ are simply repeated products, by the method of Case 2 (repeated as necessary) each of F, EF, E^2F, $\ldots$, $E^{s-1}F$ is expressible as a disjunction of terms of the two kinds. Now consider E^s; by taking the disjunction for E given by the hypothesis of the induction, and multiplying out ((6) and (7)), we obtain a sum of products of s factors each as in Case 2 we obtained a sum of products of 2 factors each. A product in which the factors are all of the first kind can be reconstituted as a unit, which will be of length $\geq s$ since the number of the factors is s, so it becomes of the second kind. All other types of products which can occur include a factor of the second kind, so by the treatment of the three types of products $E''F''$, $E''F'$ and $E'F''$ under Case 2 (repeated as necessary), each of these becomes of the second kind. So E^s (after suitable reconstitution of units) becomes of the second kind. Now by the treatment of the two types of products $E''F''$ and $E''F'$ under Case 2, $E^s(F \vee EF \vee E^2F \vee \ldots \vee E^{s-1}F)$ and hence $E^s{*}E^s(F \vee EF \vee E^2F \vee \ldots \vee E^{s-1}F)$ become of the second kind.

LEMMA 6. Each regular expression is reducible without reconstituting the units to a disjunction of one or more terms of the form E_iF_i where each E_i is a unit and F_i is empty (then E_iF_i is E_i) or regular (then E_i is non-initial).

PROOF. For a regular expression G of the second type of Lemma 4, one can see, by induction on the number n of units in G, that G can be transformed (using only (4), (6), (10)) into a disjunction of terms of the two kinds for Lemma 6.

7.3 REPRESENTABILITY OF REGULAR EVENTS. A $k \times \ell$ table is prepositive (positive), if it describes a prepositive definite event 6.4, i.e., if it is not initial and either $\ell = 1$ or there is a 1 in its lowest row (a positive definite event 5.1, 6.3, i.e., if there is a 1 in some row). A set of tables is prepositive (positive), if every table of the set is prepositive (positive).

THEOREM 3. To each regular event, there is a nerve net which represents the event by firing a certain inner neuron at time $p + 2$, when started with suitable states of the inner neurons at time 1. If the event is describable by a prepositive and positive regular set of tables, the representation can be by a net started with all inner neurons quiet.

PROOF. We begin by establishing the theorem for a regular event described by a term G of the second kind for Lemma 5 with $s = 2$. We use induction on the number n of units in G.

We will arrange that the neuron (call it the <u>output</u> <u>neuron</u>) which is to fire at $p + 2$ exactly if the event occurs ending with time p shall be of threshold 1 impinged on by only excitatory endbulbs (as in Figure 3), and shall have no axons feeding back into the net.

BASIS: $n = 1$. We construct a net to represent (the event described by) G by the method of proof of Theorem 1 Corollary 1 if G is prepositive (a fortiori positive, since $\ell \geq s > 1$), and otherwise this with the first method of 6.4 (with a neuron of Figure 16 or Figure 19) which makes the event prepositive (so positive) as an event on $k + 1$ neurons, so the representation is by firing at time $p + 2$ in both cases.

INDUCTION STEP: $n > 1$.

CASE 1: G is $E \vee F$. By the hypothesis of the induction, there are nets which represent E and F, respectively, each in the manner described, say with respective output neurons P and Q. To represent $E \vee F$, we "identify" P and Q, i.e., we replace them by a single neuron (call it P) having all the endbulbs which impinged separately on P and on Q; and of course we similarly identify the input neurons $N_1, \ldots, N_k$. The resulting net is diagramed in Figure 22. The box marked $\mathcal{E}$ stands for the net for E except for its input neurons and output neuron. The heavy line leading to P from this box represents the axons which formerly led to the output neuron $\mathcal{P}$ in the net for E.

CASE 2: G is EF. Consider the expression E' which is obtained from E by altering each unit to make it refer to one new input neuron $\mathcal{N}_{k+1}$ required to fire at the second moment of each earliest unit (but otherwise not affecting the occurrence or non-occurrence of the respective definite events); by the hypothesis that the units are of length ≥ 2, there is a second moment. Then E' is of the second kind for Lemma 5 with the same number of units as E, since the alteration gives a regular expression with the same structure in terms of its respective units under the three operations. So by the hypothesis of the induction we can represent E' and F by nets as described. However we simplify the construction by leaving out the neuron of Figure 16 in the case of each earliest non-prepositive unit of E' (this unit is non-initial, by one of the properties of G secured in Lemma 5). Now the net for EF is obtained by identifying the new input neuron $\mathcal{N}_{k+1}$ in the net for E' with the output neuron $\mathcal{Q}$ of the net for F, besides of course identifying the input neurons $\mathcal{N}_1, \ldots, \mathcal{N}_k$ for the two nets, and taking as output neuron the output neuron $\mathcal{P}$ for E' (Figure 23). The event E' is positive, since $\mathcal{N}_{k+1}$ is required to fire at its second moment. No hallucination is possible as a result of leaving out the neurons of

Figure 16 for earliest non-prepositive units of E', since $\mathcal{N}_{k+1}$ (required for those units to fire at their second moment) cannot fire until two moments after an occurrence of F, by the construction of the net for F. These omissions of neurons of Figure 16 are to give the last statement of the theorem.

CASE 3: G is E*F. The nets for E' and F are combined as in Figure 24.

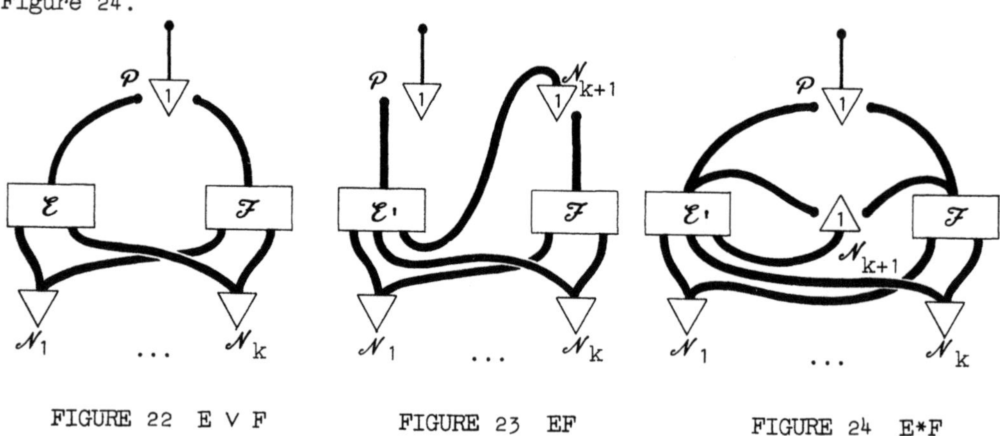

FIGURE 22 E ∨ F FIGURE 23 EF FIGURE 24 E*F

CONCLUSION. This completes the induction to show the representability of a regular event described by a term G of the second kind for Lemma 5 with s = 2. Terms of the first kind are treated as under the basis (but using Figure 16 additionally in the case with $\ell = 1$ of a prepositive non-positive term), and the disjunction of terms (if there are more than one) as under Case 1. The case the event is $\bar{I}$ has already been treated in Section 5 (Figure 8).

DISCUSSION. If the original regular expression for the event is already in terms of units each of length ≥ 2, the proof of the theorem is straightforward and yields nets of complexity corresponding very well to that of the regular expression. (For simplification of the nets representing the units, possibly at the cost of increasing the lag above 2, cf. the discussion following Theorem 1 Corollary 1.) The difficulty which calls for complicated reformulation via the proof of Lemma 5 arises when we try to combine in succession the representations of events some of them shorter than the time necessary for the net to organize a representation of the preceding event by the firing of a single neuron; the solution by Lemma 5 consists in considering grosser events before trying to combine the representations.

THEOREM 4. To each event constructed from regular events by the operations &, ∨, —, there is a nerve net which represents the event by firing a certain inner neuron at time $p + 2$, when started with suitable states of the inner neurons at time 1.

The proof will follow. By Corollary Theorem 5 below, all representable events are regular. So by Theorems 4 and 5 together, combinations of regular events by &, $\vee$ and $-$ are regular, which with Theorem 3 includes Theorem 4. We have not defined & and $-$ as operations on sets of tables, so **EF** and **E*F** cannot be used after the application of & or $-$.

PROOF OF THEOREM 4. To each of the regular events which enter in the construction by &, $\vee$ and $-$ of the given event, consider a regular expression for the regular event. Apply to this Lemma 5 with $s = 2$, and to the resulting terms of the second kind Lemma 6. Thus we obtain an expression for the given event by the operations &, $\vee$ and $-$ from components $E_1F_1, \ldots, E_mF_m$ where each E_i is an expression for a definite event and F_i is a regular expression (then the definite event expressed by E_i is non-initial and of length ≥ 2) or empty. Let E_i' come from E_i as E' came from E in the proof of Theorem 3 Case 2 if F_i is regular, and be the result of introducing an extra input neuron $\mathcal{N}_{k+i}$ to fire at the first moment of E_i' if F_i is empty. Now consider (as an event on the $k + m$ neurons $\mathcal{N}_1, \ldots, \mathcal{N}_k, \mathcal{N}_{k+1}, \ldots, \mathcal{N}_{k+m}$) the same combination of $E_1', \ldots, E_m'$ as the given event is of $E_1F_1, \ldots, E_mF_m$. If this combination of $E_1', \ldots, E_m'$ when treated as a definite event in the sense of Section 5 (not Section 6) of length equal to the greatest of the lengths of $E_1', \ldots, E_m'$ is not positive, we make it so by adding "& E'_{m+1}" where E'_{m+1} refers to the firing of a neuron $\mathcal{N}_{k+m+1}$ at time p. Now use the method of net construction for Theorem 1 Corollary 1 to construct a representing net for this event on $k + m$ or $k + m + 1$ neurons. Then for each i for which F_i is regular, identify $\mathcal{N}_{k+i}$ with the output neuron of a net given by Theorem 3 representing F_i; and for each i for which F_i is empty make $\mathcal{N}_{k+i}$ an inner neuron required to fire at time 1, as in Figure 16 if E_i is non-initial or $i = m + 1$, and as in Figure 19 if E_i is initial.

7.4 PROBLEMS. Numerous problems remain open, which the limited time we have given to this subject did not permit us to consider, although probably some of them at least can be solved quickly.

Is there an extension of Theorem 1 Corollary 2 to all regular events?

By the <u>complete set of tables</u> for an event we mean the set of tables all of them initial which describes the event. By the <u>minimal set of tables</u> for an event we mean the set of tables describing the event each of which has the property that neither a proper upper segment of it, nor itself if it is initial, as a non-initial table describes an occurrence of the event. The complete set of tables for a regular event is regular, by Theorem 3 and the proof of Theorem 5. Is the minimal set necessarily regular? If so, can a regular expression for it be obtained effectively from a regular expression for the complete set?

What kinds of events described originally in other terms are regular? We have only some examples of translation from (A) to (B) (end 7.1), and one of an indirectly established closure property of regular events (Theorem 4 with Theorem 5).

Given a regular expression for an event, it may be difficult to see of what the event consists. We know cases in which a very complicated regular expression is equivalent to a much simpler one, e.g., some arising via the proof of Theorem 5. Are there simple normal forms for regular expressions, such that any regular expression is equal, or is equivalent, to one in a normal form? Is there an effective procedure for deciding whether two regular expressions are equal, or are equivalent?

Our reason for introducing the regular events, as given by regular sets of tables described by regular expressions, is Theorem 5, which we discovered before Theorem 3. By using the notion of regular events, we thus demonstrate that a McCulloch-Pitts nerve net can represent any event which any other kind of finite digital automaton (in the sense to be developed in detail in Section 8) can represent. This of course includes a number of special results which McCulloch and Pitts obtained for alternative kinds of nerve nets, but is more general. The way is open to attempt similarly to verify the like for other kinds of "cells" in place of neurons, or to seek some characterization of the properties of the cells in order that aggregates of them have the capacity for representing all representable (i.e., all regular) events.

PART II. FINITE AUTOMATA

8. The Concept of a Finite Automaton

8.1 CELLS. Time shall consist of a succession of discrete moments numbered by the positive integers, except in Section 10 where all the integers will be used.

We shall consider automata constructed of a finite number of parts called <u>cells</u>, each being at each moment in one of a finite number ≥ 2 of states.

We shall distinguish two kinds of cells, <u>input cells</u> and <u>inner cells</u>.

An input cell admits two states, 0 and 1 (or "quiet" and "firing"), which one is assumed at a given moment being determined by the environment.

The restriction to 2 states for input cells makes the notion of an input to the automaton coincide with the notion of an input to a nerve net as formulated in Sections 4 and 6.3. But the present theory would work equally well with more than 2 states. Nothing would be gained, however, as p cells admitting each admitting 2 states could be used to replace one cell admitting any number q $(2 \leq q \leq 2^p)$ of states $0, 1, \ldots, q - 1$, where if $q < 2^p$ we could either consider only inputs in which states $q, \ldots, 2^p - 1$ do not occur or identify those states with the state $q - 1$ in all the operations of the automaton.

The number of states of an inner cell is not restricted to 2, and different inner cells may have different numbers of states.

The state of each inner cell at any time $t > 1$ is determined by the states of all the cells at time $t - 1$. Of course it may happen that we do not need to know the states of all the cells at time $t - 1$ to infer the state of a given inner cell at time t. Our formulation merely leaves it unspecified what kind of a law of determination we use, except to say that nothing else than the states of the cells at $t - 1$ can matter.

For time beginning with 1, the state of each of the inner cells at that time is to be specified (except in Section 11).

A particular example of a finite automaton is a McCulloch-Pitts nerve net (Part I). Here all the cells admit just 2 states. Another example is obtained by considering inner neurons with "alterable endbulbs" which are not effective unless at some previous time the neuron having the endbulb and the neuron on which the endbulb impinges were simultaneously fired. A neuron with r such alterable endbulbs admits 2^{r+1} states. Many other possibilities suggest themselves.

8.2 STATE. With k input cells $\mathcal{N}_1, \ldots, \mathcal{N}_k$ $(k \geq 0)$, and m inner cells $\mathcal{M}_1, \ldots, \mathcal{M}_m$ $(m \geq 1)$ with respective numbers of states $s_1, \ldots, s_m$, there are exactly $2^k \cdot s_1 \cdot \ldots \cdot s_m$ possible (<u>complete</u>) <u>states</u> of the automaton. We can consider each state as a combination of an <u>external state</u>, of which there are 2^k possible, and an <u>internal state</u>, of which there are $s_1 \cdot \ldots \cdot s_m$ possible.

The law by which the states of the inner cells at a time $t > 1$ are determined by the states of all the cells at time $t - 1$ can be given by specifying to each of the complete states at time $t - 1$ which one of the internal states at time t shall succeed it.

We could indeed consider the entire aggregate of m internal cells as replaced by a single one admitting $s_1 \cdot \ldots \cdot s_m$ states. We shall not take advantage of this possibility, because we have in view applications of the theory of finite automata in which the cells have certain simple properties and are connected in certain simple ways.

We could also (but shall not) get along with a single input cell, by scheduling the inputs on the k original input cells to come in successively in some order on the new one, which would alter the time scale so that k moments of the new scale correspond to 1 of the original. Events referring to the new time scale could then be interpreted in terms of the original.

Now let us call the states $a_1, \ldots, a_r$ where $r = 2^k \cdot s_1 \cdot \ldots \cdot s_m$ and the internal states $b_1, \ldots, b_q$ where $q = s_1 \cdot \ldots \cdot s_m$. Let the notation be arranged so that the internal state at time 1 is b_1.

With the internal state at time 1 fixed, the state at time p is a function of the input over the time 1, ..., p (including the value of p, or when k = 0 only this).

So each of the states $a_1, \ldots, a_r$ represents an event, which occurs ending with time p, if and only if the input over the time 1, ... p is one which results in that one of $a_1, \ldots, a_r$ being the state at time p. Thus the automaton can know about its past experience (inclusive of the present) only that it falls into one of r mutually exclusive classes (possibly some of them empty).

Similarly an internal state at time p + 1, or a property of the complete state at time p, or a property of the internal state at time p + 1, or a property of the internal state at time p + s for an s > 1 which does not depend on the input over the time p + 1, ..., p + s - 1, represents an event. Thus to say that the state at time p has a certain property is to say that the state then is one of a certain subclass of the r possible states, so that the past experience falls into the set sum (or disjunction) of the respective classes of past experiences which are separately represented by the states of the subclass.

What sorts of events can be represented? As the concept of input is the same as in Part I, we can use the notion of "regular event" which was introduced in Section 7. The following theorem, together with Theorem 3 referring to a special kind of finite automaton, answer the question.

9. Regularity of Representable Events

THEOREM 5. In any finite automaton (in particular, in a McCulloch-Pitts nerve net), started at time 1 in a given internal state b_1, the event represented by a given state existing at time p is regular.

PROOF. Since the initial internal state is specified, there are 2^k possible initial states (the results of combining the given initial internal state b_1 with each of the 2^k possible external states at time 1).

So if we can show that the automaton starting from a given state at time 1 will reach a given state at time p, if and only if a certain regular event occurs ending with time p, then the theorem will follow by taking the disjunction of 2^k respective regular events, which is itself a regular event.

Given any state a at time $t - 1$ $(t \geq 2)$, exactly 2^k states are possible at time t, since the internal part of the state at time t is determined by a, and the external part can happen in 2^k ways. Let us say each of these 2^k states is <u>in relation</u> R to a.

The next part of our analysis will apply to any binary relation R defined on a given set of $r \geq 1$ objects $a_1, \ldots, a_r$ (called "states"), whether or not it arises in the manner just described.

Consider any two a and $\bar{a}$ of the states, not necessarily distinct. We shall study the strings of states $d_p d_{p-1} \ldots d_1$ $(p \geq 1)$ for which d_p is a, d_1 is $\bar{a}$, and for each t $(t = 2, \ldots, p)$ d_t is in relation R to d_{t-1} (in symbols, $d_t \, R \, d_{t-1}$); say such strings <u>connect</u> a to $\bar{a}$.

We say a set of strings is "regular" under the following definition (chosen analogously to the definition of "regular" sets of tables in 7.1).

The empty set and for each i $(i = 1, \ldots, r)$ the unit set $\{a_i\}$ having as only member a_i considered as a string of length 1 are <u>regular</u>. If $\mathcal{A}$ and $\mathcal{B}$ are <u>regular</u>, so is their sum, written $\mathcal{A} \vee \mathcal{B}$. If $\mathcal{A}$ and $\mathcal{B}$ are <u>regular</u>, so is the set, written $\mathcal{A}\mathcal{B}$, of the strings obtainable by writing a string belonging to $\mathcal{A}$ just left of a string belonging to $\mathcal{B}$. If $\mathcal{A}$ and $\mathcal{B}$ are <u>regular</u>, so is the sum, written $\mathcal{A}*\mathcal{B}$, for $n = 0, 1, 2, \ldots$ of the sets $\mathcal{A} \ldots \mathcal{A}\mathcal{B}$ with n $\mathcal{A}$'s preceding the $\mathcal{B}$.

LEMMA 7. The strings $d_p \ldots d_1$ connecting a to $\bar{a}$ constitute a regular set.

PROOF OF LEMMA, by induction on r.

BASIS: $r = 1$. Then $\bar{a}$ is a. If $\overline{a \, R \, a}$ (i.e., if R is an irreflexive relation), the set of the strings connecting a to a is the unit set $\{a\}$, which is regular. If $a \, R \, a$, then the set is $\{a, aa, aaa, \ldots\}$, which is regular, since it can be written $\mathcal{A}*\mathcal{A}$ where $\mathcal{A} = \{a\}$.

INDUCTION STEP: $r > 1$.

CASE 1: $a = \bar{a}$. In this case any string connecting a to $\bar{a}$ is of the form

$$a \longrightarrow a \longrightarrow a \longrightarrow \ldots a \longrightarrow a,$$

where the number of $a \longrightarrow$'s is ≥ 0 and each $\longrightarrow$ represents independently the empty string (this being possible only if $a \, R \, a$) or a non-empty string without any a in it. Let $e_1, \ldots, e_g$ $(g \geq 0)$ be the states e such that $a \, R \, e$ but $e \neq a$, and $f_1, \ldots, f_h$ $(h \geq 0)$ the states f such that $f \, R \, a$ but $f \neq a$. Now any non-empty string represented by an $\longrightarrow$ must start with one of $e_1, \ldots, e_g$ and end with one of $f_1, \ldots, f_h$. For each pair $\mathcal{e}_i \mathcal{f}_j$, by the hypothesis of the induction the set of the strings connecting $\mathcal{e}_i$ to $\mathcal{f}_j$ without $\mathcal{a}$ in it is regular. Say $\mathcal{B}_1, \ldots, \mathcal{B}_{gh}$ are these regular sets; and let $\mathcal{A}$ be $\{\mathcal{a}\}$. Now if $\mathcal{a}$ R $\mathcal{a}$, the set of the possible strings $\mathcal{a} \longrightarrow$ is $\mathcal{A} \vee \mathcal{A}(\mathcal{B}_1 \vee \ldots \vee \mathcal{B}_{gh})$ (which reduces to $\mathcal{A}$ if gh = 0 or all the $\mathcal{B}$'s are empty); and if $\overline{\mathcal{a} R \mathcal{a}}$, it is $\mathcal{A}(\mathcal{B}_1 \vee \ldots \vee \mathcal{B}_{gh})$ (which is empty if gh = 0 or all the $\mathcal{B}$'s are empty). Let this set be $\mathcal{C}$. Then the set of the strings leading from $\mathcal{a}$ to $\mathcal{a}$ is $\mathcal{C}*\mathcal{A}$ (which reduces to $\mathcal{A}$ if $\mathcal{C}$ is empty).

CASE 2: $\mathcal{a} \neq \bar{\mathcal{a}}$. Now we have instead

$$\mathcal{a} \longrightarrow \mathcal{a} \longrightarrow \mathcal{a} \longrightarrow \ldots \mathcal{a} \longrightarrow \mathcal{a} \dashrightarrow \bar{\mathcal{a}}$$

where the number of $\mathcal{a} \longrightarrow$'s is ≥ 0 and each $\longrightarrow$ and the $\dashrightarrow$ represents independently the empty string or a non-empty string without any $\mathcal{a}$ in it. If $\mathcal{D}$ is the set of the possible strings $\mathcal{a} \dashrightarrow$, and $\mathcal{E} = \{\bar{\mathcal{a}}\}$, the set of the strings connecting $\mathcal{a}$ to $\bar{\mathcal{a}}$ is $\mathcal{C}*\mathcal{D}\mathcal{E}$, which is regular.

PROOF OF THEOREM (completed). We need to show that, for a given state $\mathcal{a}$ and each of 2^k states $\bar{\mathcal{a}}$, the state is $\mathcal{a}$ at time p and $\bar{\mathcal{a}}$ at time 1, if and only if a certain regular event occurs over the time 1, ..., p.

By the lemma, the set of the strings which can connect $\mathcal{a}$ to $\bar{\mathcal{a}}$ is regular. Consider an expression for this regular set in terms of the empty set and the sets $\{\mathcal{a}_i\}$ as the units (cf. 7.2). In this expression let us replace each unit $\{\mathcal{a}_i\}$ by the unit set consisting of the $k \times 1$ table which (if $k > 0$) describes the external part of the state $\mathcal{a}_i$, labeled initial or non-initial according to whether that unit $\{\mathcal{a}_i\}$ was earliest or not. Each empty set as unit we replace by itself (but write it $\bar{I}$). There results a regular expression. The state changes from $\bar{\mathcal{a}}$ at time 1 to $\mathcal{a}$ at time p, exactly if the event described by this regular expression occurs over the time 1, ..., p.

COROLLARY. The event represented by each of the following is likewise regular: an internal state at time p + 1, a property of the state at time p, a property of the internal state at time p + 1, a property of the internal state at time p + s for an s > 1 which does not depend on the input over the time p + 1, ..., p + s - 1.

PROOF. An event represented by a property of the state at time p is the disjunction of the events represented at time p by the states which have that property. The other modes of representation reduce to this via Lemma 1 in 5.3 (which applies here just as to McCulloch-Pitts nerve nets).

DISCUSSION. The regular expressions obtained by the proof of Theorem 5 have only initial units or $\bar{I}$ as earliest units and are built of units of length 1 (and likewise after simplification by Lemma 2). It is clear in many examples that great simplifications can be obtained by use of equivalences (7.1); but we have made no study of the possibilities for proceeding systematically with such simplifications.

The study of the structure of a set of objects $\boldsymbol{a}_1, \ldots, \boldsymbol{a}_r$ under a binary relation R, which is at the heart of the above proof, might profitably draw on some algebraic theory.

It is of course essential to our arguments that the number of cells and the number of states for each be finite, so that the number of complete states is fixed in advance. A machine of Turing [1936-7] is not a finite automaton, if the tape is considered as part of the machine, since, although only a finite number of squares of the tape are printed upon at any moment, there is no preassigned bound to this number. If the tape is considered as part of the environment, a Turing machine is a finite automaton which can in addition store information in the environment and reach for it later, so that the present input is not entirely independent of the past. Whether this comparison may lead to any useful insights into Turing machines or finite automata remains undetermined.

APPENDICES

10. Representability in a Finite Automaton with an Infinite Past

THEOREM 6. An event E is representable by a property of the state at time p of a finite automaton with an infinite past, only if E is definite.

PROOF. With $k > 0$ input cells, a complete input is generated by choosing between the finite number 2^k of possible inputs at time p, then between the same number of possible inputs at time $p - 1$, etc. ad infinitum.

By a theorem of Brouwer [1924],[4] also given by Kőnig [1927], if for each input it is determined at some finite stage (i.e., from only the part of the input occupying the time $p, \ldots, p - u$ for some $u \geq 0$) whether an event occurs or not, then there is a number $n \geq 0$ such that for any input whether or not the event occurs is determined from only the part

[4] Brouwer's treatment is intended for readers acquainted with intuitionistic set theory, and his main effort is to demonstrate the theorem intuitionistically.

of it occupying the time $p, \dots, p-n$. In this case the event would be definite of length $n+1$.

Now consider an indefinite event E. Contraposing Brouwer's theorem, there is an input $\mathcal{c}_0\mathcal{c}_1\mathcal{c}_2 \dots$ such that for every $u \geq 0$ it is not determined by the part of it $\mathcal{c}_0 \dots \mathcal{c}_u$ for the time $p, \dots, p-u$ whether E occurs or not.

CASE 1: E does not occur for the input $\mathcal{c}_0\mathcal{c}_1\mathcal{c}_2 \dots$. Then for each u there is an input $\mathcal{c}_0^u\mathcal{c}_1^u\mathcal{c}_2^u \dots$, coinciding with $\mathcal{c}_0\mathcal{c}_1\mathcal{c}_2 \dots$ over the time $p, \dots, p-u$ and diverging from it at some earlier moment, for which E occurs.

Suppose E is represented by a property of the state at time p. Say the states which have the property are $a_1, \dots, a_{r_1}$ and those which do not are $a_{r_1+1}, \dots, a_r$.

Let S be the set of all the sequences of states $d_0d_1d_2 \dots$ compatible with the present state being one of $a_1, \dots, a_{r_1}$; i.e., d_0 is one of $a_1, \dots, a_{r_1}$, and each d_i has as its internal part that which is determined by d_{i+1} being the state at the immediately preceding moment. There are r_1 choices for d_0, at most r for d_1, at most r for d_2, etc.

Any sequence of states $d_0d_1d_2 \dots$ which can be assumed for the input $\mathcal{c}_0^u\mathcal{c}_1^u\mathcal{c}_2^u \dots$ must belong to S, since E occurs for $\mathcal{c}_0^u\mathcal{c}_1^u\mathcal{c}_2^u \dots$, and must in its first $u+1$ choices $d_0 \dots d_u$ be compatible with $\mathcal{c}_0\mathcal{c}_1\mathcal{c}_2 \dots$, i.e., the external part of $d_0 \dots d_u$ must be the input $\mathcal{c}_0 \dots \mathcal{c}_u$ over the last $u+1$ moments $p, \dots, p-u$ in $\mathcal{c}_0\mathcal{c}_1\mathcal{c}_2 \dots$.

By Brouwer's theorem, if for each sequence $d_0d_1d_2 \dots$ belonging to S there were a u such that $d_0 \dots d_u$ is incompatible with $\mathcal{c}_0\mathcal{c}_1\mathcal{c}_2 \dots$, there would be an n such that for each $d_0d_1d_2 \dots$ belonging to S the part $d_0 \dots d_n$ is incompatible with $\mathcal{c}_0\mathcal{c}_1\mathcal{c}_2 \dots$, contradicting the preceding remark for $u \geq n$.

So there is an infinite sequence $d_0d_1d_2 \dots$ in S which is compatible with $\mathcal{c}_0\mathcal{c}_1\mathcal{c}_2 \dots$. But d_0 is one of the states $a_1, \dots, a_{r_1}$, although E does not occur for $\mathcal{c}_0\mathcal{c}_1\mathcal{c}_2 \dots$, contrary to our supposition that E is represented by the state at time p being one of $a_1, \dots, a_{r_1}$.

CASE 2: E occurs for the input $\mathcal{c}_0\mathcal{c}_1\mathcal{c}_2 \dots$. Applying to $\bar{E}$ the reasoning applied in Case 2 to E, it is absurd that $\bar{E}$, and hence that E, be represented by a property of the state at time p.

11. Representability with a Finite Past but an Arbitrary Initial Internal State

THEOREM 7. An event E is representable by a property of the state at time p of a finite automaton started with an arbitrary internal state at time 1, only if E is non-initial definite of length 1.

PROOF. Let E be an event not non-initial definite of length 1. Then there is some input $\mathscr{c}$ for the moment p such that whether or not E occurs is not determined by $\mathscr{c}$ alone; i.e., different choices $\mathscr{c}'_1 \ldots \mathscr{c}'_{p'-1}$ and $\mathscr{c}''_1 \ldots \mathscr{c}''_{p''-1}$ of the input over $1, \ldots, p-1$ for $p = p'$ and $p = p''$ together with $\mathscr{c}$ at p make E occur or not occur, respectively. Suppose E is represented by a certain property of the state at time p for a given initial internal state $\mathscr{b}_1$. Consider the inner states $\mathscr{b}'$ and $\mathscr{b}''$ produced at times p' and p" from the initial internal state $\mathscr{b}_1$ by the inputs $\mathscr{c}'_1 \ldots \mathscr{c}'_{p'-1}$ and $\mathscr{c}''_1 \ldots \mathscr{c}''_{p''-1}$, respectively. Now at time 1 let the input be $\mathscr{c}$ and the internal state be $\mathscr{b}'$ or $\mathscr{b}''$, respectively. Then the property of the state is possessed or not possessed, respectively. Thus the property cannot represent E for both $\mathscr{b}'$ and $\mathscr{b}''$ as initial internal state; for one of them it gives a false result for $p = 1$ and $\mathscr{c}$ as input.

12. Primitive Recursiveness of Regular Events

To illustrate that only logical and mathematical symbolism on the level of number theory is necessary to express regular events, we state the following theorem. The notion of relative primitive recursiveness is defined in Kleene [1952]. For conformity with the notation there, the time variables for this theorem shall range over 0, 1, 2, ... instead of 1, 2, 3, ...

THEOREM 8. For any regular event E referring to input neurons $\mathscr{N}_1, \ldots, \mathscr{N}_k$, the predicate E(p) (≡ E occurs ending with time p) is primitive recursive in the predicates $N_1(t), \ldots, N_k(t)$.

METHOD OF PROOF. Using Theorem 3, E(p) is equivalent to the existence of a certain kind of a string of states $\mathscr{d}_p \ldots \mathscr{d}_0$ (cf. Section 9).

13. A Simple Example of an Irregular Event

Consider the event E described as follows: $\mathscr{N}$ fired at time u^2 for every u such that $u^2 \leq p$ and only at those times. In symbols, $E(p) \equiv (t)_{t \leq p}\left[N(t) \equiv (Eu)_{u \leq p}\, t = u^2\right]$.

No finite automaton can represent E, and hence by Theorem 3 E is not regular. For suppose E is represented by a property of the state at time p of a finite automaton (admitting states $a_1, \ldots, a_r$); say the states which have this property are $a_1, \ldots, a_{r_1}$.

Consider any number s such that $2s > r_1$. Suppose $\mathcal{N}$ fires at times 1, 4, 9, ..., s^2 and never thereafter. Then E occurs for $p = 1, 2, \ldots, s^2 + 2s$ $(= (s + 1)^2 - 1)$ and for no greater p.

Consider the states $d_1, d_2, d_3, \ldots$ of the automaton at the times $s^2 + 1, s^2 + 2, s^2 + 3, \ldots$. Beginning with time $s^2 + 1$, $\mathcal{N}$ never fires, so the external state is constant. Thus each of the states $d_1, d_2, d_3, \ldots$ after the first is determined by the immediately preceding one. So, since there are only r states altogether, the sequence $d_1, d_2, d_3, \ldots$ is ultimately periodic.

However, during the time $s^2 + 1, \ldots, s^2 + 2s$ the state must be one of $a_1, \ldots, a_{r_1}$, since E occurs for these values of p. Hence, since $2s > r_1$, the period must already have become established (i.e., the first repetition in $d_1, d_2, d_3, \ldots$ must already have occurred) by the time $s^2 + 2s$. Hence the state at time $(s + 1)^2$ is one of $a_1, \ldots, a_{r_1}$, although E does not occur for $p = (s + 1)^2$.

It is not suggested that the event would be of any biological significance. The example is given to show the mathematical limitations to what events can be represented.

BIBLIOGRAPHY

[1] BROUWER, L. E. J., Beweis, dass jede volle Funktion gleichmässig stetig ist. Verhandelingen Koninklijke Nederlandsche Akademie van Wetenschappen, Amsterdam, vol. 27 (1924), pp. 189-193. Another version: Über Definitionsbereiche von Funktionen, Mathematische Annalen, vol. 97 (1927), pp. 60-75.

[2] GERARD, Ralph W., What is memory? Scientific american, vol. 189, no. 3, September 1953, pp. 118-126.

[3] HILBERT, David and ACKERMANN, Wilhelm, Grundzüge der theoretischen Logik. First ed., Berlin (Springer) 1928, viii + 120 pp. Third ed., Berlin, Göttingen, Heidelberg (Springer) 1949, viii + 155 pp. Eng. tr. of the second ed., Principles of mathematical logic, New York (Chelsea) 1950, xii + 172 pp.

[4] HOUSEHOLDER, A. S. and LANDAHL, H. D., Mathematical biophysics of the central nervous system. Mathematical biophysics monograph series, no. 1, Bloomington, Indiana (Principia press) 1945, ix + 124 pp.

[5] KLEENE, S. C., Introduction to metamathematics. Amsterdam (North Holland Pub. Co.), Groningen (Noordhoff) and New York (Van Nostrand) 1952, x + 550 pp.

[6] KÖNIG, D., Über eine Schlussweise aus dem Endlichen ins Unendliche. Acta litterarum ac scientiarum (Szeged), Sect. math., vol. III/II (1927), pp. 121-130 (particularly the appendix pp. 129-130).

[7] McCULLOCH, Warren S., The brain as a computing machine. Electrical engineering, vol. 68 (1949), pp. 492-497.

[8] McCULLOCH, Warren S. and PITTS, Walter, A logical calculus of the ideas immanent in nervous activity. Bulletin of mathematical biophysics, vol. 5 (1943), pp. 115-133.

[9] QUINE, W. V., The problem of simplifying truth functions. American mathematical monthly, vol. 59 (1952), pp. 521-531.

[10] TURING, A. M., On computable numbers, with an application to the Entscheidungsproblem. Proceedings of the London Mathematical Society, ser. 2, vol. 42 (1936-7), pp. 230-265. A correction, ibid., vol. 43 (1937), pp. 544-546.

PROBABILISTIC LOGICS AND THE SYNTHESIS OF RELIABLE ORGANISMS FROM UNRELIABLE COMPONENTS

J. von Neumann

1. INTRODUCTION

The paper that follows is based on notes taken by Dr. R. S. Pierce on five lectures given by the author at the California Institute of Technology in January 1952. They have been revised by the author but they reflect, apart from minor changes, the lectures as they were delivered.

The subject-matter, as the title suggests, is the role of error in logics, or in the physical implementation of logics — in automata-synthesis. Error is viewed, therefore, not as an extraneous and misdirected or misdirecting accident, but as an essential part of the process under consideration — its importance in the synthesis of automata being fully comparable to that of the factor which is normally considered, the intended and correct logical structure.

Our present treatment of error is unsatisfactory and ad hoc. It is the author's conviction, voiced over many years, that error should be treated by thermodynamical methods, and be the subject of a thermodynamical theory, as information has been, by the work of L. Szilard and C. E. Shannon [Cf. 5.2]. The present treatment falls far short of achieving this, but it assembles, it is hoped, some of the building materials, which will have to enter into the final structure.

The author wants to express his thanks to K. A. Brueckner and M. Gell-Mann, then at the University of Illinois, to whose discussions in 1951 he owes some important stimuli on this subject; to Dr. R. S. Pierce at the California Institute of Technology, on whose excellent notes this exposition is based; and to the California Institute of Technology, whose invitation to deliver these lectures combined with the very warm reception by the audience, caused him to write this paper in its present form, and whose cooperation in connection with the present publication is much appreciated.

2. A SCHEMATIC VIEW OF AUTOMATA

2.1 Logics and Automata

It has been pointed out by A. M. Turing [5] in 1937 and by W. S. McCulloch and W. Pitts [2] in 1943 that effectively constructive logics, that is, intuitionistic logics, can be best studied in terms of automata. Thus logical propositions can be represented as electrical networks or (idealized) nervous systems. Whereas logical propositions are built up by combining certain primitive symbols, networks are formed by connecting basic components, such as relays in electrical circuits and neurons in the nervous system. A logical proposition is then represented as a "black box" which has a finite number of inputs (wires or nerve bundles) and a finite number of outputs. The operation performed by the box is determined by the rules defining which inputs, when stimulated, cause responses in which outputs, just as a propositional function is determined by its values for all possible assignments of values to its variables.

There is one important difference between ordinary logic and the automata which represent it. Time never occurs in logic, but every network or nervous system has a definite time lag between the input signal and the output response. A definite temporal sequence is always inherent in the operation of such a real system. This is not entirely a disadvantage. For example, it prevents the occurence of various kinds of more or less overt vicious circles (related to "non-constructivity", "impredicativity", and the like) which represent a major class of dangers in modern logical systems. It should be emphasized again, however, that the representative automaton contains more than the content of the logical proposition which it symbolizes — to be precise, it embodies a definite time lag.

Before proceeding to a detailed study of a specific model of logic, it is necessary to add a word about notation. The terminology used in the following is taken from several fields of science; neurology, electrical engineering, and mathematics furnish most of the words. No attempt is made to be systematic in the application of terms, but it is hoped that the meaning will be clear in every case. It must be kept in mind that few of the terms are being used in the technical sense which is given to them in their own scientific field. Thus, in speaking of a neuron, we don't mean the animal organ, but rather one of the basic components of our network which resembles an animal neuron only superficially, and which might equally well have been called an electrical relay.

2.2 Definitions of the Fundamental Concepts

Externally an automaton is a "black box" with a finite number of inputs and a finite number of outputs. Each input and each output is capable of exactly two states, to be designated as the "stimulated" state and the "unstimulated" state, respectively. The internal functioning of such a "black box" is equivalent to a prescription that specifies which outputs will be stimulated in response to the stimulation of any given combination of the inputs, and also the time of stimulation of these outputs. As stated above, it is definitely assumed that the response occurs only after a time lag, but in the general case the complete response may consist of a succession of responses occurring at different times. This description is somewhat vague. To make it more precise it will be convenient to consider first automata of a somewhat restricted type and to discuss the synthesis of the general automaton later.

DEFINITION 1: A single output automaton with time delay δ (δ is positive) is a finite set of inputs, exactly one output, and an enumeration of certain "preferred" subsets of the set of all inputs. The automaton stimulates its output at time $t + \delta$ if and only if at time t the stimulated inputs constitute a subset which appears in the list of "preferred" subsets, describing the automaton.

In the above definition the expression "enumeration of certain subsets" is taken in its widest sense and does not exclude the extreme cases "all" and "none". If n is the number of inputs, then there exist $2^{(2^n)}$ such automata for any given δ.

Frequently several automata of this type will have to be considered simultaneously. They need not all have the same time delay, but it will be assumed that all their time lags are integral multiples of a common value δ_o. This assumption may not be correct for an actual nervous system; the model considered may apply only to an idealized nervous system. In partial justification, it can be remarked that as long as only a finite number of automata are considered, the assumption of a common value δ_o can be realized within any degree of approximation. Whatever its justification and whatever its meaning in relation to actual machines or nervous systems, this assumption will be made in our present discussions. The common value δ_o is chosen for convenience as the time unit. The time variable can now be made discrete, i.e., it need assume only integral numbers as values, and correspondingly the time delays of the automata considered are positive integers.

Single output automata with given time delays can be combined into a new automaton. The outputs of certain automata are connected by lines or wires or nerve fibers to some of the inputs of the same or other automata. The connecting lines are used only to indicate the desired connections; their function is to transmit the stimulation of an output instantaneously to all the inputs connected with that output. The network is subjected to one condition, however. Although the same output may be connected to several inputs, any one input is assumed to be connected to at most one output. It may be clearer to impose this restriction on the connecting lines, by requiring that each input and each output be attached to exactly one line, to allow lines to be split into several lines, but prohibit the merging of two or more lines. This convention makes it advisable to mention again that the activity of an output or an input, and hence of a line, is an all or nothing process. If a line is split, the stimulation is carried to all the branches in full. No energy conservation laws enter into the problem. In actual machines or neurons, the energy is supplied by the neurons themselves from some external source of energy. The stimulation acts only as a trigger device.

The most general automaton is defined to be any such network. In general it will have several inputs and several outputs and its response activity will be much more complex than that of a single output automaton with a given time delay. An intrinsic definition of the general automaton, independent of its construction as a network, can be supplied. It will not be discussed here, however.

Of equal importance to the problem of combining automata into new ones is the converse problem of representing a given automaton by a network of simpler automata, and of determining eventually a minimum number of basic types for these simpler automata. As will be shown, very few types are necessary.

2.3 Some Basic Organs

The automata to be selected as a basis for the synthesis of all automata will be called basic organs. Throughout what follows, these will be single output automata.

One type of basic organ is described by Figure 1. It has one

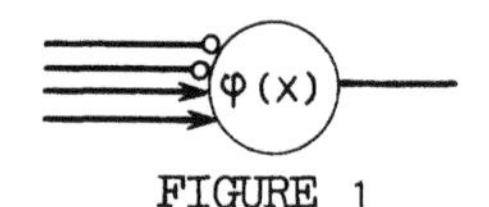

FIGURE 1

output, and may have any finite number of inputs. These are grouped into two types: Excitatory and inhibitory inputs. The excitatory inputs are distinguished from the inhibitory inputs by the addition of an arrowhead to the former and of a small circle to the latter. This distinction of inputs into two types does actually not relate to the concept of inputs, it is introduced as a means to describe the internal mechanism of the neuron. This mechanism is fully described by the so-called threshold function $\varphi(x)$ written inside the large circle symbolizing the neuron in Figure 1, according to the following convention: The output of the neuron is excited at time $t + 1$ if and only if at time t the number of stimulated excitatory inputs k and the number of stimulated inhibitory inputs ℓ satisfy the relation $k \geqq \varphi(\ell)$. (It is reasonable to require that the function $\varphi(x)$ be monotone non-decreasing.) For the purposes of our discussion of this subject it suffices to use only certain special classes of threshold functions $\varphi(x)$. E. g.

$$(1) \qquad \varphi(x) = \psi_h(x) \begin{cases} = 0 & \text{for } x < h \\ = \infty & \text{for } x \geqq h \end{cases}$$

(i.e., $< h$ inhibitions are absolutely ineffective, $\geqq h$ inhibitions are absolutely effective), or

$$(2) \qquad \varphi(x) = \chi_h(x) = x + h$$

(i.e., the excess of stimulations over inhibitions must be $\geqq h$). We will use χ_h, and write the inhibition number h (instead of χ_h) inside the large circle symbolizing the neuron. Special cases of this type are the three basic organs shown in Figure 2. These are, respectively, a threshold two neuron with two excitatory inputs, a threshold one neuron with two excitatory inputs, and finally a threshold one neuron with one excitatory input and one inhibitory input.

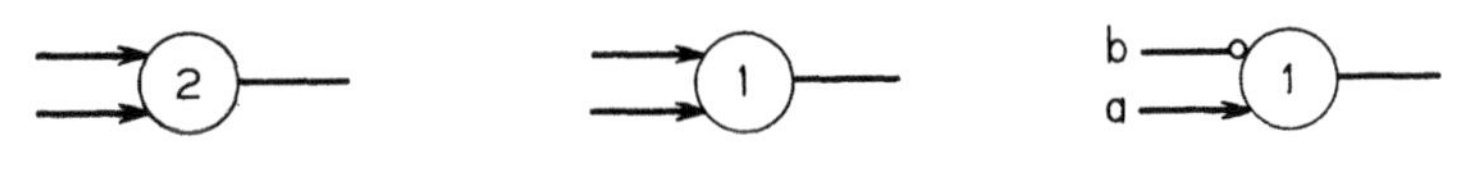

FIGURE 2

The automata with one output and one input described by the networks shown in Figure 3 have simple properties: The first one's output is never stimulated, the second one's output is stimulated at all times if its input has been ever (previously) stimulated. Rather than add these automata to a network, we shall permit lines leading to an input to be either always non-stimulated, or always stimulated. We call the latter "grounded" and designate it by the symbol ||⊢ and we call the former "live" and designate it by the symbol |ı|⊢

FIGURE 3

3. AUTOMATA AND THE PROPOSITIONAL CALCULUS

3.1 The Propositional Calculus

The propositional calculus deals with propositions irrespective of their truth. The set of propositions is closed under the operations of negation, conjunction and disjunction. If a is a proposition, then "not a", denoted by a^{-1} (we prefer this designation to the more conventional ones $-a$ and $\sim a$), is also a proposition. If a, b are two propositions, then "a and b", "a or b", denoted respectively by ab, $a + b$, are also propositions. Propositions fall into two sets, T and F, depending whether they are true or false. The proposition a^{-1} is in T if and only if a is in F. The proposition ab is in T if and only if a and b are both in T, and a + b is in T if and only if either a or b is in T. Mathematically speaking the set of propositions, closed under the three fundamental operations, is mapped by a homomorphism onto the Boolean algebra of the two elements $\underline{1}$ and $\underline{0}$. A proposition is true if and only if it is mapped onto the element $\underline{1}$. For convenience, denote by $\underline{1}$ the proposition $\bar{a} + \bar{a}^{-1}$, by $\underline{0}$ the proposition $\bar{a}\bar{a}^{-1}$, where $\bar{a}$ is a fixed but otherwise arbitrary proposition. Of course, $\underline{0}$ is false and $\underline{1}$ is true.

A polynomial P in n variables, $n \geq 1$, is any formal expression obtained from $x_1, \ldots, x_n$ by applying the fundamental operations to them a finite number of times, for example $[(x_1 + x_2^{-1})\, x_3]^{-1}$ is a polynomial. In the propositional calculus two polynomials in the same variables are considered equal if and only if for any choice of the propositions $x_1, \ldots, x_n$ the resulting two propositions are always either both true or both false. A fundamental theorem of the propositional calculus states that every polynomial P is equal to

$$\sum_{i_1=\pm 1} \cdots \sum_{i_n=\pm 1} f_{i_1 \ldots i_n}\, x_1^{i_1} \cdots x_n^{i_n},$$

where each of the $f_{i_1 \ldots i_n}$ is equal to $\underline{0}$ or $\underline{1}$. Two polynomials are equal if and only if their f's are equal. In particular, for each n, there exist exactly $2^{(2^n)}$ polynomials.

3.2 Propositions, Automata and Delays

These remarks enable us to describe the relationship between automata and the propositional calculus. Given a time delay s, there exists a one-to-one correspondence between single output automata with time delay s and the polynomials of the propositional calculus. The number n of inputs (to be designated $\nu = 1, \ldots, n$) is equal to the number of variables. For every combination $i_1 = \pm 1, \ldots, i_n = \pm 1$, the coefficient $f_{i_1 \ldots i_n} = \underline{1}$, if and only if a stimulation at time t of exactly those inputs ν for which $i_\nu = 1$, produces a stimulation of the output at time t + s.

DEFINITION 2: Given a polynomial $P = P(x_1, \ldots, xn)$ and a time delay s, we mean by a P,s-network a network built from the three basic organs of Figure 2, which as an automaton represents P with time delay s.

THEOREM 1: Given any P, there exists a (unique) $s^* = s^*(P)$, such that a P,s-network exists if and only if $s \geq s^*$.

PROOF: Consider a given P. Let S(P) be the set of those s for which a P,s -network exists. If $s' \geq s$, then tying $s'-s$ unit-delays, as shown in Figure 4, in series to the output of a P,s -network produces a P,s' -network. Hence $S(P)$ contains with an s all $s' \geq s$. Hence if $S(P)$ is not empty, then it is precisely the set of all $s \geq s^*$, where $s^* = s^*(P)$ is its smallest element. Thus the theorem holds for P if $S(P)$ is not empty, i.e., if the existence of at least one P,s -network (for some s!) is established.

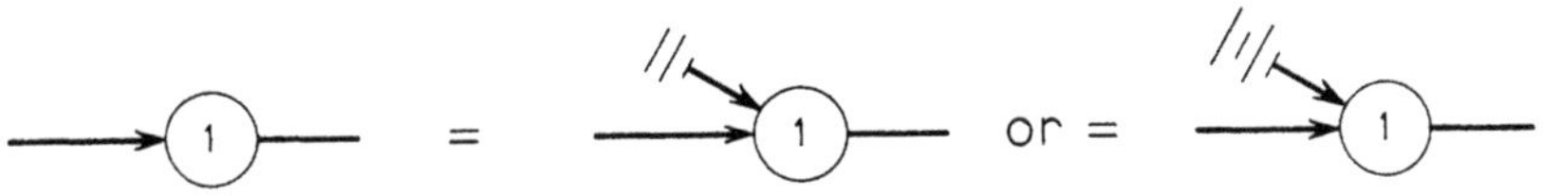

FIGURE 4

Now the proof can be effected by induction over the number $p = p(P)$ of symbols used in the definitory expression for P (counting each occurrence of each symbol separately).

If $p(P) = 1$, then $P(x_1, \ldots, x_n) \equiv x_\nu$ (for one of the $\nu = 1, \ldots, n$). The "trivial" network which obtains by breaking off all input lines other than ν, and taking the input line ν directly to the output, solves the problem with $s = 0$. Hence $s^*(P) = 0$.

If $p(P) > 1$, then $P \equiv Q^{-1}$ or $P \equiv QR$ or $P \equiv Q + R$, where $p(Q)$, $p(R) < p(P)$. For $P \equiv Q^{-1}$ let the box $\boxed{Q}$ represent a Q, s' -network, with $s' = s^*(Q)$. Then the network shown in Figure 5 is clearly a P,s -network, with $s = s' + 1$. Hence $s^*(P) \leq s^*(Q) + 1$. For $P \equiv QR$ or $Q + R$ let the boxes $\boxed{Q}$, $\boxed{R}$ represent a Q,s'' -network and an R,s'' -network, respectively, with $s'' = \mathrm{Max}(s^*(Q), s^*(R))$. Then the network shown in Figure 6 is clearly a P,s -network, with $P \equiv QR$ or $Q + R$ for $h = 2$ or 1, respectively, and with $s = s'' + 1$. Hence $s^*(P) \leq \mathrm{Max}(s^*(Q), s^*(R)) + 1$.

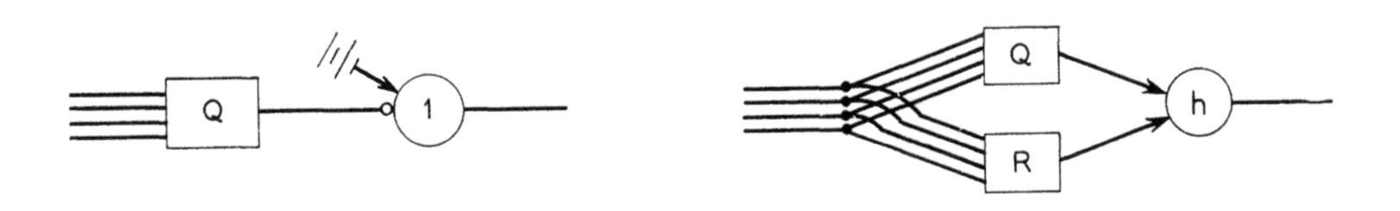

FIGURE 5

FIGURE 6

Combine the above theorem with the fact that every single output automaton can be equivalently described — apart from its time delay s — by a polynomial P, and that the basic operations ab, $a + b$, a^{-1} of the propositional calculus are represented (with unit delay) by the basic organs of Figure 2. (For the last one, which represents ab^{-1}, cf. the remark at the beginning of 4.1.1.) This gives:

DEFINITION 3: Two single output automata are equivalent in the wider sense, if they differ only in their time delays — but otherwise the same input stimuli produce the same output stimulus (or non-stimulus) in both.

THEOREM 2 (Reduction Theorem): Any single output automaton r is equivalent in the wider sense to a network of basic organs of Figure 2. There exists a (unique) $s^* = s^*(r)$, such that the latter network exists if and only if its prescribed time delay s satisfies $s \geq s^*$.

3.3 Universality. General Logical Considerations

Now networks of arbitrary single output automata can be replaced by networks of basic organs of Figure 2: It suffices to replace the unit delay in the former system by $\bar{s}$ unit delays in the latter, where $\bar{s}$ is the maximum of the $s^*(r)$ of all the single output automata that occur in the former system. Then all delays that will have to be matched will be multiples of $\bar{s}$, hence $\geqq \bar{s}$, hence $\geqq s^*(r)$ for all r that can occur in this situation, and so the Reduction Theorem will be applicable throughout.

Thus this system of basic organs is universal: It permits the construction of essentially equivalent networks to any network that can be constructed from any system of single output automata. I.e., no redefinition of the system of basic organs can extend the logical domain covered by the derived networks.

The general automaton is any network of single output automata in the above sense. It must be emphasized, that, in particular, feedbacks, i.e., arrangements of lines which may allow cyclical stimulation sequences, are allowed. (I.e., configurations like those shown in Figure 7. There will be various, non-trivial, examples of this later.) The above arguments have shown, that a limitation of the underlying single output automata to our original basic organs causes no essential loss of generality. The question, as to which logical operations can be equivalently represented (with suitable, but not a priori specified, delays) is nevertheless not without difficulties.

These general automata are, in particular, not immediately equivalent to all of effectively constructive (intuitionistic) logics. I.e., given a problem involving (a finite number of) variables, which can be solved (identically in these variables) by effective construction, it is not always possible to construct a general automaton that will produce this solution identically (i.e., under all conditions). The reason for this is essentially, that the memory requirements of such a problem may depend on (actual values assumed by) the variables (i.e., they must be finite for any specific system of values of the variables, but they may be unbounded for the totality of all possible systems of values), while a general automaton in the above sense necessarily has a fixed memory capacity. I.e., a fixed general automaton can only handle (identically, i.e., generally) a problem with fixed (bounded) memory requirements.

We need not go here into the details of this question. Very simple addenda can be introduced to provide for a (finite but) unlimited memory capacity. How this can be done has been shown by A. M. Turing [5]. Turing's analysis loc. cit. also shows, that with such addenda general automata become strictly equivalent to effectively constructive (intuitionistic) logics. Our system in its present form (i.e., general automata with limited memory capacity) is still adequate for the treatment of all problems with neurological analogies, as our subsequent examples will show. (Cf. also W. S. McCulloch and W. Pitts [2].) The exact logical domain that they cover has been recently characterized by Kleene [1]. We will return to some of these questions in 5.1.

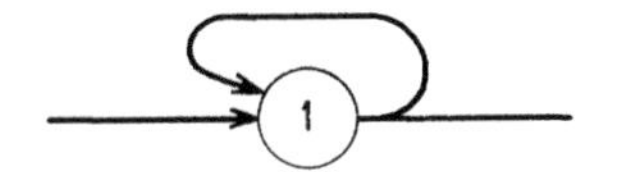

FIGURE 7

4. BASIC ORGANS

4.1 Reduction of the Basic Components

4.1.1 THE SIMPLEST REDUCTIONS. The previous section makes clear the way in which the elementary neurons should be interpreted logically. Thus the ones shown in Figure 2 respectively represent the logical functions ab, $a + b$, and ab^{-1}. In order to get b^{-1}, it suffices to make the a-terminal of the third organ, as shown in Figure 8, live. This will be abbreviated in the following, as shown in Figure 8.

FIGURE 8

Now since $ab \equiv ((a^{-1}) + (b^{-1}))^{-1}$ and $a + b \equiv ((a^{-1})(b^{-1}))^{-1}$, it is clear that the first organ among the three basic organs shown in Figure 2 is equivalent to a system built of the remaining two organs there, and that the same is true for the second organ there. Thus the first and second organs shown in Figure 2 are respectively equivalent (in the wider sense) to the two networks shown in Figure 9. This tempts one to consider

FIGURE 9

a new system, in which —o○— (viewed as a basic entity in its own right, and not an abbreviation for a composite, as in Figure 8), and either the first or the second basic organ in Figure 2, are the basic organs. They permit forming the second or the first basic organ in Figure 2, respectively, as shown above, as (composite) networks. The third basic organ in Figure 2 is easily seen to be also equivalent (in the wider sense) to a composite of the above, but, as was observed at the beginning of 4.1.1 the necessary organ is in any case not this, but —o○— (cf. also the remarks concerning Figure 8), respectively. Thus either system of new basic organs permits reconstructing (as composite networks) all the (basic) organs of the original system. It is true, that these constructs have delays varying from 1 to 3, but since unit delays, as shown in Figure 4, are available in either new system, all these delays can be brought up to the value 3. Then a trebling of the unit delay time obliterates all differences.

To restate: Instead of the three original basic organs, shown again in Figure 10, we can also (essentially equivalently) use the two basic organs Nos. one and three or Nos. two and three in Figure 10.

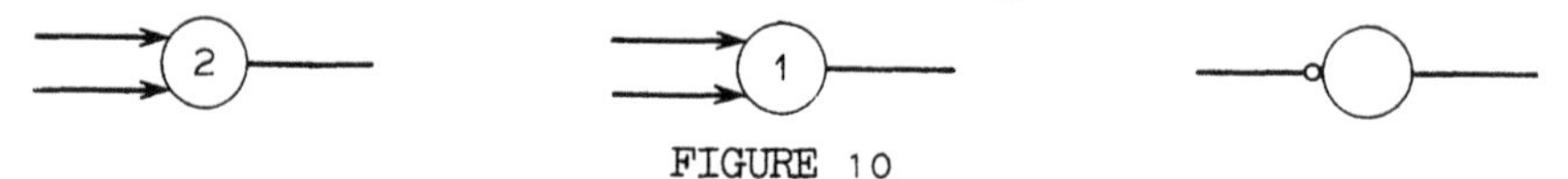

FIGURE 10

4.1.2 THE DOUBLE LINE TRICK. This result suggests strongly that one consider the one remaining combination, too: The two basic organs Nos. one and two in Figure 10, as the basis of an essentially equivalent system.

One would be inclined to infer that the answer must be negative: No network built out of the two first basic organs of Figure 10 can be equivalent (in the wider sense) to the last one. Indeed, let us attribute to T and F, i.e., to the stimulated or non-stimulated state of a line, respectively, the "truth values" $\underline{1}$ or $\underline{0}$, respectively. Keeping the ordering $\underline{0} < \underline{1}$ in mind, the state of the output is a monotone non-decreasing function of the states of the inputs for both basic organs Nos. one and two in Figure 10, and hence for all networks built from these organs exclusively as well. This, however, is not the case for the last organ of Figure 10 (nor for the last organ of Figure 2), irrespectively of delays.

Nevertheless a slight change of the underlying definitions permits one to circumvent this difficulty, and to get rid of the negation (the last organ of Figure 10) entirely. The device which effects this is of additional methodological interest, because it may be regarded as the prototype of one that we will use later on in a more complicated situation. The trick in question is to represent propositions on a double line instead of a single one. One assumes that of the two lines, at all times precisely one is stimulated. Thus there will always be two possible states of the line pair: The first line stimulated, the second non-stimulated; and the second line stimulated, the first non-stimulated. We let one of these states correspond to the stimulated single line of the original system —

that is, to a true proposition — and the other state to the unstimulated single line — that is, to a false proposition. Then the three fundamental Boolean operations can be represented by the three first schemes shown in Figure 11. (The last scheme shown in Figure 11 relates to the original system of Figure 2.)

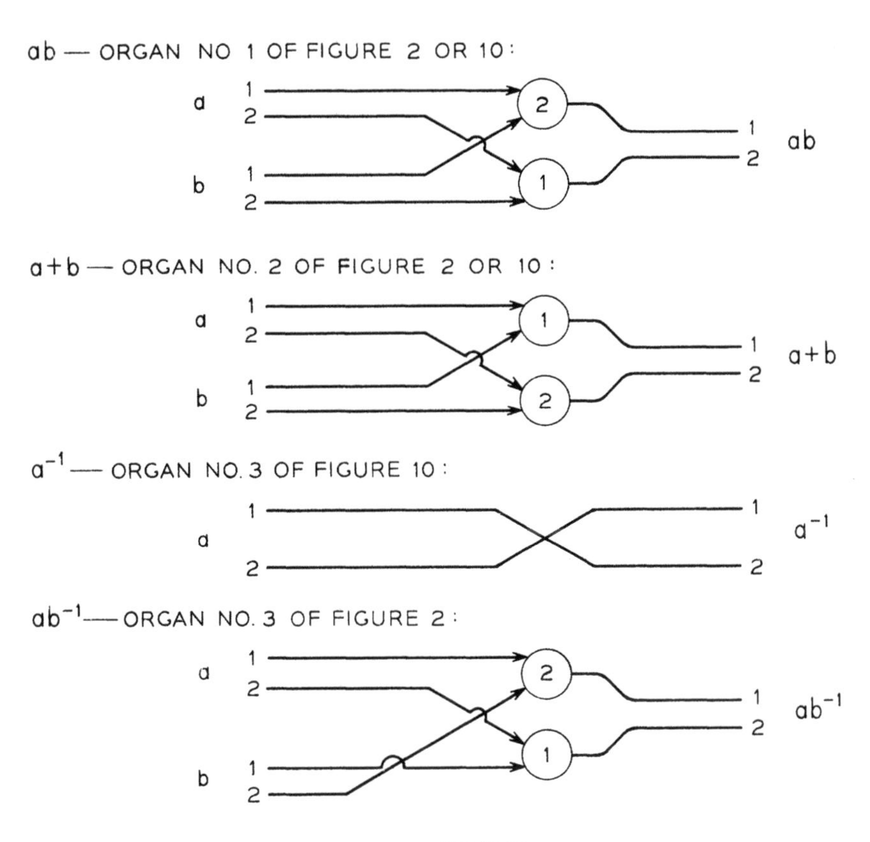

FIGURE 11

In these diagrams, a true proposition corresponds to 1 stimulated, 2 unstimulated, while a false proposition corresponds to 1 unstimulated, 2 stimulated. The networks of Figure 11, with the exception of the third one, have also the correct delays: Unit delay. The third one has zero delay, but whenever this is not wanted, it can be replaced by unit delay, by replacing the third network by the fourth one, making its a1 -line live, its a2 -line grounded, and then writing a for its b.

Summing up: Any two of the three (single delay) organs of Figure 10 — which may simply be designated ab, a + b, a^{-1} — can be stipulated to be the basic organs, and yield a system that is essentially equivalent to the original one.

4.2 Single Basic Organs

4.2.1 THE SHEFFER STROKE. It is even possible to reduce the number of basic organs to one, although it cannot be done with any of the three organs enumerated above. We will, however, introduce two new organs, either of which suffices by itself.

The first universal organ corresponds to the well-known "Sheffer stroke" function. Its use in this context was suggested by K. A. Brueckner and G. Gell-Mann. In symbols, it can be represented (and abbreviated) as shown on Figure 12. The three fundamental Boolean operations can now be performed as shown in Figure 13.

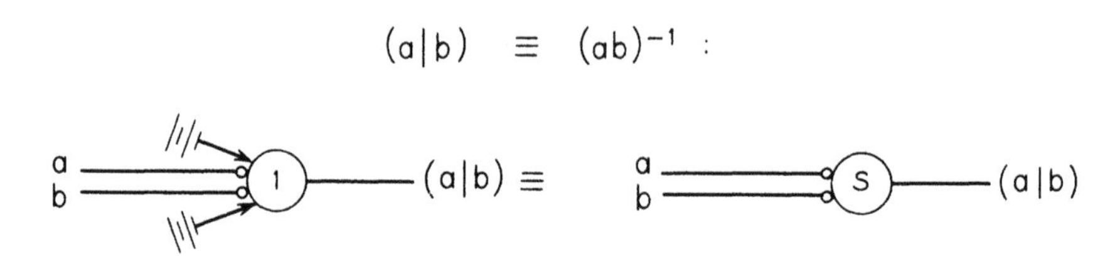

FIGURE 12

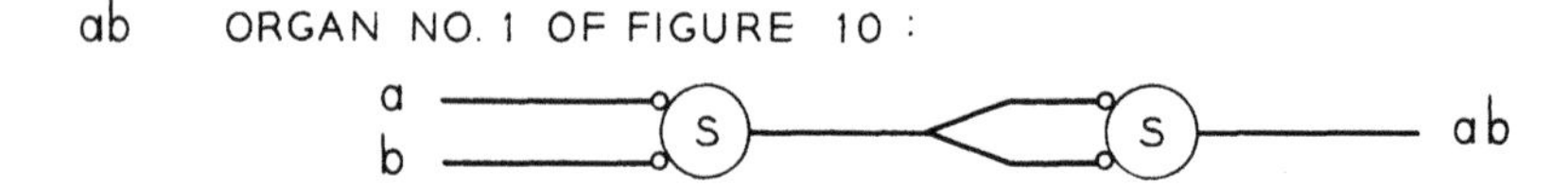

a+b ORGAN NO. 2 OF FIGURE 10 :

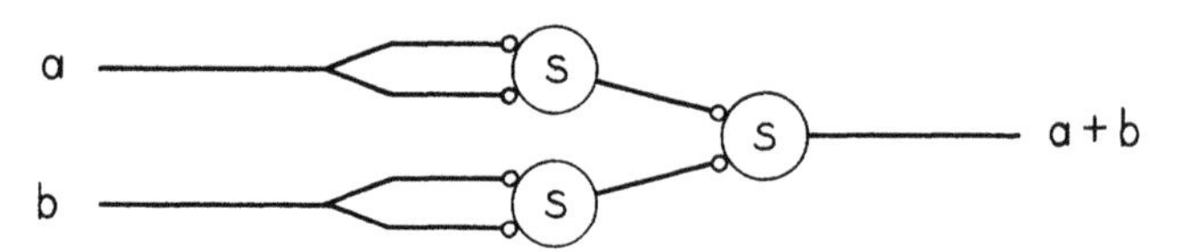

a^{-1} ORGAN NO. 3 OF FIGURE 10 :

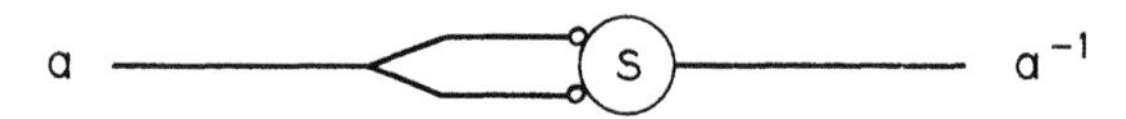

FIGURE 13

The delays are 2, 2, 1, respectively, and in this case the complication caused by these delay-relationships is essential. Indeed, the output of the Sheffer-stroke is an antimonotone function of its inputs. Hence in every network derived from it, even-delay outputs will be monotone functions of its inputs, and odd-delay outputs will be antimonotone ones. Now ab and a + b are not antimonotone, and ab^{-1} and a^{-1} are not monotone. Hence no delay-value can simultaneously accomodate in this set up one of the two first organs and one of the two last organs.

The difficulty can, however, be overcome as follows: ab and a + b are represented in Figure 13, both with the same delay, namely 2. Hence our earlier result (in 4.1.2), securing the adequacy of the system of the two basic organs ab and a + b applies: Doubling the unit delay time reduces the present set up (Sheffer stroke only !) to the one referred to above.

4.2.2 THE MAJORITY ORGAN. The second universal organ is the "majority organ". In symbols, it is shown (and alternatively designated) in Figure 14. To get conjunction and disjunction, is a simple matter, as shown in Figure 15. Both delays are 1. Thus ab and a + b (according to Figure 10) are correctly represented, and the new system (majority organ only !) is adequate because the system based on those two organs is known to be adequate (cf. 4.1.2).

$$m(a,b,c) \equiv ab + ac + bc \equiv (a+b)(a+c)(b+c) :$$

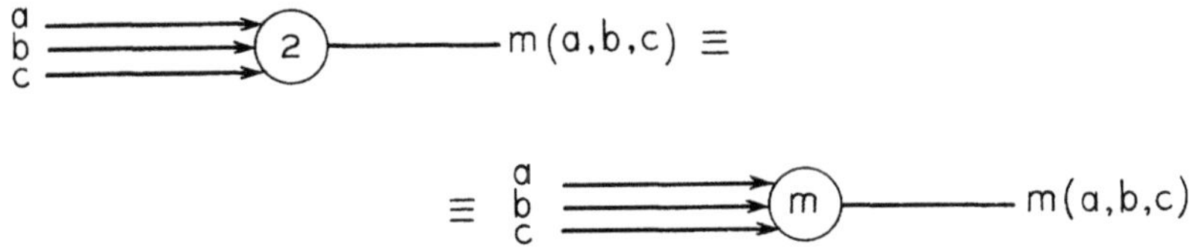

FIGURE 14

ab:

a+b:

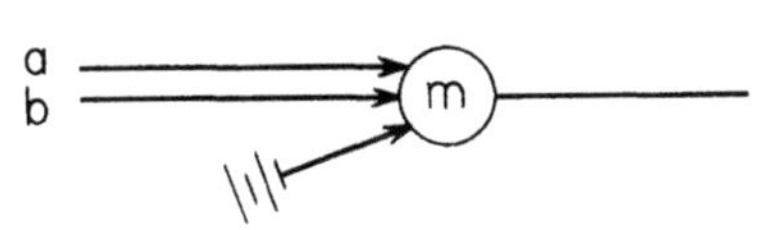

FIGURE 15

5. LOGICS AND INFORMATION

5.1 Intuitionistic Logics

All of the examples which have been described in the last two sections have had a certain property in common; in each, a stimulus of one of the inputs at the left could be traced through the machine until at a certain time later it came out as a stimulus of the output on the right. To be specific, no pulse could ever return to a neuron through which it had once passed. A system with this property is called circle-free by W. S. McCulloch and W. Pitts [2]. While the theory of circle-free machines is attractive because of its simplicity, it is not hard to see that these machines are very limited in their scope.

When the assumption of no circles in the network is dropped, the situation is radically altered. In this far more complicated case, the output of the machine at any time may depend on the state of the inputs in the indefinitely remote past. For example, the simplest kind of cyclic circuit, as shown in Figure 16, is a kind of memory machine. Once this organ has been stimulated by a, it remains stimulated and sends forth a pulse in b at all times thereafter. With more complicated networks, we can construct machines which will count, which will do simple arithmetic, and which will even perform certain unlimited inductive processes. Some of these will be illustrated by examples in 6. The use of cycles or feedback in automata

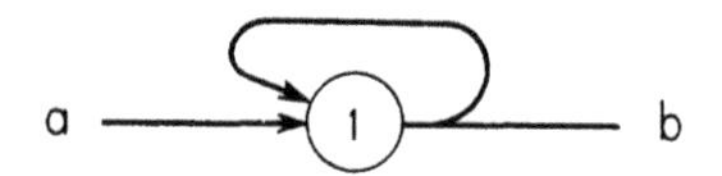

FIGURE 16

extends the logic of constructable machines to a large portion of intuitionistic logic. Not all of intuitionistic logic is so obtained, however, since these machines are limited by their fixed size. (For this, and for the remainder of this chapter cf. also the remarks at the end of 3.3.) Yet, if our automata are furnished with an unlimited memory — for example, an infinite tape, and scanners connected to afferent organs, along with suitable efferent organs to perform motor operations and/or print on the tape — the logic of constructable machines becomes precisely equivalent to intuitionistic logic (see A. M. Turing [5]). In particular, all numbers computable in the sense of Turing can be computed by some such network.

5.2 Information

5.2.1 GENERAL OBSERVATIONS. Our considerations deal with varying situations, each of which contains a certain amount of information. It is desirable to have a means of measuring that amount. In most cases of importance, this is possible. Suppose an event is one selected from a finite set of possible events. Then the number of possible events can be regarded as a measure of the information content of knowing which event occurred, provided all events are a priori equally probable. However, instead of using the number n of possible events as the measure of information, it is advantageous to use a certain function of n, namely the logarithm. This step can be (heuristically) justified as follows: If two physical systems I and II represent n and m (a priori equally probable) alternatives, respectively, then union I + II represents nm such alternatives. Now it is desirable that the (numerical) measure of information be (numerically) additive under this (substantively) additive composition I + II. Hence some function $f(n)$ should be used instead of n, such that

$$f(nm) = f(n) + f(m). \tag{3}$$

In addition, for $n > m$ I represents more information than II, hence it is reasonable to require

$$n > m \text{ implies } f(n) > f(m). \tag{4}$$

Note, that $f(n)$ is defined for $n = 1,2, \ldots$ only. From (3), (4) one concludes easily, that

$$f(n) \equiv C \ln n \tag{5}$$

for some constant $C > 0$. (Since $f(n)$ is defined for $n = 1,2, \ldots$ only, (3) alone does not imply this, even not with a constant $C \gtreqless 0$!) Next, it is conventional to let the minimum non-vanishing amount of information, i.e., that which corresponds to $n = 2$, be the unit of information — the "bit". This means that $f(2) = 1$, i.e., $C = 1/\ln 2$, and so

$$f(n) \equiv {}^2\log n. \tag{6}$$

This concept of information was successively developed by several authors in the late 1920's and early 1930's, and finally integrated into a broader system by C. E. Shannon [3].

5.2.2 EXAMPLES. The following simple examples give some illustration: The outcome of the flip of a coin is one bit. That of the roll of a die is ${}^2\log 6 = 2.5$ bits. A decimal digit represents ${}^2\log 10 = 3.3$ bits, a letter of the alphabet represents ${}^2\log 26 = 4.7$ bits, a single character from a 44-key, 2-setting typewriter represents ${}^2\log(44 \times 2) = 6.5$ bits. (In all these we assume, for the sake of the argument, although actually unrealistically, a priori equal probability of all possible choices.) It follows that any line or nerve fibre which can be classified as either stimulated or non-stimulated carries precisely one bit of information, while a bundle of n such lines can communicate n bits. It is important to observe that this definition is possible only on the assumption that a background of a priori knowledge exists, namely, the knowledge of a system of a priori equally probable events.

This definition can be generalized to the case where the possible events are not all equally probable. Suppose the events are known to have probabilities $p_1, p_2, \ldots, p_n$. Then the information contained in the knowledge of which of these events actually occurs, is defined to be

$$H = -\sum_{i=1}^{n} p_i \, {}^2\log p_i \text{ (bits)}. \tag{7}$$

In case $p_1 = p_2 = \cdots = p_n = 1/n$, this definition is the same as the previous one. This result, too, was obtained by C. E. Shannon [3], although it is implicit in the earlier work of L. Szilard [4].

An important observation about this definition is that it bears close resemblance to the statistical definition of the entropy of a thermodynamical system. If the possible events are just the known possible states of the system with their corresponding probabilities, then the two definitions are identical. Pursuing this, one can construct a mathematical theory of the communication of information patterned after statistical mechanics. (See L. Szilard [4] and C. E. Shannon [3].) That information theory should thus reveal itself as an essentially thermodynamical discipline, is not at all surprising: The closeness and the nature of the connection between information and entropy is inherent in L. Boltzman's classical definition of entropy (apart from a constant, dimensional factor) as the logarithm of the "configuration number." The "configuration number" is the number of a priori equally probable states that are compatible with the macroscopic description of the state — i.e., it corresponds to the amount of (miscroscopic) information that is missing in the (macroscopic) description.

6. TYPICAL SYNTHESES OF AUTOMATA

6.1 The Memory Unit

One of the best ways to become familiar with the ideas which have been introduced, is to study some concrete examples of simple networks. This section is devoted to a consideration of a few of them.

The first example will be constructed with the help of the three basic organs of Figure 10. It is shown in Figure 18. It is a slight refinement of the primitive memory network of Figure 16.

This network has two inputs a and b and one output x. At time t, x is stimulated if and only if a has been stimulated at an earlier time, and no stimulation of b has occurred since then. Roughly speaking, the machine remembers whether a or b was the last input to be stimulated. Thus x is stimulated, if it has been stimulated immediately before — to be designated by x' — or if a has been stimulated immediately before, but b has not been stimulated immediately before. This is expressed by the formula $x = (x' + a)b^{-1}$, i.e., by the network shown in Figure 17. Now x should be fed back into x' (since x' is the

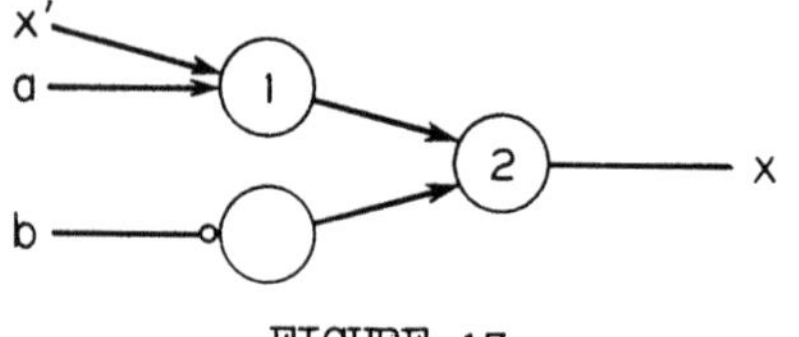

FIGURE 17

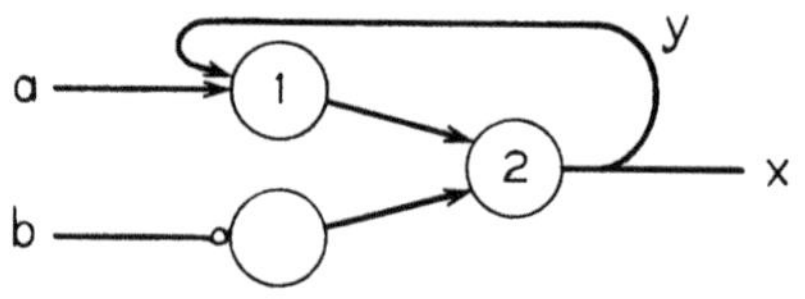

FIGURE 18

immediately preceding state of x). This gives the network shown in Figure 18, where this branch of x is designated by y. However, the delay of the first network is 2, hence the second network's memory extends over past events that lie an even number of time (delay) units back. I.e., the output x is stimulated if and only if a has been stimulated at an earlier time, an even number of units before, and no stimulation of b has occurred since then, also an even number of units before. Enumerating the time units by an integer t, it is thus seen, that this network represents a separate memory for even and for odd t. For each case it is a simple "off-on", i.e., one bit, memory. Thus it is in its entirety a two bit memory.

6.2 Scalers

In the examples that follow, free use will be made of the general family of basic organs considered in 2.3, at least for all $\varphi = x_h$ (cf. (2) there). The reduction thence to elementary organs in the original sense is secured by the Reduction Theorem in 3.2, and in the subsequently developed interpretations, according to section 4, by our considerations there. It is therefore unnecessary to concern ourselves here with these reductions.

The second example is a machine which counts input stimuli by two's. It will be called a "scaler by two". Its diagram is shown in Figure 19.

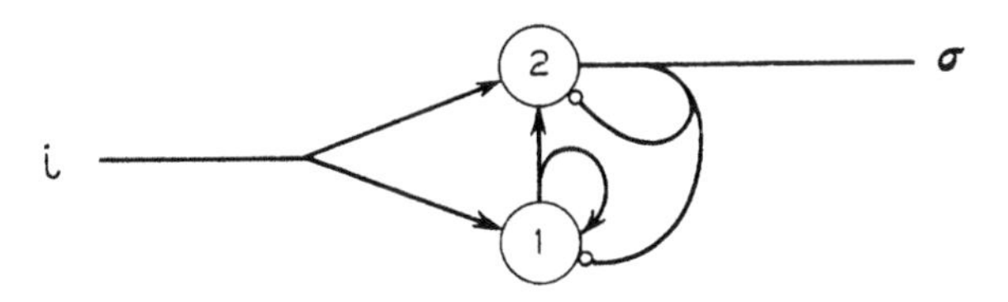

FIGURE 19

By adding another input, the repressor, the above mechanism can be turned off at will. The diagram becomes as shown in Figure 20. The result will be called a "scaler by two" with a repressor and denoted as indicated by Figure 20.

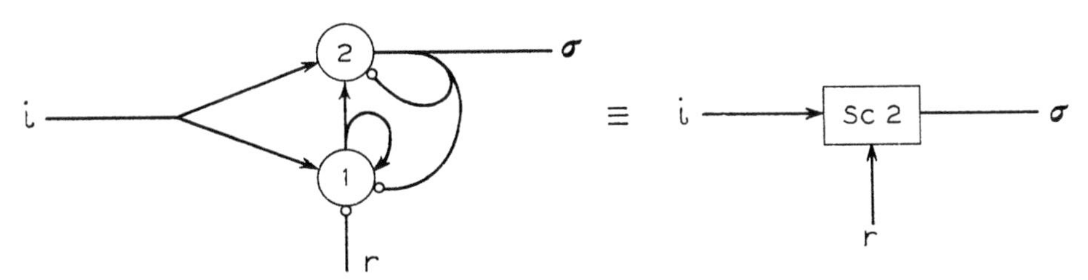

FIGURE 20

In order to obtain larger counts, the "scaler by two" networks can be hooked in series. Thus a "scaler by 2^n" is shown in Figure 21. The use of the repressor is of course optional here. "Scalers by m", where m is not necessarily of the form 2^n, can also be constructed with little difficulty, but we will not go into this here.

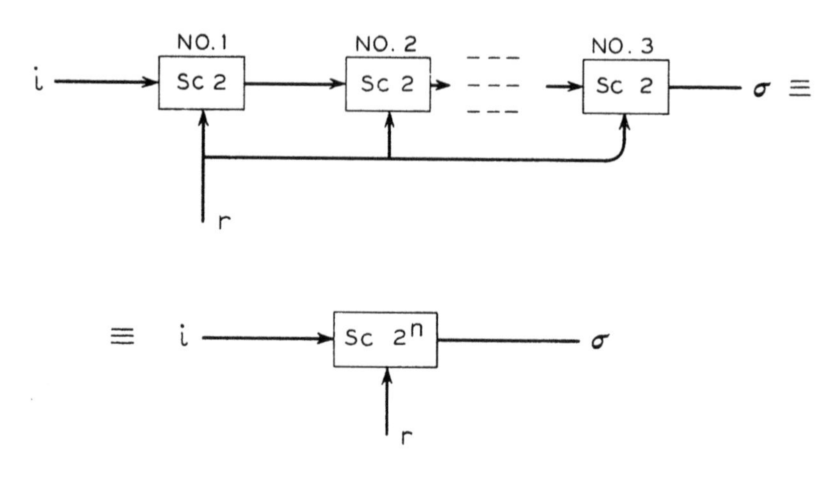

FIGURE 21

6.3 Learning

Using these "scalers by 2^n" (i.e., n-stage counters), it is possible to construct the following sort of "learning device". This network has two inputs a and b. It is designed to learn that whenever a is stimulated, then, in the next instant, b will be stimulated. If this occurs 256 times (not necessarily consecutively and possibly with many exceptions to the rule), the machine learns to anticipate a pulse from b one unit of time after a has been active, and expresses this by being stimulated at its b output after every stimulation of a. The diagram is shown in Figure 22. (The "expression" described above will be made effective in the desired sense by the network of Figure 24, cf. its discussion below).

This is clearly learning in the crudest and most inefficient way, only. With some effort, it is possible to refine the machine so that, first, it will learn only if it receives no counter-instances of the pattern "b follows a" during the time when it is collecting these 256 instances; and, second, having once learned, the machine can unlearn by the occurrence of 64 counter-examples to "b follows a" if no (positive) instances of this pattern interrupt the (negative) series. Otherwise, the behaviour is as before. The diagram is shown in Figure 23. To make this learning effective, one has to use x to gate a so as to replace b at its normal functions. Let these be represented by an output c. Then this process is mediated by the network shown in Figure 24. This network must then be attached to the lines a,b and to the output x of the preceding network (according to Figure 22,23).

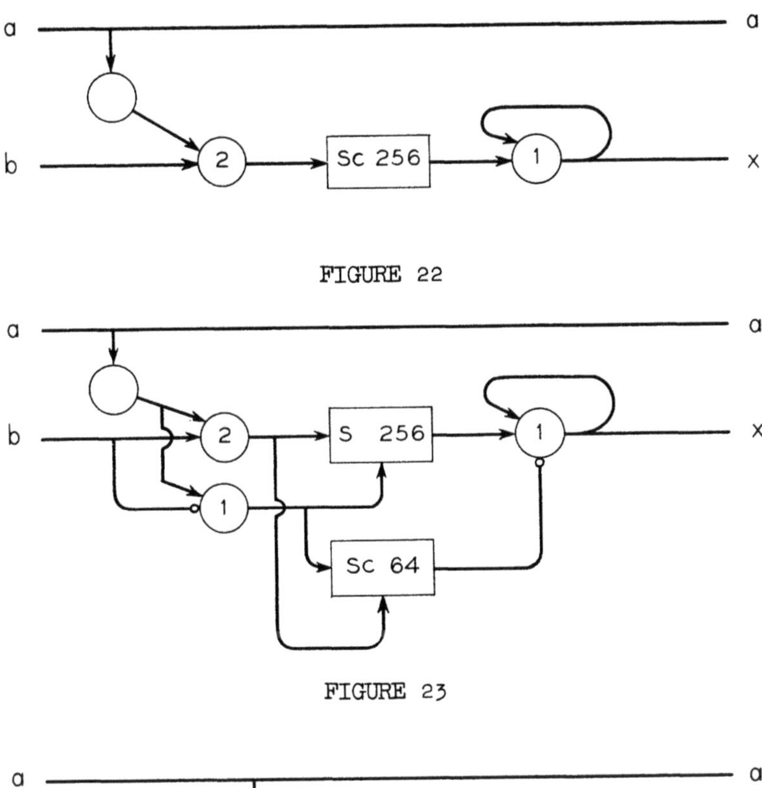

FIGURE 22

FIGURE 23

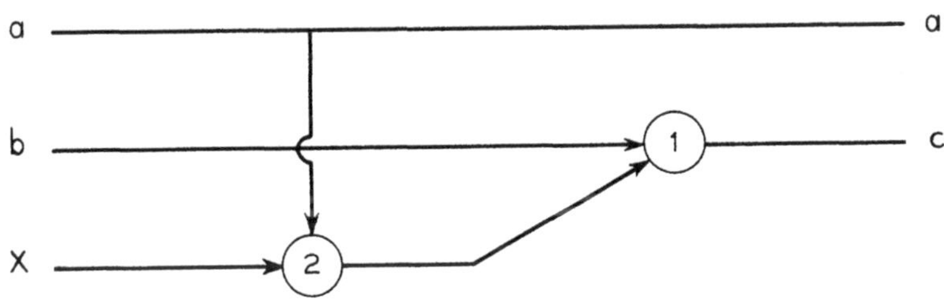

FIGURE 24

7. THE ROLE OF ERROR

7.1 Exemplification with the Help of the Memory Unit

In all the previous considerations, it has been assumed that the basic components were faultless in their performance. This assumption is clearly not a very realistic one. Mechanical devices as well as electrical ones are statistically subject to failure, and the same is probably true for animal neurons too. Hence it is desirable to find a closer approximation to reality as a basis for our constructions, and to study this revised situation. The simplest assumption concerning errors is this: With every basic organ is associated a positive number ϵ such that in any operation, the organ will fail to function correctly with the (precise) probability ϵ. This malfunctioning is assumed to occur statistically independently of the general state of the network and of the occurrence of other malfunctions. A more general assumption, which is a good deal more realistic, is this: The malfunctions are statistically dependent on the general state of the network and on each other. In any particular state, however, a malfunction of the basic organ in question has a probability of malfunctioning which is $\leqq \epsilon$. For the present occasion, we make the first (narrower and simpler) assumption, and that with a single ϵ: Every neuron has statistically independently of all else exactly the probability ϵ of misfiring. Evidently, it might as well be supposed $\epsilon \leqq 1/2$, since an organ which consistently misbehaves with a probability $> 1/2$, is just behaving with the negative of its attributed function, and a (complementary) probability of error $< 1/2$. Indeed, if the organ is thus redefined as its own opposite, its ϵ $(> 1/2)$ goes then over into $1 - \epsilon$ $(< 1/2)$. In practice it will be found necessary to have ϵ a rather small number, and one of the objectives of this investigation is to find the limits of this smallness, such that useful results can still be achieved.

It is important to emphasize, that the difficulty introduced by allowing error is not so much that incorrect information will be obtained, but rather that irrelevant results will be produced. As a simple example, consider the memory organ of Figure 16. Once stimulated, this network should continue to emit pulses at all later times; but suppose it has the probability ϵ of making an error. Suppose the organ receives a stimulation at time t and no later ones. Let the probability that the organ is still excited after s cycles be denoted ρ_s. Then the recursion formula

$$\rho_{s+1} = (1 - \epsilon)\, \rho_s + \epsilon\, (1 - \rho_s)$$

is clearly satisfied. This can be written

$$\rho_{s+1} - 1/2 = (1 - 2\epsilon)\, (\rho_s - 1/2)$$

and so

$$\text{(8)} \qquad \rho_s - 1/2 = (1 - 2\epsilon)^s\, (\rho_o - 1/2) \sim e^{-2\epsilon s}\, (\rho_o - 1/2)$$

for small ϵ. The quantity $\rho_s - 1/2$ can be taken as a rough measure of the amount of discrimination in the system after the s -th cycle. According to the above formula, $\rho_s \to 1/2$ as $s \to \infty$ — a fact which is expressed by saying that, after a long time, the memory content of the machine disappears, since it tends to equal likelihood of being right or wrong, i.e., to irrelevancy.

7.2 The General Definition

This example is typical of many. In a complicated network, with long stimulus-response chains, the probability of errors in the basic organs makes the response of the final outputs unreliable, i.e., irrelevant, unless some control mechanism prevents the accumulation of these basic errors. We will consider two aspects of this problem. Let the data be these: The function which the automaton is to perform is given; a basic organ is given (Sheffer stroke, for example); a number ϵ ($< 1/2$), which is the probability of malfunctioning of this basic organ, is prescribed. The first question is: Given $\delta > 0$, can a corresponding automaton be constructed from the given organs, which will perform the desired function and will commit an error (in the final result, i.e., output) with probability $\leqq \delta$? How small can δ be prescribed? The second question is: Are there other ways to interpret the problem which will allow us to improve the accuracy of the result?

7.3 An Apparent Limitation

In partial answer to the first question, we notice now that δ, the prescribed maximum allowable (final) error of the machine, must not be less than ϵ. For any output of the automaton is the immediate result of the operation of a single final neuron and the reliability of the whole system cannot be better than the reliability of this last neuron.

7.4 The Multiple Line Trick

In answer to the second question, a method will be analyzed by which this threshold restriction $\delta \geqq \epsilon$ can be removed. In fact we will be able to prescribe δ arbitrarily small (for suitable, but fixed, ϵ). The trick consists in carrying all the messages simultaneously on a bundle of N lines (N is a large integer) instead of just a single or double strand as in the automata described up to now. An automaton would then be represented by a black box with several bundles of inputs and outputs, as shown in Figure 25. Instead of requiring that all or none of the lines of

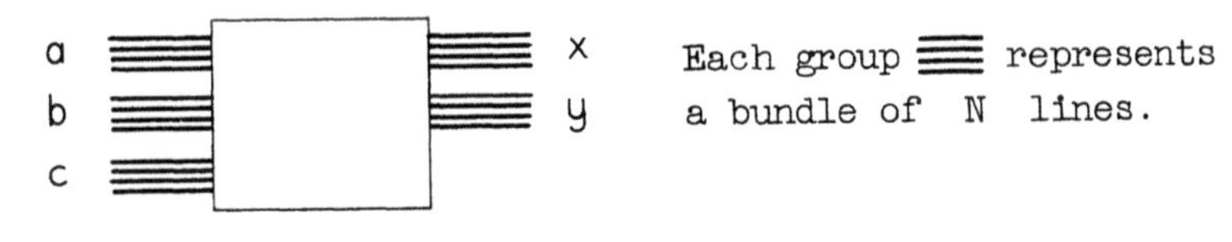

FIGURE 25

the bundle be stimulated, a certain critical (or fiduciary) level Δ is set: $0 < \Delta < 1/2$. The stimulation of $\geq (1 - \Delta)N$ lines of a bundle is interpreted as a positive state of the bundle. The stimulation of $\leq \Delta N$ lines is considered as a negative state. All levels of stimulation between these values are intermediate or undecided. It will be shown that by suitably constructing the automaton, the number of lines deviating from the "correctly functioning" majorities of their bundles can be kept at or below the critical level ΔN (with arbitrarily high probability). Such a system of construction is referred to as "multiplexing". Before turning to the multiplexed automata, however, it is well to consider the ways in which error can be controlled in our customary single line networks.

8. CONTROL OF ERROR IN SINGLE LINE AUTOMATA

8.1 The Simplified Probability Assumption

In 7.3 it was indicated that when dealing with an automaton in which messages are carried on a single (or even a double) line, and in which the components have a definite probability ϵ of making an error, there is a lower bound to the accuracy of the operation of the machine. It will now be shown that it is nevertheless possible to keep the accuracy within reasonable bounds by suitably designing the network. For the sake of simplicity only circle-free automata (cf. 5.1) will be considered in this section, although the conclusions could be extended, with proper safeguards, to all automata. Of the various essentially equivalent systems of basic organs (cf. section 4) it is, in the present instance, most convenient to select the majority organ, which is shown in Figure 14, as the basic organ for our networks. The number ϵ $(0 < \epsilon < 1/2)$ will denote the probability each majority organ has for malfunctioning.

8.2 The Majority Organ

We first investigate upper bounds for the probability of errors as impulses pass through a single majority organ of a network. Three lines constitute the inputs of the majority organ. They come from other organs or are external inputs of the network. Let η_1, η_2, η_3 be three numbers $(0 < \eta_i \leq 1)$, which are respectively upper bounds for the probabilities that these lines will be carrying the wrong impulses. Then $\epsilon + \eta_1 + \eta_2 + \eta_3$ is an upper bound for the probability that the output line of the majority organ will act improperly. This upper bound is valid in all cases. Under proper circumstances it can be improved. In particular, assume: (i) The probabilities of errors in the input lines are independent, (ii) under proper functioning of the network, these lines should always be in the same state of excitation (either all stimulated, or all unstimulated). In this latter case

$$\theta = \eta_1 \eta_2 + \eta_1 \eta_3 + \eta_2 \eta_3 - 2\eta_1\eta_2\eta_3$$

is an upper bound for at least two of the input lines carrying the wrong impulses, and thence

$$\epsilon' = (1 - \epsilon)\,\theta + \epsilon(1 - \theta) = \epsilon + (1 - 2\epsilon)\theta$$

is a smaller upper bound for the probability of failure in the output line. If all $\eta_i \leq \eta$, then $\epsilon + 3\eta$ is a general upper bound, and $\epsilon + (1 - 2\epsilon)\,(3\eta^2 - 2\eta^3) \leqq \epsilon + 3\eta^2$ is an upper bound for the special case. Thus it appears that in the general case each operation of the automaton increases the probability of error, since $\epsilon + 3\eta > \eta$, so that if the serial depth of the machine (or rather of the process to be performed) is very great, it will be impractical or impossible to obtain any kind of accuracy. In the special case, on the other hand, this is not necessarily so — $\epsilon + 3\eta^2 < \eta$ is possible. Hence, the chance of keeping the error under control lies in maintaining the conditions of the special case throughout the construction. We will now exhibit a method which achieves this.

8.3 Synthesis of Automata

8.3.1 THE HEURISTIC ARGUMENT. The basic idea in this procedure is very simple. Instead of running the incoming data into a single machine, the same information is simultaneously fed into a number of identical machines, and the result that comes out of a majority of these machines is assumed to be true. It must be shown that this technique can really be used to control error.

Denote by O the given network (assume two outputs in the specific instance picture in Figure 26). Construct O in triplicate, labeling the copies O^1, O^2, O^3 respectively. Consider the system shown in Figure 26.

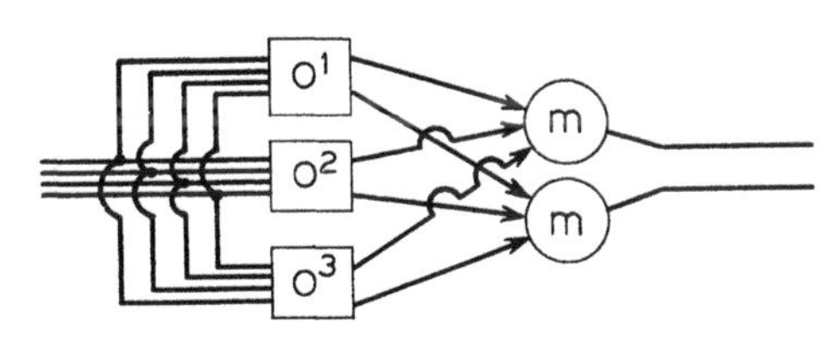

FIGURE 26

For each of the final majority organs the conditions of the special case considered above obtain. Consequently, if η is an upper bound for the probability of error at any output of the original network O, then

$$\eta^* = \epsilon + (1 - 2\epsilon)\,(3\eta^2 - 2\eta^3) \equiv f_\epsilon(\eta) \tag{9}$$

is an upper bound for the probability of error at any output of the new network O^*. The graph is the curve $\eta^* = f_\epsilon(\eta)$, shown in Figure 27.

Consider the intersections of the curve with the diagonal $\eta^* = \eta$: First, $\eta = 1/2$ is at any rate such an intersection. Dividing $\eta - f_\epsilon(\eta)$ by $\eta - 1/2$ gives $2((1 - 2\epsilon)\eta^2 - (1 - 2\epsilon)\eta + \epsilon)$, hence the other intersections are the roots of $(1 - 2\epsilon)\eta^2 - (1 - 2\epsilon)\eta + \epsilon = 0$, i.e.,

$$\eta = \frac{1}{2}\left(1 \pm \sqrt{\frac{1 - 6\epsilon}{1 - 2\epsilon}}\right)$$

I.e., for $\epsilon \geq 1/6$ they do not exist (being complex (for $\epsilon > 1/6$) or $= 1/2$ (for $\epsilon = 1/6$)); while for $\epsilon < 1/6$ they are $\eta = \eta_o$, $1 - \eta_o$, where

$$\eta_o = \frac{1}{2}\left(1 - \sqrt{\frac{1 - 6\epsilon}{1 - 2\epsilon}}\right) = \epsilon + 3\epsilon^2 + \ldots \tag{10}$$

For $\eta = 0$; $\eta^* = \epsilon > \eta$. This, and the monotony and continuity of $\eta^* = f_\epsilon(\eta)$ therefore imply:

First case, $\epsilon \geq 1/6$: $0 \leq \eta < 1/2$ implies $\eta < \eta^* < 1/2$; $\frac{1}{2} < \eta \leq 1$ implies $1/2 < \eta^* < \eta$.

Second case, $\epsilon < 1/6$: $0 \leq \eta < \eta_o$ implies $\eta < \eta^* < \eta_o$: $\eta_o < \eta < 1/2$ implies $\eta_o < \eta^* < \eta$; $1/2 < \eta < 1 - \eta_o$ implies $\eta < \eta^* < 1 - \eta_o$; $1 - \eta_o < \eta < 1$ implies $1 - \eta_o < \eta^* < \eta$.

Now we must expect numerous successive occurrences of the situation under consideration, if it is to be used as a basic procedure. Hence the iterative behavior of the operation $\eta \to \eta^* = f_\epsilon(\eta)$ is relevant. Now it is clear from the above, that in the first case the successive iterates of the process in question always converge to $1/2$, no matter what the original η; while in the second case these iterates converge to η_o if the original $\eta < 1/2$, and to $1 - \eta_o$ if the original $\eta > 1/2$.

In other words: In the first case no error level other than $\eta \sim 1/2$ can maintain itself in the long run. I.e., the process asymptotically degenerates to total irrelevance, like the one discussed in 7.1. In the second case the error-levels $\eta \sim \eta_o$ and $\eta \sim 1 - \eta_o$ will not only maintain themselves in the long run, but they represent the asymptotic behavior for any original $\eta < 1/2$ or $\eta > 1/2$, respectively.

These arguments, although heuristic, make it clear that the second case alone can be used for the desired error-level control. I.e., we must require $\epsilon < 1/6$, i.e., the error-level for a single basic organ function must be less than $\sim 16\%$. The stable, ultimate error-level should then be η_o (we postulate, of course, that the start be made with an error-level $\eta < 1/2$). η_o is small if ϵ is, hence ϵ must be small, and so

$$\eta_o = \epsilon + 3\epsilon^2 + \ldots \tag{11}$$

This would therefore give an ultimate error-level of $\sim 10\%$ (i.e., $\eta_o \sim .1$)

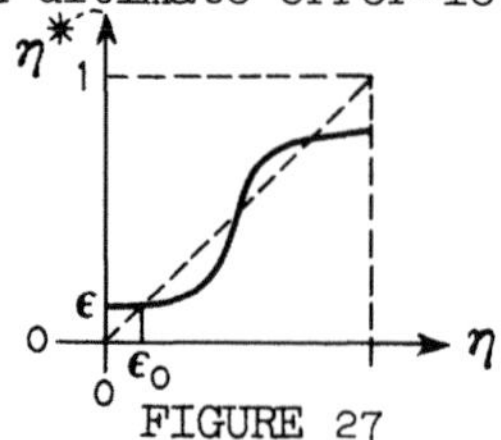

FIGURE 27

for a single basic organ function error-level of $\sim 8\%$ (i.e. $\epsilon \sim .08$).

8.3.2 THE RIGOROUS ARGUMENT. To make this heuristic argument binding, it would be necessary to construct an error controlling network P* for any given network P, so that all basic organs in P* are so connected as to put them into the special case for a majority organ, as discussed above. This will not be uniformly possible, and it will therefore be necessary to modify the above heuristic argument, although its general pattern will be maintained.

It is, then desired, to find for any given network P an essentially equivalent network P*, which is error-safe in some suitable sense, that conforms with the ideas expressed so far. We will define this as meaning, that for each output line of P* (corresponding to one of P) the (separate) probability of an incorrect message (over this line) is $\leq \eta_1$. The value of η_1 will result from the subsequent discussion.

The construction will be an induction over the longest serial chain of basic organs in P, say $\mu = \mu(P)$.

Consider the structure of P. The number of its inputs i and outputs σ is arbitrary, but every output of P must either come from a basic organ in P, or directly from an input, or from a ground or live source. Omit the first mentioned basic organs from P, as well as the outputs other than the first mentioned ones, and designate the network that is left over by Q. This is schematically shown in Figure 28. (Some of the apparently separate outputs of Q may be split lines coming from a single one, but this is irrelevant for what follows.)

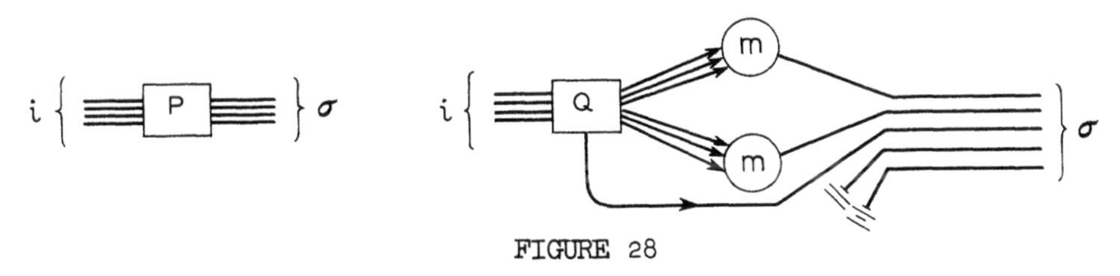

FIGURE 28

If Q is void, then there is nothing to prove; let therefore Q be non-void. Then clearly $\mu(Q) = \mu(P) - 1$.

Hence the induction permits us to assume the existence of a network Q* which is essentially equivalent to Q, and has for each output a (separate) error-probability $\leq \eta_1$.

We now provide three copies of Q*: Q^{*1}, Q^{*2}, Q^{*3}, and construct P* as shown in Figure 29. (Instead of drawing the, rather complicated, connections across the two dotted areas, they are indicated by attaching identical markings to endings that should be connected.)

Now the (separate) output error-probabilities of Q* are (by inductive assumption) $\leq \eta_1$. The majority organs in the first column in the above figure (those without a $\square$) are so connected as to belong into the special case for a majority organ (cf. 8.2), hence their outputs have (separate) error-probabilities $\leq f_\epsilon(\eta_1)$. The majority organs in the second column in the above figure (those with a $\square$) are in the general case, hence their (separate) error-probabilities are $\leq \epsilon + 3\, f_\epsilon(\eta_1)$.

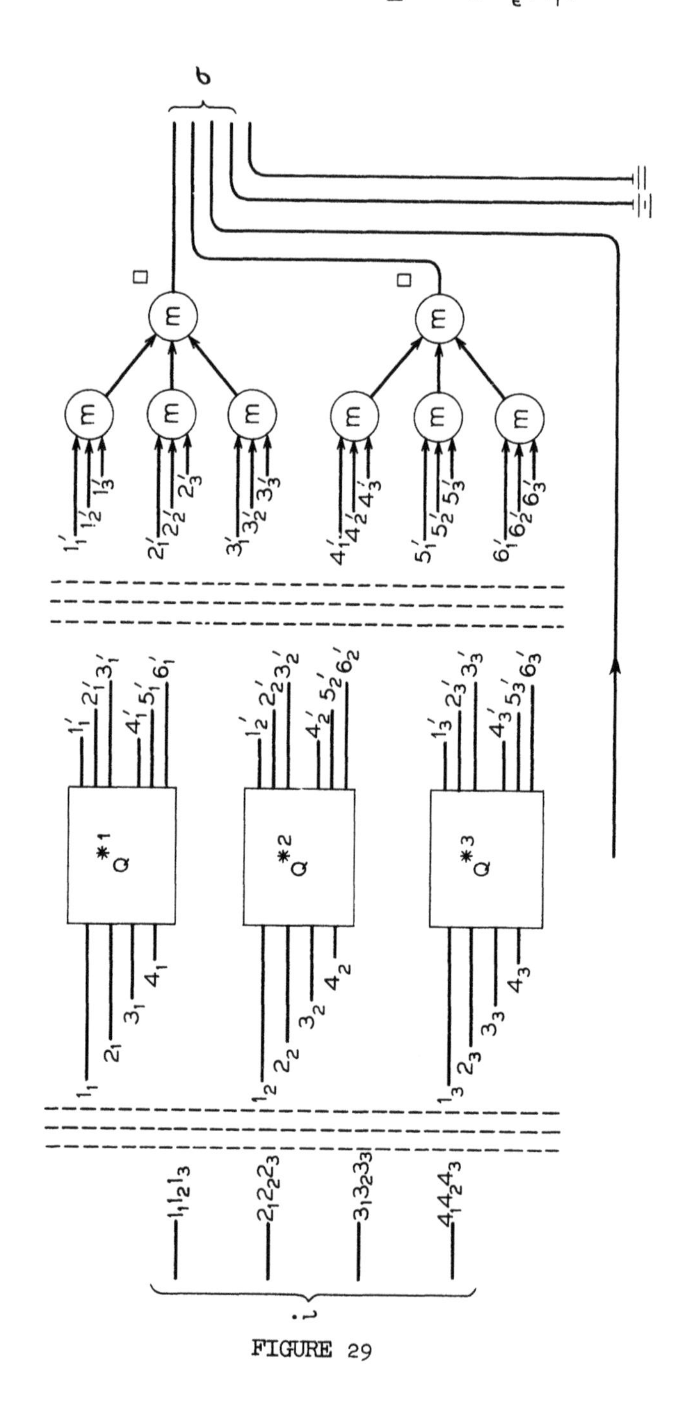

FIGURE 29

Consequently the inductive step succeeds, and therefore the attempted inductive proof is binding, if

$$\epsilon + 3f_\epsilon(\eta_1) \leq \eta_1 . \tag{12}$$

8.4 Numerical Evaluation

Substituting the expression (9) for $f_\epsilon(\eta)$ into condition (12) gives

$$4\epsilon + 3(1 - 2\epsilon)(3\eta_1^2 - 2\eta_1^3) \leq \eta_1 ,$$

i.e.,

$$\eta_1^3 - \frac{3}{2}\eta_1^2 + \frac{1}{6(1 - 2\epsilon)}\ \eta_1 - \frac{2\epsilon}{3(1 - 2\epsilon)} \geq 0 .$$

Clearly the smallest $\eta_1 > 0$ fulfilling this condition is wanted. Since the left hand side is < 0 for $\eta_1 \leq 0$, this means the smallest (real, and hence, by the above, positive) root of

$$\eta_1^3 - \frac{3}{2}\ \eta_1^2 + \frac{1}{6(1 - 2\epsilon)}\ \eta_1 - \frac{2\epsilon}{3(1 - 2\epsilon)} = 0 . \tag{13}$$

We know from the preceding heuristic argument, that $\epsilon \leq 1/6$ will be necessary — but actually even more must be required. Indeed, for $\eta_1 = 1/2$ the left hand side of (13) is $= -(1 + \epsilon)/(6 - 12\epsilon) < 0$, hence a significant and acceptable η_1 (i.e., an $\eta_1 < 1/2$), can be obtained from (13) only if it has three real roots. A simple calculation shows, that for $\epsilon = 1/6$ only one real root exists $\eta_1 = 1.42_5$. Hence the limiting ϵ calls for the existence of a double root. Further calculation shows, that the double root in question occurs for $\epsilon = .0073$, and that its value is $\eta_1 = .060$. Consequently $\epsilon < .0073$ is the actual requirement, i.e., the error-level of a single basic organ function must be $< .73\%$. The stable, ultimate error-level is then the smallest positive root η_1 of (13). η_1 is small if ϵ is, hence ϵ must be small, and so (from (13))

$$\eta_1 = 4\epsilon + 152\epsilon^2 + \ldots$$

It is easily seen, that e.g. an ultimate error level of 2% (i.e., $\eta_1 = .02$) calls for a single basic organ function error-level of .41% (i.e., $\epsilon = .0041$).

This result shows that errors can be controlled. But the method of construction used in the proof about threefolds the number of basic organs in P^* for an increase of $\mu(P)$ by 1, hence P^* has to contain about $3^{\mu(P)}$ such organs. Consequently the procedure is impractical.

The restriction $\epsilon < .0073$ has no absolute significance. It could be relaxed by iterating the process of triplication at each step. The inequality $\epsilon < 1/6$ is essential, however, since our first argument showed, that for $\epsilon \geq 1/6$ even for a basic organ in the most favorable situation (namely in the "special" one) no interval of improvement exists.

9. THE TECHNIQUE OF MULTIPLEXING

9.1 General Remarks on Multiplexing

The general process of multiplexing in order to control error was already referred to in 7.4. The messages are carried on N lines. A positive number $\Delta (< 1/2)$ is chosen and the stimulation of $\geq (1 - \Delta)N$ lines of the bundle is interpreted as a positive message, the stimulation of $\leq \Delta N$ lines as a negative message. Any other number of stimulated lines is interpreted as malfunction. The complete system must be organized in such a manner, that a malfunction of the whole automaton cannot be caused by the malfunctioning of a single component, or of a small number of components, but only by the malfunctioning of a large number of them. As we will see later, the probability of such occurrences can be made arbitrarily small provided the number of lines in each bundle is made sufficiently great. All of section 9 will be devoted to a description of the method of constructing multiplexed automata and its discussion, without considering the possibility of error in the basic components. In section 10 we will then introduce errors in the basic components, and estimate their effects.

9.2 The Majority Organ

9.2.1 THE BASIC EXECUTIVE ORGAN. The first thing to consider is the method of constructing networks which will perform the tasks of the basic organs for bundles of inputs and outputs instead of single lines.

A simple example will make the process clear. Consider the problem of constructing the analog of the majority organ which will accomodate bundles of five lines. This is easily done using the ordinary majority organ of Figure 12, as shown in Figure 30. (The connections are replaced by suitable markings, in the same way as in Figure 29.)

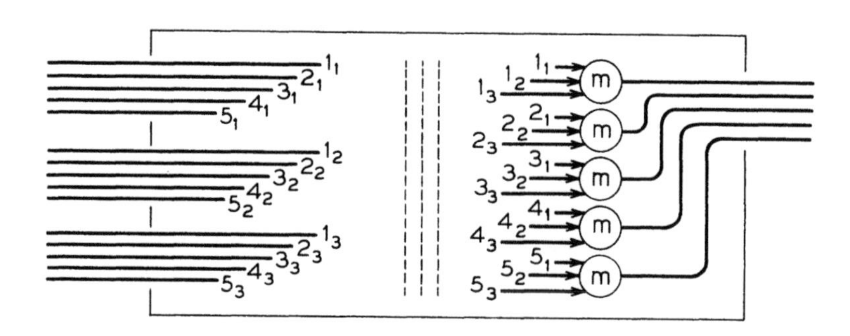

FIGURE 30

9.2.2 THE NEED FOR A RESTORING ORGAN. It is intuitively clear that if almost all lines of two of the input bundles are stimulated, then almost all lines of the output bundle will be stimulated. Similarly if almost none of the lines of two of the input bundles are stimulated, then the mechanism will stimulate almost none of its output lines. However, another fact is brought to light. Suppose that a critical level $\Delta = 1/5$ is set for the bundles. Then if two of the input bundles have 4 lines stimulated while the other has none, the output may have only 3 lines stimulated. The same effect prevails in the negative case. If two bundles have just one input each stimulated, while the third bundle has all of its inputs stimulated, then the resulting output may be the stimulation of two lines. In other words, the relative number of lines in the bundle, which are not in the majority state, can double in passing through the generalized majority system. A more careful analysis (similar to the one that will be gone into in more detail for the case of the Sheffer organ in 10) shows the following: If, in some situation, the operation of the organ should be governed by a two-to-one majority of the input bundles (i.e., if two of these bundles are both prevalently stimulated or both prevalently non-stimulated, while the third one is in the opposite condition), then the most probable level of the output error will be (approximately) the sum of the errors in the two governing input bundles; on the other hand, in an operation in which the organ is governed by a unanimous behavior of its input bundles (i.e., if all three of these bundles are prevalently stimulated or all three are prevalently non-stimulated), then the output error will generally be smaller than the (maximum of the) input errors. Thus in the significant case of two-to-one majorization, two significant inputs may combine to produce a result lying in the intermediate region of uncertain information. What is needed therefore, is a new type of organ which will restore the original stimulation level. In other words, we need a network having the property that, with a fairly high degree of probability, it transforms an input bundle with a stimulation level which is near to zero or to one into an output bundle with stimulation level which is even closer to the corresponding extreme.

Thus the multiplexed systems must contain two types of organs. The first type is the executive organ which performs the desired basic operations on the bundles. The second type is an organ which restores the stimulation level of the bundles, and hence erases the degradation caused by the executive organs. This situation has its analog in many of the real automata which perform logically complicated tasks. For example in electrical circuits, some of the vacuum tubes perform executive functions, such as detection or rectification or gateing or coincidence-sensing, while the remainder are assigned the task of amplification, which is a restorative operation.

9.2.3 THE RESTORING ORGAN

9.2.3.1 CONSTRUCTION. The construction of a restoring organ is quite simple in principle, and in fact contained in the second remark made in 9.2.2. In a crude way, the ordinary majority organ already performs this task. Indeed in the simplest case, for a bundle of three lines, the majority organ has precisely the right characteristics: It suppresses a single incoming impulse as well as a single incoming non-impulse, i.e., it amplifies the prevalence of the presence as well as of the absence of impulses. To display this trait most clearly, it suffices to split its output line into three lines, as shown in Figure 31.

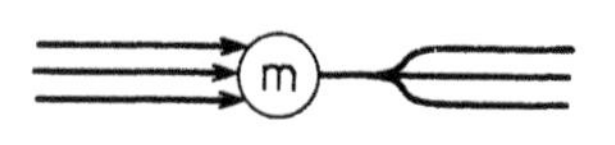

FIGURE 31

Now for large bundles, in the sense of the remark referred to above, concerning the reduction of errors in the case of a response induced by a unanimous behavior of the input bundles, it is possible to connect up majority organs in parallel and thereby produce the desired restoration. However, it is necessary to assume that the stimulated (or non-stimulated) lines are distributed at random in the bundle. This randomness must then be maintained at all times. The principle is illustrated by Figure 32. The "black box" U is supposed to permute the lines of the input bundle that pass through it, so as to restore the randomness of the pulses in its lines. This is necessary, since to the left of U the input bundle consists of a set of triads, where the lines of each triad originate in the splitting of a single line, and hence are always all three in the same condition. Yet, to the right of U the lines of the corresponding triad must be statistically independent, in order to permit the application of the statistical formula to be given below for the functioning of the majority organ into which they feed. The way to select such a "randomizing" permutation will not be considered here – it is intuitively plausible that most "complicated" permutations will be suited for this "randomizing" role. (Cf. 11.2.)

9.2.3.2 NUMERICAL EVALUATION. If αN of the N incoming lines are stimulated, then the probability of any majority organ being stimulated (by two or three stimulated inputs) is

$$\alpha^* = 3\alpha^2 - 2\alpha^3 = g(\alpha). \tag{14}$$

Thus approximately (i.e., with high probability, provided N is large) $\alpha^* N$ outputs will be excited. Plotting the curve of α^* against α, as shown in Figure 33, indicates clearly that this organ will have the desired characteristics:

This curve intersects the diagonal $\alpha^* = \alpha$ three times: For $\alpha = 0, 1/2, 1$. $0 < \alpha < 1/2$ implies $0 < \alpha^* < \alpha$; $1/2 < \alpha < 1$ implies $\alpha < \alpha^* < 1$. I.e., successive iterates of this process converge to 0 if the original $\alpha < 1/2$ and to 1 if the original $\alpha > 1/2$.

In other words: The error levels $\alpha \sim 0$ and $\alpha \sim 1$ will not only maintain themselves in the long run, but they represent the asymptotic behavior for any original $\alpha < 1/2$ or $\alpha > 1/2$, respectively. Note, that because of $g(1 - \alpha) \equiv 1 - g(\alpha)$ there is complete symmetry between the $\alpha < 1/2$ region and the $\alpha > 1/2$ region.

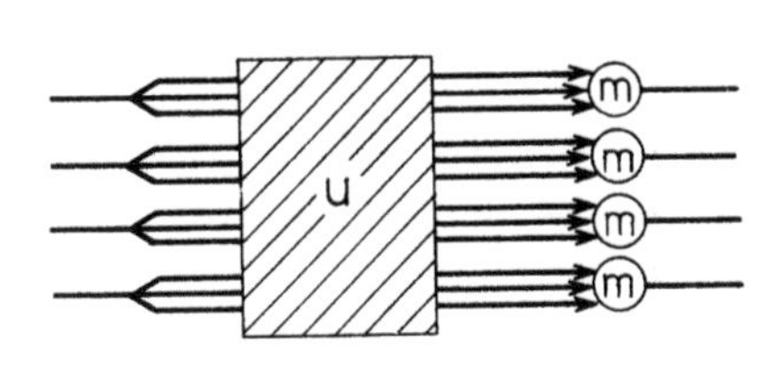

FIGURE 32

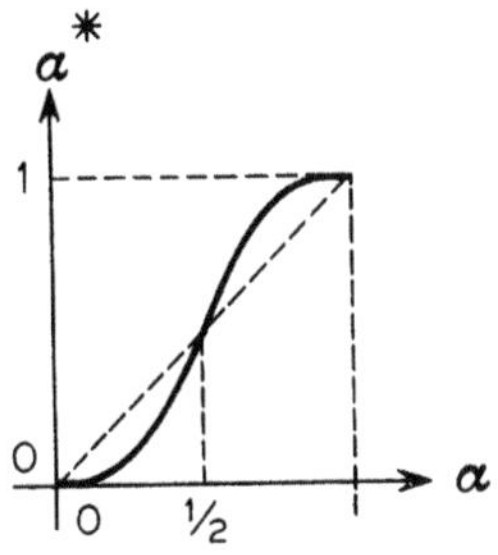

FIGURE 33

The process $\alpha \to \alpha^*$ thus brings every α nearer to that one of 0 and 1, to which it was nearer originally. This is precisely that process of restoration, which was seen in 9.2.2 to be necessary. I.e., one or more (successive) applications of this process will have the required restoring effect.

Note, that this process of restoration is most effective when $\alpha - \alpha^* = 2\alpha^3 - 3\alpha^2 + \alpha$ has its minimum or maximum, i.e., for $6\alpha^2 - 6\alpha + 1 = 0$, i.e., for $\alpha = (3 \pm \sqrt{3})/6 = .788, .212$. Then $\alpha - \alpha^* = \mp .096$. I.e., the maximum restoration is effected on error levels at the distance of 21.2% from 0% or 100% – these are improved (brought nearer) by 9.6%.

9.3 Other Basic Organs

We have so far assumed that the basic components of the construction are majority organs. From these, an analog of the majority organ – one which picked out a majority of bundles instead of a majority of single lines – was constructed. Since this, when viewed as a basic organ, is a universal organ, these considerations show that it is at least theoretically possible to construct any network with bundles instead of single lines. However there was no necessity for starting from majority organs. Indeed, any other basic system whose universality was established in section 4 can be used instead. The simplest procedure in such a case is to construct an (essential) equivalent of the (single line) majority organ from the given basic system (cf. 4.2.2), and then proceed with this composite majority organ in the same way, as was done above with the basic majority organ.

Thus, if the basic organs are those Nos. one and two in Figure 10 (cf. the relevant discussion in 4.1.2), then the basic synthesis (that of the majority organ, cf. above) is immediately derivable from the introductory formula of Figure 14.

9.4 The Sheffer Stroke

9.4.1 THE EXECUTIVE ORGAN. Similarly, it is possible to construct the entire mechanism starting from the Sheffer organ of Figure 12. In this case, however, it is simpler not to effect the passage to an (essential) equivalent of the majority organ (as suggested above), but to start de novo. Actually, the same procedure, which was seen above to work for the majority organ, works mutatis mutandis for the Sheffer organ, too. A brief description of the direct procedure in this case is given in what follows:

Again, one begins by constructing a network which will perform the task of the Sheffer organ for bundles of inputs and outputs instead of single lines. This is shown in Figure 34 for bundles of five wires. (The connections are replaced by suitable markings, as in Figures 29 and 30.)

It is intuitively clear that if almost all lines of both input bundles are stimulated, then almost none of the lines of the output bundle will be stimulated. Similarly, if almost none of the lines of one input bundle are stimulated, then almost all lines of the output bundle will be stimulated. In addition to this overall behavior, the following detailed behavior is found (cf. the detailed consideration in 10.4). If the condition of the organ is one of prevalent non-stimulation of the output bundle, and hence is governed by (prevalent stimulation of) both input bundles, then the most probable level of the output error will be (approximately) the sum of the errors in the two governing input bundles; if on the other hand the condition of the organ is one of prevalent stimulation of the output bundle, and hence is governed by (prevalent non-stimulation of) one or of both input bundles, then the output error will be on (approximately) the same level as the input error, if (only) one input bundle is governing (i.e., prevalently non-stimulated), and it will be generally smaller than the input error, if both input bundles are governing (i.e., prevalently non-stimulated). Thus two significant inputs may produce a result lying in the intermediate zone of uncertain information. Hence a restoring organ (for the error level) is again needed, in addition to the executive organ.

9.4.2 THE RESTORING ORGAN. Again, the above indicates that the restoring organ can be obtained from a special case functioning of the standard executive organ, namely by obtaining all inputs from a single input bundle, and seeing to it that the output bundle has the same size as the original input bundle. The principle is illustrated by Figure 35. The "black box" U is again supposed to effect a suitable permutation of the lines that pass through it, for the same reasons and in the same manner as in the corresponding situation for the majority organ (cf. Figure 32). I.e., it must have a "randomizing" effect.

If αN of the N incoming lines are stimulated, then the probability of any Sheffer organ being stimulated (by at least one non-stimulated input) is

(15) $$\alpha^+ = 1 - \alpha^2 \equiv h(\alpha).$$

Thus approximately (i.e., with high probability provided N is large) $\sim \alpha^+ N$ outputs will be excited. Plotting the curve of α^+ against α discloses some characteristic differences against the previous case (that one of the majority organs, i.e., $\alpha^* = 3\alpha^2 - 2\alpha^3 \equiv g(\alpha)$, cf. 9.2.3), which require further discussion. This curve is shown in Figure 36. Clearly α^+ is an antimonotone function of α, i.e., instead of restoring an excitation level (i.e., bringing it closer to 0 or to 1, respectively), it transforms it into its opposite (i.e., it brings the neighborhood of 0 close to 1, and the neighborhood of 1 close to 0). In addition it produces for α near to 1 an α^+ less near to 0 (about twice farther), but for α near to 0 an α^+ much nearer to 1 (second order !). All these circumstances suggest, that the operation should be iterated.

Let the restoring organ therefore consist of two of the previously pictured organs in series, as shown in Figure 37. (The "black boxes" U_1, U_2 play the same role as their analog U plays in Figure 35.) This organ transforms an input excitation level αN into an output excitation level of approximately (cf. above) $\sim \alpha^{++}$ where

$$\alpha^{++} = 1 - (1 - \alpha^2)^2 \equiv h(h(\alpha)) \equiv k(\alpha),$$

i.e.,

(16) $$\alpha^{++} = 2\alpha^2 - \alpha^4 \equiv k(\alpha).$$

This curve of α^{++} against α is shown in Figure 38. This curve is very similar to that one obtained for the majority organ (i.e., $\alpha^* = 3\alpha^2 - 2\alpha^3 \equiv g(\alpha)$, cf. 9.2.3). Indeed: The curve intersects the diagonal $\alpha^{++} = \alpha$ in the interval $0 \leq \alpha \leq 1$ three times: For $\alpha = 0, \alpha_0, 1$, where $\alpha_0 = (-1 + \sqrt{5})/2 = .618$. (There is a fourth intersection $\alpha = -1 - \alpha_0 = -1.618$, but this is irrelevant, since it is not in the interval $0 \leq \alpha \leq 1$.) $0 < \alpha < \alpha_0$ implies $0 < \alpha^{++} < \alpha$; $\alpha_0 < \alpha < 1$ implies $\alpha < \alpha^{++} < 1$.

In other words: The role of the error levels $\alpha \sim 0$ and $\alpha \sim 1$ is precisely the same as for the majority organ (cf. 9.2.3), except that the limit between their respective areas of control lies at $\alpha = \alpha_0$ instead of at $\alpha = 1/2$. I.e., the process $\alpha \rightarrow \alpha^{++}$ brings every α nearer to either 0 or to 1, but the preference to 0 or to 1 is settled at a discrimination level of 61.8% (i.e., α_0) instead of one of 50% (i.e., $1/2$). Thus, apart from a certain asymmetric distortion, the organ behaves like its counterpart considered for the majority organ — i.e., it is an effective restoring mechanism.

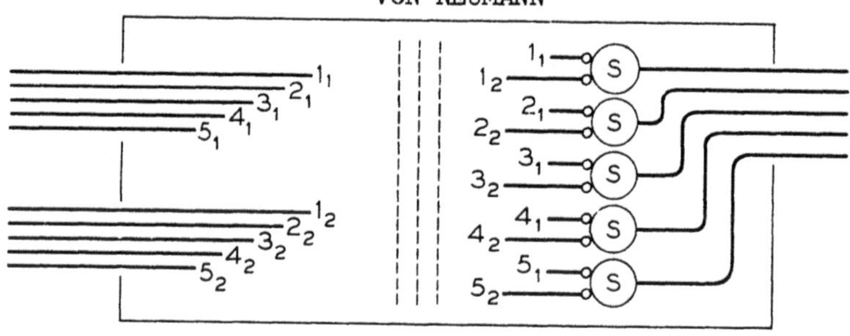

FIGURE 34

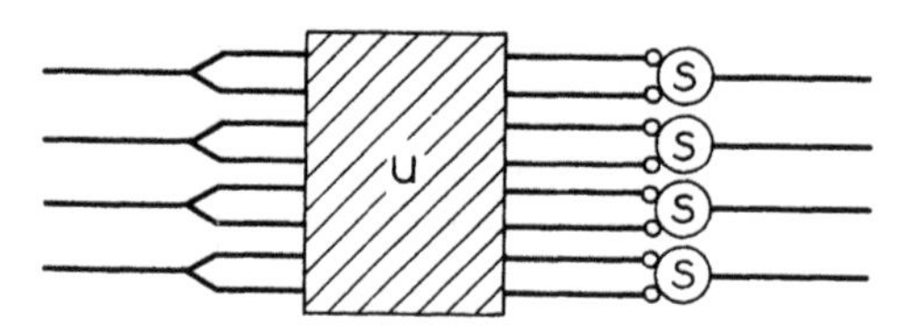

FIGURE 35

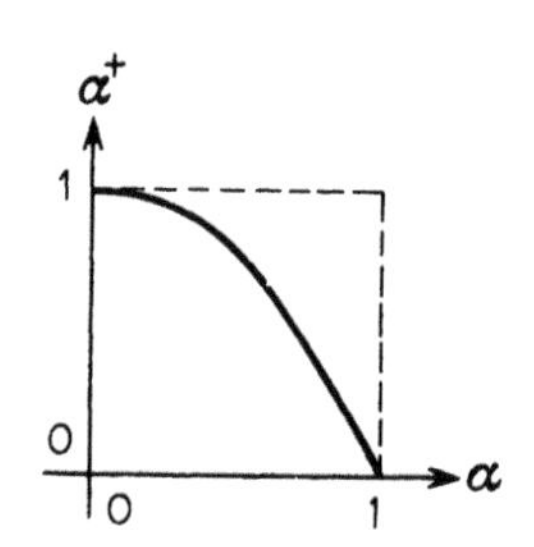

FIGURE 36

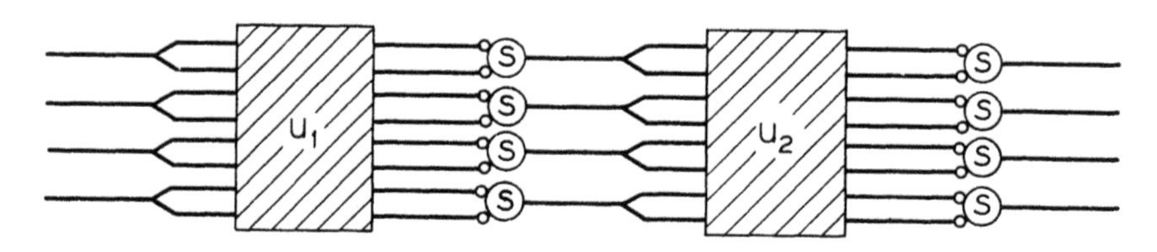

FIGURE 37

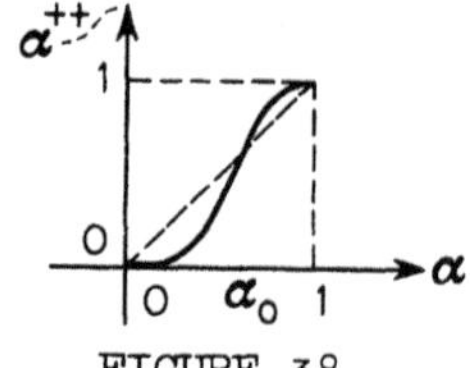

FIGURE 38

10. ERROR IN MULTIPLEX SYSTEMS

10.1 General Remarks

In section 9 the technique for constructing multiplexed automata was described. However, the role of errors entered at best intuitively and summarily, and therefore it has still not been proved that these systems will do what is claimed for them — namely control error. Section 10 is devoted to a sketch of the statistical analysis necessary to show that, by using large enough bundles of lines, any desired degree of accuracy (i.e., as small a probability of malfunction of the ultimate output of the network as desired) can be obtained with a multiplexed automaton.

For simplicity, we will only consider automata which are constructed from the Sheffer organs. These are easier to analyze since they involve only two inputs. At the same time, the Sheffer organ is (by itself) universal (cf. 4.2.1), hence every automaton is essentially equivalent to a network of Sheffer organs.

Errors in the operation of an automaton arise from two sources. First, the individual basic organs can make mistakes. It will be assumed as before, that, under any circumstance, the probability of this happening is just ϵ. Any operation on the bundle can be considered as a random sampling of size N (N being the size of the bundle). The number of errors committed by the individual basic organs in any operation on the bundle is then a random variable, distributed approximately normally with mean ϵN and standard deviation $\sqrt{\epsilon(1 - \epsilon)N}$. A second source of failures arises because in operating with bundles which are not all in the same state of stimulation or non-stimulation, the possibility of multiplying error by unfortunate combinations of lines into the basic (single line) organs is always present. This interacts with the statistical effects, and in particular with the processes of degeneration and of restoration of which we spoke in 9.2.2, 9.2.3 and 9.4.2.

10.2 The Distribution of the Response Set Size

10.2.1 EXACT THEORY. In order to give a statistical treatment of the problem, consider the Figure 34, showing a network of Sheffer organs, which was discussed in 9.4.1. Let again N be the number of lines in each (input or output) bundle. Let X be the set of those $i = 1, \ldots, N$ for which line No. i in the first input bundle is stimulated at time t; let Y be the corresponding set for the second input bundle and time t; and let Z be the corresponding set for the output bundle, assuming the correct functioning of all the Sheffer organs involved, and time $t + 1$. Let X, Y have ξN, ηN elements, respectively, but otherwise be random — i.e., equidistributed over all pairs of sets with these numbers of elements. What can then be said about the number of elements ζN of Z? Clearly ξ, η, ζ, are the relative levels of excitation of the two input bundles and of the output bundle, respectively, of the network under consideration. The question is then: What is the distribution of the (stochastic) variable ζ in terms of the (given) ξ, η ?

Let W be the complementary set of Z. Let p, q, r be the numbers of elements of X, Y, W, respectively, so that $p = \xi N$, $q = \eta N$, $r = (1-\zeta)N$. Then the problem is to determine the distribution of the (stochastic) variable r in terms of the (given) p, q — i.e., the probability of any given r in combination with any given p, q.

W is clearly the intersection of the sets X, Y: $W = X \cdot Y$. Let U, V be the (relative) complements of W in X, Y, respectively: $U = X - W$, $V = Y - W$, and let S be the (absolute, i.e., in the set $(1, \ldots, N)$) complement of the sum of X and Y: $S = -(X + Y)$. Then W, U, V, S are pairwise disjoint sets making up together precisely the entire set $(1, \ldots, N)$, with r, $p - r$, $q - r$, $N - p - q + r$ elements, respectively. Apart from this they are unrestricted. Thus they offer together $N!/[r!(p - r)!(q - r)!(N - p - q + r)!]$ possible choices. Since there are a priori $N!/[p!(N - p)!]$ possible choices of an X with p elements and a priori $N!/[q!(N - q)!]$ possible choices of a Y with q elements, this means that the looked for probability of W having r elements is

$$\rho = \left(\frac{N!}{r!(p-r)!(q-r)!(N-p-q+r)!} \bigg/ \frac{N!}{p!(N-p)!} \; \frac{N!}{q!(N-q)!} \right)$$

$$= \frac{p!(N-p)!q!(N-q)!}{r!(p-r)!(q-r)!(N-p-q+r)!N!}$$

Note, that this formula also shows that $\rho = 0$ when $r < 0$ or $p - r < 0$ or $q - r < 0$ or $N - p - q + r < 0$, i.e., when r violates the conditions

$$\mathrm{Max}(0, p + q - N) \leq r \leq \mathrm{Min}(p, q).$$

This is clear combinatorially, in view of the meaning of X, Y and W. In terms of ξ, η, ζ the above conditions become

$$(17) \qquad 1 - \mathrm{Max}(0, \xi + \eta - 1) \geqq \zeta \geqq 1 - \mathrm{Min}(\xi, \eta).$$

Returning to the expression for ρ, substituting the ξ, η, ζ expressions for p, q, r and using Stirling's formula for the factorials involved, gives

$$(18) \qquad \rho \sim \frac{1}{\sqrt{2\pi N}} \sqrt{a} \;\; e^{-\theta N},$$

where

$$a = \frac{\xi(1-\xi)\eta(1-\eta)}{(\zeta+\xi-1)(\zeta+\eta-1)(1-\zeta)(2-\xi-\eta-\zeta)}$$

$$\theta = (\zeta + \xi - 1) \ln (\zeta + \xi - 1) + (\zeta + \eta - 1) \ln (\zeta + \eta - 1) + (1 - \zeta) \ln (1 - \zeta) + (2 - \xi - \eta - \zeta) \ln (2 - \xi - \eta - \zeta) - \xi \ln \xi - (1 - \xi) \ln (1 - \xi) - \eta \ln \eta - (1 - \eta) \ln (1 - \eta).$$

From this

$$\frac{\partial\theta}{\partial\zeta} = \ln \frac{(\zeta + \xi - 1)(\zeta + \eta - 1)}{(1 - \zeta)(2 - \xi - \eta - \zeta)} ,$$

$$\frac{\partial^2\theta}{\partial\zeta^2} = \frac{1}{\zeta + \xi - 1} + \frac{1}{\zeta + \eta - 1} + \frac{1}{1 - \zeta} + \frac{1}{2 - \xi - \eta - \zeta}$$

Hence $\theta = 0$, $\frac{\partial\theta}{\partial\zeta} = 0$ for $\zeta = 1 - \xi\eta$, and $\partial^2\theta/\partial\zeta^2 > 0$ for all ζ (in its entire interval of variability according to (17)). Consequently $\theta > 0$ for all $\zeta \neq 1 - \xi\eta$ (within the interval (17)). This implies, in view of (18) that for all ζ which are significantly $\neq 1 - \xi\eta$, ρ tends to 0 very rapidly as N gets large. It suffices therefore to evaluate (18) for $\zeta \sim 1 - \xi\eta$. Now $a = 1/[\xi (1 - \xi) \eta(1 - \eta)]$, $\partial^2\theta/\partial\zeta^2 = 1/[\xi (1 - \xi) \eta(1 - \eta)]$ for $\zeta = 1 - \xi\eta$. Hence

$$a \sim \frac{1}{\xi(1 - \xi)\eta(1 - \eta)} ,$$

$$\theta \sim \frac{(\zeta - (1 - \xi\eta))^2}{2\xi(1 - \xi)\eta(1 - \eta)}$$

for $\zeta \sim 1 - \xi\eta$. Therefore

$$(19) \qquad \rho \sim \frac{1}{\sqrt{2\pi \xi (1 - \xi) \eta (1 - \eta) N}} \; e^{-\frac{(\zeta - (1 - \xi\eta))^2}{2\xi(1 - \xi)\eta(1 - \eta)} N}$$

is an acceptable approximation for ρ.

r is an integer-valued variable, hence $\zeta = 1 - \frac{r}{N}$ is a rational-valued variable, with the fixed denominator N. Since N is assumed to be very large, the range of ζ is very dense. It is therefore permissible to replace it by a continuous one, and to describe the distribution of ζ by a probability-density σ. ρ is the probability of a single value of ζ, and since the values of ζ are equidistant, with a separation $d\zeta = 1/N$, the relation between σ and ρ is best defined by $\sigma d\zeta = \rho$, i.e., $\sigma = \rho N$. Therefore (19) becomes

$$(20) \qquad \sigma \sim \frac{1}{\sqrt{2\pi}\sqrt{\xi(1 - \xi)\eta(1 - \eta)/N}} \; e^{-\frac{1}{2}\left(\frac{\zeta - (1 - \xi\eta)}{\sqrt{\xi(1 - \xi)\eta(1 - \eta)/N}}\right)^2}$$

This formula means obviously the following:

ζ is approximately normally distributed, with the mean $1 - \xi\eta$ and the dispersion $\sqrt{\xi(1 - \xi)\eta(1 - \eta)/N}$. Note, that the rapid decrease of the normal distribution function (i.e., the right hand side of (20)) with N (which is exponential !) is valid as long as ζ is near to $1 - \xi\eta$, only the coefficient of N (in the exponent, i. e., $-\frac{1}{2}\left([\zeta - (1 - \xi\eta)] / \sqrt{\xi (1 - \xi) \eta (1 - \eta) / N}\right)^2$ is somewhat altered as ζ deviates from $1 - \xi$. (This follows from the discussion of θ given above.)

The simple statistical discussion of 9.4 amounted to attributing to ζ the unique value $1 - \xi\eta$. We see now that this is approximately true:

$$(21)\quad \begin{cases} \zeta = (1 - \xi\eta) + \sqrt{\xi (1 - \xi) \eta (1 - \eta) / N}\ \delta, \\ \delta \text{ is a stochastic variable, normally distributed, with the} \\ \text{mean } 0 \text{ and the dispersion } 1. \end{cases}$$

10.2.2 THEORY WITH ERRORS. We must now pass from r, ζ, which postulate faultless functioning of all Sheffer organs in the network, to r', ζ' which correspond to the actual functioning of all these organs — i.e., to a probability ϵ of error on each functioning. Among the r organs each of which should correctly stimulate its output, each error reduces r' by one unit. The number of errors here is approximately normally distributed, with the mean ϵr and the dispersion $\sqrt{\epsilon (1 - \epsilon) r}$ (cf. the remark made in 10.1). Among the $N - r$ organs, each of which should correctly not stimulate its output, each error increases r' by one unit. The number of errors here is again approximately normally distributed, with the mean $\epsilon (N - r)$, and the dispersion $\sqrt{\epsilon (1 - \epsilon) (N - r)}$ (cf. as above). Thus $r' - r$ is the difference of these two (independent) stochastic variables. Hence it, too, is approximately normally distributed, with the mean $-\epsilon r + \epsilon(N - r) = \epsilon(N - 2r)$, and the dispersion

$$\sqrt{(\sqrt{\epsilon (1 - \epsilon) r})^2 + (\sqrt{\epsilon (1 - \epsilon) (N - r)})^2} = \sqrt{\epsilon (1 - \epsilon) N}.$$

I. e., (approximately)

$$r' = r + 2\epsilon \left(\frac{N}{2} - r\right) + \sqrt{\epsilon (1 - \epsilon) N}\ \delta',$$

where δ' is normally distributed, with the mean 0 and the dispersion 1. From this

$$\zeta' = \zeta + 2\epsilon \left(\frac{1}{2} - \zeta\right) - \sqrt{\epsilon (1 - \epsilon) / N}\ \delta',$$

and then by (21)

$$\begin{aligned} \zeta' = {} & (1 - \xi\eta) + 2\epsilon \left(\xi\eta - \frac{1}{2}\right) + \\ & + (1 - 2\epsilon) \sqrt{\xi (1 - \xi) \eta (1 - \eta) / N}\ \delta - \\ & - \sqrt{\epsilon (1 - \epsilon) / N}\ \delta'. \end{aligned}$$

Clearly $(1 - 2\epsilon) \sqrt{\xi (1 - \xi) \eta (1 - \eta) / N}\ \delta - \sqrt{\epsilon (1 - \epsilon) / N}\ \delta'$, too, is normally distributed, with the mean 0 and the dispersion

$$\begin{aligned} & \sqrt{((1 - 2\epsilon) \sqrt{\xi (1 - \xi) \eta (1 - \eta) / N})^2 + (\sqrt{\epsilon (1 - \epsilon) / N})^2} = \\ & = \sqrt{((1 - 2\epsilon)^2 \xi (1 - \xi) \eta (1 - \eta) + \epsilon (1 - \epsilon)) / N}. \end{aligned}$$

Hence (21) becomes at last (we write again ζ in place of ζ'):

(22)
$$\begin{cases} \zeta = (1 - \xi\eta) + 2\epsilon\,(\xi\eta - \tfrac{1}{2}) + \\ + \sqrt{((1-2\epsilon)^2\,\xi\,(1-\xi)\,\eta\,(1-\eta) + \epsilon\,(1-\epsilon))/N}\;\delta^*, \\ \delta^* \text{ is a stochastic variable, normally distributed, with the mean} \\ 0 \text{ and the dispersion } 1. \end{cases}$$

10.3 The Restoring Organ

This discussion equally covers the situations that are dealt with in Figures 35 and 37, showing networks of Sheffer organs in 9.4.2.

Consider first Figure 35. We have here a single input bundle of N lines, and an output bundle of N lines. However, the two-way split and the subsequent "randomizing" permutation produce an input bundle of $2N$ lines and (to the right of U) the even lines of this bundle on one hand, and its odd lines on the other hand, may be viewed as two input bundles of N lines each. Beyond this point the network is the same as that one of Figure 34, discussed in 9.4.1. If the original input bundle had ξN stimulated lines, then each one of the two derived input bundles will also have ξN stimulated lines. (To be sure of this, it is necessary to choose the "randomizing" permutation U of Figure 35 in such a manner, that it permutes the even lines among each other, and the odd lines among each other. This is compatible with its "randomizing" the relationship of the family of all even lines to the family of all odd lines. Hence it is reasonable to expect, that this requirement does not conflict with the desired "randomizing" character of the permutation.) Let the output bundle have ζN stimulated lines. Then we are clearly dealing with the same case as in (22), except that it is specialized to $\xi = \eta$.
Hence (22) becomes:

(23)
$$\begin{cases} \zeta = (1 - \xi^2) + 2\epsilon\,(\xi^2 - \tfrac{1}{2}) + \\ + \sqrt{((1-2\epsilon)^2\,(\xi\,(1-\xi))^2 + \epsilon\,(1-\epsilon))/N}\;\delta^* \\ \delta^* \text{ is a stochastic variable, normally distributed, with the} \\ \text{mean } 0 \text{ and the dispersion } 1. \end{cases}$$

Consider next Figure 37. Three bundles are relevant here: The input bundle at the extreme left, the intermediate bundle issuing directly from the first tier of Sheffer organs, and the output bundle, issuing directly from the second tier of Sheffer organs, i.e., at the extreme right. Each one of these three bundles consists of N lines. Let the number of stimulated lines in each bundle be ζN, ωN, ψN, respectively. Then (23) above applies, with its ξ, ζ replaced first by ζ, ω, and second by ω, ψ:

$$(24) \quad \begin{cases} \omega = (1 - \zeta^2) + 2\epsilon\ (\zeta^2 - \frac{1}{2}) + \\ \quad + \sqrt{((1 - 2\epsilon)^2\ (\zeta(1 - \zeta))^2 + \epsilon\ (1 - \epsilon))/N}\ \delta^{**}\ , \\ \psi = (1 - \omega^2) + 2\epsilon\ (\omega^2 - \frac{1}{2}) + \\ \quad + \sqrt{((1 - 2\epsilon)^2\ (\omega(1 - \omega))^2 + \epsilon\ (1 - \epsilon))/N}\ \delta^{***}\ , \end{cases}$$

δ^{**}, δ^{***} are stochastic variables, independently and normally distributed, with the mean 0 and the dispersion 1.

10.4 Qualitative Evaluation of the Results

In what follows, (22) and (24) will be relevant — i.e., the Sheffer organ networks of Figures 34 and 37.

Before going into these considerations, however, we have to make an observation concerning (22). (22) shows that the (relative) excitation levels ξ, η on the input bundles of its network generate approximately (i.e., for large N and small ϵ) the (relative) excitation level $\zeta_o = 1 - \xi\eta$ on the output bundle of that network. This justifies the statements made in 9.4.1 about the detailed functioning of the network. Indeed: If the two input bundles are both prevalently stimulated, i.e., if $\xi \sim 1$, $\eta \sim 1$ then the distance of ζ_o from 0 is about the sum of the distances of ξ and of η from 1: $\zeta_o = (1 - \xi) + \xi\ (1 - \eta)$. If one of the two input bundles, say the first one, is prevalently non-stimulated, while the other one is prevalently stimulated, i.e., if $\xi \sim 0$, $\eta \sim 1$, then the distance of ζ_o from 1 is about the distance of ξ from 0: $1 - \zeta_o = \xi\eta$. If both input bundles are prevalently non-stimulated, i.e., if $\xi \sim 0$, $\eta \sim 0$, then the distance of ζ_o from 1 is small compared to the distances of both ξ and η from 0: $1 - \zeta_o = \xi\eta$.

10.5 Complete Quantitative Theory

10.5.1 GENERAL RESULTS. We can now pass to the complete statistical analysis of the Sheffer stroke operation on bundles. In order to do this, we must agree on a systematic way to handle this operation by a network. The system to be adopted will be the following: The necessary executive organ will be followed in series by a restoring organ. I.e., the Sheffer organ network of Figure 34 will be followed in series by the Sheffer organ network of Figure 37. This means that the formulas of (22) are to be followed by those of (24). Thus ξ,η are the excitation levels of the two input bundles, ψ is the excitation level of the output bundle, and we have:

(25)
$$\begin{cases} \zeta = (1 - \xi\eta) + 2\epsilon\,(\xi\eta - \frac{1}{2}) + \\ \quad + \sqrt{((1-2\epsilon)^2\,\xi(1-\xi)\eta\,(1-\eta) + \epsilon(1-\epsilon))/N}\;\delta^*, \\ \omega = (1 - \zeta^2) + 2\epsilon\,(\zeta^2 - \frac{1}{2}) + \\ \quad + \sqrt{((1-2\epsilon)^2\,(\zeta(1-\zeta))^2 + \epsilon(1-\epsilon))/N}\;\delta^{**}, \\ \psi = (1 - \omega^2) + 2\epsilon(\omega^2 - \frac{1}{2}) + \\ \quad + \sqrt{((1-2\epsilon)^2\,(\omega(1-\omega))^2 + \epsilon(1-\epsilon))/N}\;\delta^{***}, \end{cases}$$

δ^*, δ^{**}, δ^{***} are stochastic variables, independently and normally distributed, with the mean 0 and the dispersion 1.

Consider now a given fiduciary level Δ. Then we need a behavior, like the "correct" one of the Sheffer stroke, with an overwhelming probability. This means: The implication of $\psi \leq \Delta$ by $\xi \geq 1 - \Delta$, $\eta \geq 1 - \Delta$; the implication of $\psi \geq 1 - \Delta$ by $\xi \leq \Delta$, $\eta \geq 1 - \Delta$; the implication of $\psi \geq 1 - \Delta$ by $\xi \leq \Delta$, $\eta \leq \Delta$. (We are, of course, using the symmetry in ξ, η.)

This may, of course, only be expected for N sufficiently large and ϵ sufficiently small. In addition, it will be necessary to make an appropriate choice of the fiduciary level Δ.

If N is so large and ϵ is so small, that all terms in (25) containing factors $1/\sqrt{N}$ and ϵ can be neglected, then the above desired "overwhelmingly probable" inferences become even strictly true, if Δ is small enough. Indeed, then (25) gives $\zeta = \zeta_o = 1 - \xi\eta$, $\omega = \omega_o = 1 - \zeta^2$, $\psi = \psi_o = 1 - \omega^2$, i.e., $\psi = 1 - (2\xi\eta - (\xi\eta)^2)^2$. Now it is easy to verify $\psi = 0\,(\Delta^2)$ for $\xi \geq 1 - \Delta$, $\eta \geq 1 - \Delta$; $\psi = 1 - 0\,(\Delta^2)$ for $\xi \leq \Delta$, $\eta \geq 1 - \Delta$; $\psi = 1 - 0\,(\Delta^4)$ for $\xi \leq \Delta$, $\eta \leq \Delta$. Hence sufficiently small Δ will guarantee the desiderata stated further above.

10.5.2 NUMERICAL EVALUATION. Consider next the case of a fixed, finite N and a fixed, positive ϵ. Then a more elaborate calculation must be based on the complete formulae of (25). This calculation will not be carried out here, but its results will be described.

The most favorable fiduciary level Δ, from the point of view of this calculation turns out to be $\Delta = .07$. I.e., stimulation of at least 93% of the lines of a bundle represents a positive message; stimulation of at most 7% of the lines of a bundle represents a negative message; the interval between 7% and 93% is a zone of uncertainty, indicating an effective malfunction of the network.

Having established this fiduciary level, there exists also an upper bound for the allowable values of ϵ. This is $\epsilon = .0107$. In other words, if $\epsilon \geq .0107$, the risk of effective malfunction of the network will be above a fixed, positive lower bound, no matter how large a bundle size N is used. The calculations were therefore continued with a specific $\epsilon < .0107$, namely, with $\epsilon = .005$.

With these assumptions, then, the calculation yields an estimate for the probability of malfunction of the network, i.e., of the violation of the desiderata stated further above. As is to be expected, this estimate is given by an error integral. This is

$$(26)\quad \begin{cases} \rho(N) = \dfrac{1}{\sqrt{2\pi}} \displaystyle\int_{\kappa}^{\infty} e^{-\frac{1}{2}x^2}\, dx \sim \dfrac{1}{\sqrt{2\pi}\,\kappa}\, e^{-\frac{1}{2}\kappa^2} \\ \text{where} \\ \quad k = .062\sqrt{N} \end{cases}$$

expresses, in a certain sense, the total allowable error divided by a composite standard deviation. The approximation is of course valid only for large N. It can also be written in the form

$$(27)\qquad \rho(N) \sim \frac{6.4}{\sqrt{N}}\, 10^{-\frac{8.6\,N}{10,000}}$$

The following table gives a better idea of the dependency expressed by the formula:

N = number of lines in a bundle	$\rho(N)$ = probability of malfunction
1,000	2.7×10^{-2}
2,000	2.6×10^{-3}
3,000	2.5×10^{-4}
5,000	4×10^{-6}
10,000	1.6×10^{-10}
20,000	2.8×10^{-19}
25,000	1.2×10^{-23}

Notice that for as many as 1000 lines in a bundle, the reliability (about 3%) is rather poor. (Indeed, it is inferior to the $\epsilon = .005$, i.e., 1/2%, that we started with.) However, a 25 fold increase in this size gives very good reliability.

10.5.3 EXAMPLES.

10.5.3.1 FIRST EXAMPLE. To get an idea of the significance of these sizes and the corresponding approximations, consider the two following examples.

Consider first a computing machine with 2500 vacuum tubes, each of which is actuated on the average once every 5 microseconds. Assume that a mean free path of 8 hours between errors is desired. In this period of time there will have been $\frac{1}{5} \times 2{,}500 \times 8 \times 3{,}600 \times 10^6 = 1.4 \times 10^{13}$ actuations, hence the above specification calls for $\delta \sim 1/[1.4 \times 10^{13}] = 7 \times 10^{-14}$. According to the above table this calls for an N between 10,000 and 20,000 — interpolating linearly on $-{}^{10}\log \delta$ gives $N = 14{,}000$. I.e., the system should be multiplexed 14,000 times.

It is characteristic for the steepness of statistical curves in this domain of large numbers of events, that a 25 percent increase of N, i.e., $N = 17{,}500$, gives (again by interpolation) $\delta = 4.5 \times 10^{-17}$, i.e., a reliability which is 1,600 times better.

10.5.3.2 SECOND EXAMPLE. Consider second a plausible quantitative picture for the functioning of the human nervous system. The number of neurons involved is usually given as 10^{10}, but this number may be somewhat low, also the synaptic end-bulbs and other possible autonomous sub-units may increase it significantly, perhaps a few hundred times. Let us therefore use the figure 10^{13} for the number of basic organs that are present. A neuron may be actuated up to 200 times per second, but this is an abnormally high rate of actuation. The average neuron will probable be actuated a good deal less frequently, in the absence of better information 10 actuations per second may be taken as an average figure of at least the right order. It is hard to tell what the mean free path between errors should be. Let us take the view that errors properly defined are to be quite serious errors, and since they are not ordinarily observed, let us take a mean free path which is long compared to an ordinary human life — say 10,000 years. This means $10^{13} \times 10{,}000 \times 31{,}536{,}000 \times 10 = 3.2 \times 10^{25}$ actuations, hence it calls for $\delta \sim 1/(3.2 \times 10^{25}) = 3.2 \times 10^{-26}$. According to the table this lies somewhat beyond $N = 25{,}000$ — extrapolating linearly on $-{}^{10}\log \delta$ gives $N = 28{,}000$.

Note, that if this interpretation of the functioning of the human nervous system were a valid one (for this cf. the remark of 11.1), the number of basic organs involved would have to be reduced by a factor 28,000. This reduces the number of relevant actuations and increases the value of the necessary δ by the same factor. I.e., $\delta = 9 \times 10^{-22}$, and hence $N = 23{,}000$. The reduction of N is remarkably small — only 20%! This makes a reevaluation of the reduced N with the new N, δ unnecessary: In fact the new factor, i.e., 23,000, gives $\delta = 7.4 \times 10^{-22}$ and this with the approximation used above, again $N = 23{,}000$. (Actually the change of N is ~ 120, i.e., only 1/2%!)

Replacing the 10,000 years, used above rather arbitrarily, by 6 months, introduces another factor 20,000, and therefore a change of about the same size as the above one — now the value is easily seen to be $N = 23{,}000$ (uncorrected) or $N = 19{,}000$ (corrected).

10.6 Conclusions

All this shows, that the order of magnitude of N is remarkably insensitive to variations in the requirements, as long as these requirements are rather exacting ones, but not wholly outside the range of our (industrial or natural) experience. Indeed, the N obtained above were all $\sim 20,000$, to within variations lying between - 30% and + 40%.

10.7 The General Scheme of Multiplexing

This is an opportune place to summarize our results concerning multiplexing, i.e., the sections 9 and 10. Suppose it is desired to build a machine to perform the logical function $f(x, y, \ldots)$ with a given accuracy (probability of malfunction on the final result of the entire operation) η, using Sheffer neurons whose reliability (or accuracy , i.e., probability of malfunction on a single operation) is ϵ. We assume $\epsilon = .005$. The procedure is then as follows.

First, design a network R for this function $f(x, y, \ldots)$ as though the basic (Sheffer) organs had perfect accuracy. Second, estimate the maximum number of single (perfect) Sheffer organ reactions (summed over all successive operations of all the Sheffer organs actually involved) that occur in the network R in evaluating $f(x, y, \ldots)$ — say m such reactions. Put $\delta = \eta/m$. Third, estimate the bundle size N that is needed to give the multiplexed Sheffer organ like network (cf. 10.5.2) an error probability of at most δ. Fourth, replace each single line of the network R by a bundle of size N, and each Sheffer neuron of the network R by the multiplexed Sheffer organ network that goes with this N (cf. 10.5.1) — this gives a network $R^{(N)}$. A "yes" will then be transmitted by the stimulation of more than 93% of the strands in a bundle, a "no" by the stimulation of less than 7%, and intermediate values will signify the occurrence of an essential malfunction of the total system.

It should be noticed that this construction multiplies the number of lines by N and the number of basic organs by $3N$. (In 10.5.3 we used a uniform factor of multiplication N. In view of the insensitivity of N to moderate changes in δ, that we observed in 10.5.3.2, this difference is irrelevant.) Our above considerations show, that the size of N is $\sim 20,000$ in all cases that interest us immediately. This implies, that such techniques are impractical for present technologies of componentry (although this may perhaps not be true for certain conceivable technologies of the future), but they are not necessarily unreasonable (at least not on grounds of size alone) for the micro-componentry of the human nervous system.

Note, that the conditions are significantly less favorable for the non-multiplexing procedure to control error described in section 8. That process multiplied the number of basic organs by about 3^{μ}, μ being the number of consecutive steps (i.e., basic organ actuations) from input to output, (cf. the end of 8.4). (In this way of counting, iterative processes must be counted as many times as iterations occur.) Thus for $\mu = 160$, which is not an excessive "logical depth," even for a conventional calculation, $3^{160} \sim 2 \times 10^{76}$, i.e., somewhat above the putative order of the number of electrons in the universe. For $\mu = 200$ (only 25 percent more!) then $3^{200} \sim 2.5 \times 10^{95}$, i.e., 1.2×10^{19} times more — in view of the above this requires no comment.

11. GENERAL COMMENTS ON DIGITALIZATION AND MULTIPLEXING

11.1 Plausibility of Various Assumptions Regarding the Digital vs. Analog Character of the Nervous System

We now pass to some remarks of a more general character.

The question of the number of basic neurons required to build a multiplexed automaton serves as an introduction for the first remark. The above discussion shows, that the multiplexing technique is impractical on the level of present technology, but quite practical for a perfectly conceivable, more advanced technology, and for the natural relay-organs (neurons). I.e., it merely calls for micro-componentry which is not at all unnatural as a concept on this level. It is therefore quite reasonable to ask specifically, whether it, or something more or less like it, is a feature of the actually existing human (or rather: animal) nervous system.

The answer is not clear cut. The main trouble with the multiplexing systems, as described in the preceding section, is that they follow too slavishly a fixed plan of construction — and specifically one, that is inspired by the conventional procedures of mathematics and mathematical logics. It is true, that the animal nervous systems, too, obey some rigid "architectural" patterns in their large-scale construction, and that those variations, which make one suspect a merely statistical design, seem to occur only in finer detail and on the micro-level. (It is characteristic of this duality, that most investigators believe in the existence of overall laws of large-scale nerve- stimulation and composite action that have only a statistical character, and yet occasionally a single neuron is known to control a whole reflex-arc.) It is true, that our multiplexing scheme, too, is rigid only in its large-scale pattern (the prototype network R, as a pattern, and the general layout of the executive-plus-restoring organ, as discussed in 10.7 and in 10.5.1), while the "random" permutation "black boxes" (cf. the relevant Figures 32, 35, 37 in 9.2.3 and 9.4.2) are typical of a "merely statistical design." Yet the nervous system seems to be somewhat more flexibly designed. Also, its "digital" (neural) operations are rather freely alternating with "analog" (humoral) processes in their complete chains of causation. Finally the whole logical pattern of the nervous system seems to deviate in certain important traits qualitatively and significantly from our ordinary mathematical and mathematical-logical modes of operation: The pulse-trains that carry "quantitative" messages along the nerve fibres do not seem to be coded digital expressions (like a binary or a [Morse or binary coded] decimal digitalization) of a number, but rather "analog" expressions of one, by way of their pulse-density, or something similar – although much more than ordinary care should be exercised in passing judgments in this field, where we have so little factual information. Also, the "logical depth" of our neural operations – i.e., the total number of basic operations from (sensory) input to (memory) storage or (motor) output seems to be much less than it would be in any artificial automaton (e.g. a computing machine) dealing with problems of anywhere nearly comparable complexity. Thus deep differences in the basic organizational principles are probably present.

Some similarities, in addition to the one referred to above, are nevertheless undeniable. The nerves are bundles of fibres – like our bundles. The nervous system contains numerous "neural pools" whose function may well be that of organs devoted to the restoring of excitation levels. (At least of the two [extreme] levels, e.g. one near to 0 and one near to 1, as in the case discussed in section 9, especially in 9.2.2 and 9.2.3, 9.4.2. Restoring one level only – by exciting or quenching or establishing some intermediate stationary level – destroys rather than restores information, since a system with a single stable state has a memory capacity 0 [cf. the definition given in 5.2]. For systems which can stabilize [i.e., restore] more than two excitation levels, cf. 12.6.)

11.2 Remarks Concerning the Concept of a Random Permutation

The second remark on the subject of multiplexed systems concerns the problem (which was so carefully sidestepped in section 9) of maintaining randomness of stimulation. For all statistical analyses, it is necessary to assume that this randomness exists. In networks which allow feedback, however, when a pulse from an organ gets back to the same organ at some later time, there is danger of strong statistical correlation. Moreover, without randomness, situations may arise where errors tend to be amplified instead of cancelled out. E.g. it is possible, that the machine remembers its mistakes, so to speak, and thereafter perpetuates them. A simplified example of this effect is furnished by the elementary memory organ of Figure 16, or by a similar one, based on the Sheffer stroke, shown in Figure 39. We will discuss the latter. This system, provided it makes no mistakes, fires on alternate moments of time. Thus it has two possible states: Either it fires at even times or at odd times. (For a quantitative discussion of Figure 16, cf. 7.1.) However, once

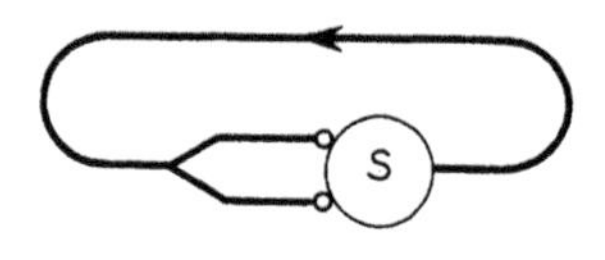

FIGURE 39

the mechanism makes a mistake, i.e., if it fails to fire at the right parity, or if it fires at the wrong parity, that error will be remembered, i.e., the parity is now lastingly altered, until there occurs a new mistake. A single mistake thus destroys the memory of this particular machine for all earlier events. In multiplex systems, single errors are not necessarily disastrous: But without the "random" permutations introduced in section 9, accumulated mistakes can be still dangerous.

To be more specific: Consider the network shown in Figure 35, but without the line-permuting "black box" U. If each output line is now fed back into its input line (i.e., into the one with the same number from above), then pulses return to the identical organ from which they started, and so the whole organ is in fact a sum of separate organs according to Figure 39, and hence it is just as subject to error as a single one of those organs acting independently. However, if a permutation of the bundle is interposed, as shown, in principle, by U in Figure 35, then the accuracy of the system may be (statistically) improved. This is, of course, the trait which is being looked for by the insertion of U, i.e., of a "random" permutation in the sense of section 9. But how is it possible to perform a "random" permutation?

The problem is not immediately rigorously defined. It is, however, quite proper to reinterpret it as a problem that can be stated in a rigorous form, namely: It is desired to find one or more permutations which can be used in the "black boxes" marked with U or U_1, U_2 in the relevant Figures 35, 37, so that the essential statistical properties that are asserted there are truly present. Let us consider the simpler one of these two, i.e., the multiplexed version of the simple memory organ of Figure 39 — i.e., a specific embodiment of Figure 35. The discussion given in 10.3 shows that it is desirable, that the permutation U of Figure 35 permute the even lines among each other, and the odd lines among each other. A possible rigorous variant of the question that should now be asked is this.

Find a fiduciary level $\Delta > 0$ and a probability $\epsilon > 0$, such that for any $\eta > 0$ and any $s = 1, 2, \ldots$ there exists an $N = N(\eta, s)$ and a permutation $U = U^{(N)}$, satisfying the following requirement: Assume that the probability of error in a single operation of any given Sheffer organ is ϵ. Assume that at the time t all lines of the above network are stimulated, or that all are not stimulated. Then the number of lines stimulated at the time $t + s$ will be $\geq (1 - \Delta)N$ or $\leq \Delta N$, respectively, with a probability $\geq 1 - \delta$. In addition $N(\eta, s) \leq C \ln(s/\eta)$, with a constant C (which should not be excessively great).

Note, that the results of section 10 make the surmise seem plausible, that $\Delta = .07$, $\epsilon = .005$ and $C \sim 10{,}000/[8.6 \times \ln 10] \sim 500$ are suitable choices for the above purpose.

The following surmise concerning the nature of the permutation

$U^{(N)}$ has a certain plausibility: Let $N = 2^\ell$. Consider the 2^ℓ complexes $(d_1, d_2, \ldots, d_\ell)$ ($d_\lambda = 0, 1$ for $\lambda = 1, \ldots, \ell$). Let these correspond in some one to one way to the 2^ℓ integers $i = 1, \ldots, N$:

$$i \rightleftarrows (d_1, d_2, \ldots, d_\ell). \tag{28}$$

Now let the mapping

$$i \longrightarrow i' = U^{(N)} i \tag{29}$$

be induced, under the correspondence (28), by the mapping

$$(d_1, d_2, \ldots, d_\ell) \longrightarrow (d_\ell, d_1, \ldots, d_{\ell-1}). \tag{30}$$

Obviously, the validity of our assertion is independent of the choice of the correspondence (28). Now (30) does not change the parity of

$$\sum_{\lambda=1}^{\ell} d_\lambda$$

hence the desideratum that $U^{(N)}$, i.e., (29), should not change the parity of i (cf. above) is certainly fulfilled, if the correspondence (28) is so chosen as to let i have the same parity as

$$\sum_{\lambda=1}^{\ell} d_\lambda .$$

This is clearly possible, since on either side each parity occurs precisely $2^{\ell-1}$ times. This $U^{(N)}$ should fulfill the above requirements.

11.3 Remarks Concerning the Simplified Probability Assumption

The third remark on multiplexed automata concerns the assumption made in defining the unreliability of an individual neuron. It was assumed that the probability of the neuron failing to react correctly was a constant ϵ, independent of time and of all previous inputs. This is an unrealistic assumption. For example, the probability of failure for the Sheffer organ of Figure 12 may well be different when the inputs a and b are both stimulated, from the probability of failure when a and not b is stimulated. In addition, these probabilities may change with previous history, or simply with time and other environmental conditions. Also, they are quite likely to be different from neuron to neuron. Attacking the problem with these more realistic assumptions means finding the domains of operability of individual neurons, finding the intersection of these domains (even when drift with time is allowed) and finally, carrying out the statistical estimates for this far more complicated situation. This will not be attempted here.

12. ANALOG POSSIBILITIES

12.1 Further Remarks Concerning Analog Procedures

There is no valid reason for thinking that the system which has been developed in the past pages is the only or the best model of any existing nervous system or of any potential error-safe computing machine or logical machine. Indeed, the form of our model-system is due largely to the influence of the techniques developed for digital computing and to the trends of the last sixty years in mathematical logics. Now, speaking specifically of the human nervous system, this is an enormous mechanism — at leat 10^6 times larger than any artifact with which we are familiar — and its activities are correspondingly varied and complex. Its duties include the interpretation of external sensory stimuli, of reports of physical and chemical conditions, the control of motor activities and of internal chemical levels, the memory function with its very complicated procedures for the transformation of and the search for information, and of course, the continuous relaying of coded orders and of more or less quantitative messages. It is possible to handle all these processes by digital methods (i.e., by using numbers and expressing them in the binary system — or, with some additional coding tricks, in the decimal or some other system), and to process the digitalized, and usually numericized, information by algebraical (i.e., basically arithmetical) methods. This is probably the way a human designer would at present approach such a problem. It was pointed out in the discussion in 11.1, that the available evidence, though scanty and inadequate, rather tends to indicate that the human nervous system uses different principles and procedures. Thus message pulse trains seem to convey meaning by certain analogic traits (within the pulse notation — i.e., this seems to be a mixed, part digital, part analog system), like the time density of pulses in one line, correlations of the pulse time series between different lines in a bundle, etc.

Hence our multiplexed system might come to resemble the basic traits of the human nervous system more closely, if we attenuated its rigidly discrete and digital character in some respects. The simplest step in this direction, which is rather directly suggested by the above remarks about the human nervous system, would seem to be this.

12.2 A Possible Analog Procedure

12.2.1 THE SET UP. In our prototype network R each line carries a "yes" (i.e., stimulation) or a "no" (i.e., non-stimulation) message — these are interpreted as digits 1 and 0, respectively. Correspondingly, in the final (multiplexed) network $R^{(N)}$ (which is derived from R) each bundle carries a "yes" = 1 (i.e., prevalent stimulation) or a "no" = 0 (i.e., prevalent non-stimulation) message. Thus only two meaningful states, i.e., average levels of excitation ξ, are allowed for a bundle — actually for one of these $\xi \sim 1$ and for the other $\xi \sim 0$.

Now for large bundle sizes N the average excitation level ξ is an approximately continuous quantity (in the interval $0 \leq \xi \leq 1$) — the larger N, the better the approximation. It is therefore not unreasonable to try to evolve a system in which ξ is treated as a continuous quantity in $0 \leq \xi \leq 1$. This means an analog procedure (or rather, in the sense discussed above, a mixed, part digital, part analog procedure). The possibility of developing such a system depends, of course, on finding suitable algebraic procedures that fit into it, and being able to assure its stability in the mathematical sense (i.e., adequate precision) and in the logical sense (i.e., adequate control of errors). To this subject we will now devote a few remarks.

12.2.2 THE OPERATIONS. Consider a multiplex automaton of the type which has just been considered in 12.2.1, with bundle size N. Let ξ denote the level of excitation of the bundle at any point, that is, the relative number of excited lines. With this interpretation, the automaton is a mechanism which performs certain numerical operations on a set of numbers to give a new number (or numbers). This method of interpreting a computer has some advantages, as well as some disadvantages in comparison with the digital, "all or nothing", interpretation. The conspicuous advantage is that such an interpretation allows the machine to carry more information with fewer components than a corresponding digital automaton. A second advantage is that it is very easy to construct an automaton which will perform the elementary operations of arithmetics. (Or, to be more precise: An adequate subset of these. Cf. the discussion in 12.3.) For example, given ξ and η, it is possible to obtain $\frac{1}{2}(\xi + \eta)$ as shown in Figure 40. Similarly, it is possible to obtain $\alpha\xi + (1 - \alpha)\eta$ for any constant α with $0 \leq \alpha \leq 1$. (Of course, there must be $\alpha = M/N$, $M = 0, 1, \ldots, N$, but this range for α is the same "approximate continuum" as that one for ξ, hence we may treat the former as a continuum just as properly as the latter.) We need only choose αN lines from the first bundle and combine them with $(1 - \alpha)N$ lines from the second. To obtain the quantity $1 - \xi\eta$ requires the set-up shown in Figure 41. Finally we can produce any constant excitation level α $(0 \leq \alpha \leq 1)$, by originating a bundle so that αN lines come from a live source and $(1 - \alpha)N$ from ground.

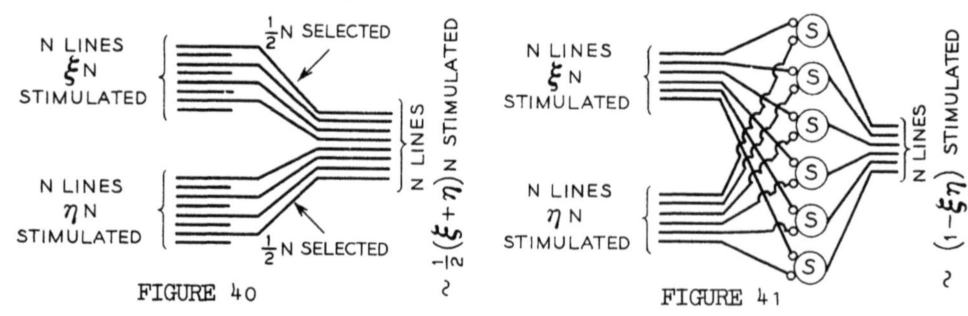

FIGURE 40 FIGURE 41

12.3 Discussion of the Algebraical Calculus Resulting from the Above Operations

Thus our present analog system can be used to build up a system of algebra where the fundamental operations are

$$(31) \qquad \left\{ \begin{array}{l} \alpha \\ \alpha\xi + (1-\alpha)\eta \\ 1 - \xi\eta \end{array} \right. \qquad \left\{ \begin{array}{l} \text{(for any constant } \alpha \\ \text{in } 0 \leq \alpha \leq 1), \end{array} \right.$$

All these are to be viewed as functions of ξ, η. They lead to a system, in which one can operate freely with all those functions $f(\xi_1, \xi_2, \ldots, \xi_k)$ of any k variables $\xi_1, \xi_2, \ldots, \xi_k$, that the functions of (31) generate. I.e., with all functions that can be obtained by any succession of the following processes:

(A) In the functions of (31) replace ξ, η by any variables ξ_i, ξ_j.

(B) In a function $f(\xi_1^*, \ldots, \xi_\ell^*)$, that has already been obtained, replace the variables $\xi_1^*, \ldots, \xi_\ell^*$, by any functions $g_1(\xi_1, \ldots, \xi_k), \ldots, g_\ell(\xi_1, \ldots, \xi_k)$, respectively, that have already been obtained.

To these, purely algebraical-combinatorial processes we add a properly analytical one, which seems justified, since we have been dealing with approximative procedures, anyway:

(C) If a sequence of functions $f_u(\xi_1, \ldots, \xi_k)$, $u = 1, 2, \ldots$, that have already been obtained, converges uniformly (in the domain $0 \leq \xi_1 \leq 1, \ldots, 0 \leq \xi_k \leq 1$) for $u \to \infty$ to $f(\xi_1, \ldots, \xi_k)$, then form this $f(\xi_1, \ldots, \xi_k)$.

Note, that in order to have the freedom of operation as expressed by (A), (B), the same "randomness" conditions must be postulated as in the corresponding parts of sections 9 and 10. Hence "randomizing" permutations U must be interposed between consecutive executive organs (i.e., those described above and reenumerated in (A)), just as in the sections referred to above.

In ordinary algebra the basic functions are different ones, namely:

$$(32) \qquad \left\{ \begin{array}{l} \alpha \\ \xi + \eta \\ \xi\eta \end{array} \right. \qquad \left\{ \begin{array}{l} \text{(for any constant } \alpha \\ \text{in } 0 \leq \alpha \leq 1), \end{array} \right.$$

It is easily seen, that the system (31) can be generated (by (A), (B)) from the system (32), while the reverse is not obvious (not even with (C) added). In fact (31) is intrinsically more special than (32), i.e., the functions that (31) generates are fewer than those that (32) generates (this is true for (A), (B), and also for (A), (B), (C)) — the former do not even include $\xi + \eta$. Indeed all functions of (31), i.e., of (A) based on (31), have this property: If all variables lie in the interval $0 \leq \xi \leq 1$, then the function, too, lies in that interval. This property is conserved under the applications of (B), (C). On the other hand $\xi + \eta$ does not possess this property — hence it cannot be generated by (A), (B), (C) from (31). (Note, that the above property of the functions of (31), and of all those that they generate, is a quite natural one: They are all dealing with excitation levels, and excitation levels must, by their nature, be numbers ξ with $0 \leq \xi \leq 1$.)

In spite of this limitation, which seems to mark it as essentially narrower than conventional algebra, the system of functions generated (by (A), (B), (C)) from (31) is broad enough for all reasonable purposes. Indeed, it can be shown that the functions so generated comprise precisely the following class of functions:

All functions $f(\xi_1, \xi_2, \ldots, \xi_k)$ which, as long as their variables $\xi_1, \ldots, \xi_k$ lie in the interval $0 \leq \xi \leq 1$, are continuous and have their value lying in that interval, too.

We will not give the proof here, it runs along quite conventional lines.

12.4 Limitations of this System

This result makes it clear, that the above analog system, i.e., the system of (31), guarantees for numbers ξ with $0 \leq \xi \leq 1$ (i.e., for the numbers that it deals with, namely excitation levels) the full freedom of algebra and of analysis.

In view of these facts, this analog system would seem to have clear superiority over the digital one. Unfortunately, the difficulty of maintaining accuracy levels counterbalances the advantages to a large extent. The accuracy can never be expected to exceed $1/N$. In other words, there is an intrinsic noise level of the order $1/N$, i.e., for the N considered in 10.5.2 and 10.5.3 (up to $\sim 20{,}000$) at best 10^{-4}. Moreover, in its effects on the operations of (31), this noise level rises from $1/N$ to $1/\sqrt{N}$. (E.g., for the operation $1 - \xi\eta$, cf. the result (21) and the argument that leads to it.) With the above assumptions, this is at best $\sim 10^{-2}$, i.e., 1%! Hence after a moderate number of operations, the excitation levels are more likely to resemble a random sampling of numbers than mathematics.

It should be emphasized, however, that this is not a conclusive argument that the human nervous system does not utilize the analog system. As was pointed out earlier, it is in fact known for at least some nervous processes that they are of an analog nature, and that the explanation of this may, at least in part, lie in the fact that the "logical depth" of the nervous network is quite shallow in some relevant places. To be more specific: The number of synapses of neurons from the peripheral sensory organs, down the afferent nerve fibres, through the brain, back through the efferent nerves to the motor system may not be more than ~ 10. Of course the parallel complexity of the network of neurons is indisputable. "Depth" introduced by feedback in the human brain may be overcome by some kind of self-stabilization. At the same time, a good argument can be put up that the animal nervous system uses analog methods (as they are interpreted above) only in the crudest way, accuracy being a very minor consideration.

12.5 A Plausible Analog Mechanism: Density Modulation by Fatigue

Two more remarks should be made at this point. The first one deals with some more specific aspects of the analog element in the organization and functioning of the human nervous system. The second relates to the possibility of stabilizing the precision level of the analog procedure that was outlined above.

This is the first remark. As we have mentioned earlier, many neurons of the nervous system transmit intensities (i.e., quantitative data) by analog methods, but, in a way entirely different from the method described in 12.2, 12.3 and 12.4. Instead of the level of excitation of a nerve (i.e., of a bundle of nerve fibres) varying, as described in 12.2, the single nerve fibres fire repetitiously, but with varying frequency in time. For example, the nerves transmitting a pressure stimulus may vary in frequency between, say, 6 firings per second and, say, 60 firings per second. This frequency is a monotone function of the pressure. Another example is the optic nerve, where a certain set of fibres responds in a similar manner to the intensity of the incoming light. This kind of behavior is explained by the mechanism of neuron operation, and in particular with the phenomena of threshold and of fatigue. With any peripheral neuron at any time can be associated a threshold intensity: A stimulus will make the neuron fire if and only if its magnitude exceeds the threshold intensity. The behavior of the threshold intensity as a function of the time after a typical neuron fires is qualitatively pictured in Figure 42. After firing, there is an "absolute refractory period" of about 5 milliseconds, during which no stimulus can make the neuron fire again. During this period, the threshold value is infinite. Next comes a "relative refractory period" of about 10 milliseconds, during which time the threshold level drops back to its equilibrium value (it may even oscillate about this value a few times at the end). This decrease is for the most part monotonic. Now the nerve will fire again as soon as it is stimulated with an intensity greater than its excitation threshold. Thus if the neuron is subjected to continual excitation of constant intensity (above the equilibrium intensity), it will fire periodically with a period between 5 and 15 milliseconds, depending on the intensity of the stimulus.

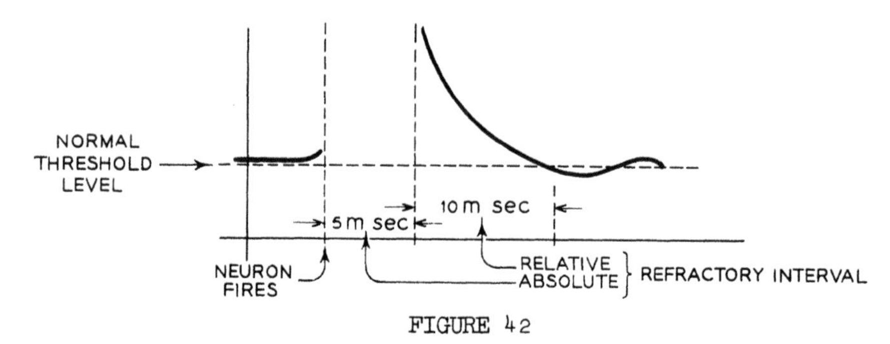

FIGURE 42

Another interesting example of a nerve network which transmits intensity by this means is the human acoustic system. The ear analyzes a sound wave into its component frequencies. These are transmitted to the brain through different nerve fibres with the intensity variations of the corresponding component represented by the frequency modulation of nerve firing.

The chief purpose of all this discussion of nervous systems is to point up the fact that it is dangerous to identify the real physical (or biological) world with the models which are constructed to explain it. The problem of understanding the animal nervous action is far deeper than the problem of understanding the mechanism of a computing machine. Even plausible explanations of nervous reaction should be taken with a very large grain of salt.

12.6 Stabilization of the Analog System

We now come to the second remark. It was pointed out earlier, that the analog mechanism that we discussed may have a way of stabilizing excitation levels to a certain precision for its computing operations. This can be done in the following way.

For the digital computer, the problem was to stabilize the excitation level at (or near) the two values 0 and 1. This was accomplished by repeatedly passing the bundle through a simple mechanism which changed an excitation level ξ into the level $f(\xi)$, where the function $f(\xi)$ had the general form shown in Figure 43. The reason that such a mechanism is a restoring organ for the excitation levels $\xi \sim 0$ and $\xi \sim 1$ (i.e., that it stabilizes at — or near — 0 and 1) is that $f(\xi)$ has this property: For some suitable $b (0 \leq b \leq 1)$ $0 < \xi < b$ implies $0 < f(\xi) < \xi$; $b < \xi < 1$ implies $\xi < f(\xi) < 1$. Thus $\xi = 0, 1$ are the only stable fixpoints of $f(\xi)$. (Cf. the discussion in 9.2.3 and 9.4.2.)

Now consider another $f(\xi)$, which has the form shown in Figure 44. I.e., we have:

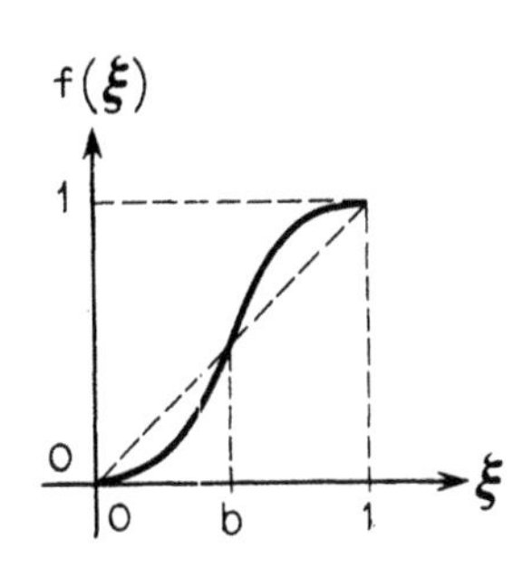

FIGURE 43

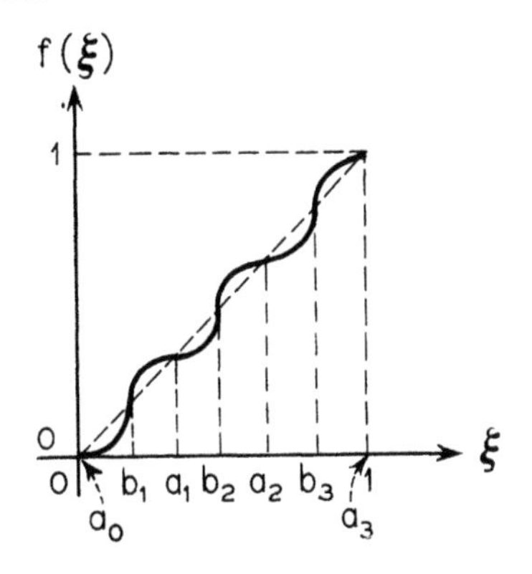

FIGURE 44

$$0 = a_0 < b_1 < a_1, < \dots < a_{\nu-1} < b_\nu < a_\nu = 1,$$
$$\text{for } i = 1, \dots, \nu: \quad a_{i-1} < \xi < b_i \text{ implies } a_{i-1} < f(\xi) < \xi$$
$$b_i < \xi < a_i \text{ implies } \xi < f(\xi) < a_i.$$

Here $a_0(= 0), a_1, \dots, a_{\nu-1}, a_\nu(= 1)$ are $f(\xi)$'s only stable fixpoints, and such a mechanism is a restoring organ for the excitation levels $\xi \sim a_0(= 0), a_1, \dots, a_{\nu-1}, a_\nu(= 1)$. Choose, e.g., $a_i = i/\nu$ $(i = 0, 1, \dots, \nu)$, with $\nu^{-1} < \delta$, or more generally, just $a_i - a_{i-1} < \delta$ $(i = 1, \dots, \nu)$ with some suitable ν. Then this restoring organ clearly conserves precisions of the order δ (with the same prevalent probability with which it restores).

13. CONCLUDING REMARK

13.1 A Possible Neurological Interpretation

There remains the question, whether such a mechanism is possible, with the means that we are now envisaging. We have seen further above, that this is the case, if a function $f(\xi)$ with the properties just described can be generated from (31). Such a function can indeed be so generated. Indeed, this follows immediately from the general characterization of the class of functions that can be generated from (31), discussed in 12.3. However, we will not go here into this matter any further.

It is not inconceivable that some "neural pools" in the human nervous system may be such restoring organs, to maintain accuracy in those parts of the network where the analog princple is used, and where there is enough "logical depth" (cf. 12.4) to make this type of stabilization necessary.

BIBLIOGRAPHY

[1] KLEENE, S. C., "Representation of events in nerve nets and finite automata," pp. 3-42.

[2] McCULLOCH, W. S. and PITTS, W., "A logical calculus of the ideas immanent in nervous activity," Bull. of Math. Biophysics, 5 (1943), pp. 115-133

[3] SHANNON, C. E., "A mathematical theory of communication," The Bell System Tech. Jour., 27 (1948), pp. 379-423.

[4] SZILARD, L., "Über die Entropieverminderung in einem thermodynamischen System bei Eingriffen intelligenter Wesen," Zeitschrift für Physik, 53 (1929), pp. 840-856.

[5] TURING, A. M., "On computable numbers," Proc. of the London Math. Soc. 2-42 (1936), pp. 230-265.

SOME UNECONOMICAL ROBOTS

James T. Culbertson

Our neurons are modeled in certain respects after biological neurons but are defined so as to have simpler and more uniform properties. Our postulates about these neurons and how they interact differ only slightly from the customary ones [3;4]. We will, however, explain the necessary concepts as clearly as we can and will not assume that the reader is already familiar with the literature in this field.

Each neuron has one entrance and one or more exits. An impulse entering the entrance of a neuron travels to all exits of that neuron.

We will use three kinds of neurons:

(A) Receptor neurons receive impulses from the environment and send impulses to neurons. Certain changes (stimuli) in the environment cause impulses in receptor neurons. "Environment" means everything except the neurons.

(B) Central neurons receive impulses from and send impulses to neurons.

(C) Effector neurons receive impulses from neurons and send impulses to the environment. Impulses coming out the exits of effector neurons cause changes in the environment. (Effector nerve cells in actual organisms cause muscles to contract.)

These neurons are represented, as in Figure 2, by lines; branching lines in any case where the neuron has more than one exit.

There are two kinds of neuron entrances — receptors or the entrances of receptor neurons and synapses or the entrances of other neurons. Each receptor is indicated by a square, □ , and each synapse by a simple line ending, sometimes enlarged.

There are two kinds of neuron exits — effectors or the exits of effector neurons and endbulbs or the exits of other neurons. Each effector is indicated by a diagonalized square ⧄ and each endbulb by a small circle, ● or ○ , depending on whether it is excitatory or inhibitory.

Receptor, central, and effector neurons may be connected together to form nerve nets in many different ways [1]. The anatomical rule concerning the way neurons may be assembled is as follows: All neuron contacts are contacts between endbulbs and synapses, no endbulb can contact more than one synapse.

Thus, in Figure 2, receptor neuron r_2 has two excitatory endbulbs on the synapse of effector neuron e_2, and receptor neuron r_1 has one excitatory on effector neuron e_1 and one inhibitory bulb on e_2, etc.

If at any time, t, an impulse enters a receptor (from the environment) or enters a synapse (from one or more bulbs contacting it) then the neuron having that receptor or synapse is said to _fire_ at t. If a neuron fires at t, the impulse passes to each exit, each exit becomes active, and then, finally, at $t + 1$ each exit ceases its activity or _dies_.

If a neuron fires at t, then all its endbulbs die at $t + 1$. It is assumed that this action time, or the interval between when a neuron fires and when its endbulbs die, is the same for all neurons. For convenience it is taken as the unit of time.

To simplify nerve net theory, it is customary to make an assumption, called the _quantized-time assumption_, which is clearly recognized by all to be contrary-to-fact for any sort of relays like our neurons. The quantized-time assumption is that neurons can fire only at integral values of time. That is, setting our clock appropriately, neurons can fire when t equals 1, or 2, or 3, or any other integer, but they cannot fire when t equals 1 1/2, or 2 1/4, or any other non-integer. This idealizes any nerve net into a discrete state machine, like a synchronous type digital or relay computer, clicking from each state to the next. It keeps the neuron firings all in phase, and hence simplifies the analysis.

Neurons influence each other as follows. Associated with each synapse there is a number, θ, called its _threshold_. The threshold (an integer either positive, negative, or zero) is constant for any one synapse but may vary from synapse to synapse. A neuron fires at time, t, if and only if $E - I \geqslant \theta$, where E is the number of excitatory bulbs contacting it and dying at t, and I is the number of inhibitory bulbs contacting it and dying at t.

Using suitable receptors and effectors we can connect them together via central cells. _If we could get enough central cells and if they were small enough and if each cell had enough endbulbs and if we could put enough bulbs at each synapse and if we had time enough to assemble them_, then we could construct robots to satisfy _any_ given input-output specifications; i.e., we could construct robots that would behave in any way we desired under any environmental circumstances. We will illustrate this point later in the paper. There would be no difficulty in constructing a robot with behavioral properties just like John Jones or Henry Smith or in constructing a robot with any desired behavioral improvements over Jones and Smith. Actually, of course, we will never have many billions of cells available. Even if we did have many billions of them, the brain resulting when they were put together would be of an absurdly large size; and since the time of life is short, we would not have time enough to assemble them. The robots discussed in this paper are totally impractical for these reasons. They are of theoretical interest only, illustrating a general method of robot design. We don't care here about the practical difficulties of where to get all the neurons and how to find room and time to assemble them. In fact, we don't care just now whether we need more neurons than there are atoms in the whole universe. We can always discuss these practical neuro-economical matters later. In the following we will assume that we can assemble as many cells as we want into a brain as small as we want and do it within a time commensurate with that of other human projects.

On the other hand, of course, some robots with very complex behavior could actually be built — just how complex is an interesting question.

A nerve net robot contains some receptor neurons $r_1, r_2, \dots r_n$, some effector neurons $e_1, e_2, \dots e_\mu$ and generally some, usually many, central neurons. The input-output properties of the robot depend on how these neurons are put together and how thresholds are assigned. These neurons are all embedded in a supporting and protecting structure or body.

Contained in or connected with this body are power devices which may be of various kinds. These are activated by the effectors. Which ones are active depends upon which effectors are or have been active. Some of these power devices may change the shape of the body, either grossly or in relatively local parts, while others may transmit energy, like radio signals, for example, to the environment without any apparent mechanical bodily changes, as in the emitting of light by a firefly or electricity by an electric eel.

Those power devices which change the shape of the body or move its parts relative to each other could be constructed of contractile material like the muscles of animals, if that becomes technically possible, or of course they could involve contemporary engines like electric motors with accessory wheels, pulleys and levers.

Some of these power devices could move the robot through its environment just as in the locomotion of animals, except that the robot could be built with wheels instead of legs if so desired. Others could be used for offensive or defensive action against other robots or animals or for manipulating any sort of things in any sort of way. Appropriate power devices for moving body parts or appendages could be used by the robot for fixing its own flat tires or repairing various other damages external or internal — except of course disabling damage to some part required in fixing that same part. Thus, if a bad fall or hostile blow disabled some of its snyapses, the robot could not make the necessary repairs in any case where doing so would require the use of those same synapses.

MEMORYLESS ROBOTS

In this section we will limit the analysis to robots without memory. In the strictest sense of the word, a memoryless robot has no "memory" whatsoever. The activity of its effectors at any time t depends only on the activity of its receptors at one previous time, $t - \Delta$, and the interval Δ, called the reaction time, is the same for all values of t, i.e., this reaction time is the same for all inputs throughout the life of the individual. We will make the reaction time equal to three time units for a reason which will become clear, namely because if we let $\Delta = 3$ we can construct the robot so that it will satisfy any given input-output specifications. For $\Delta < 3$ this is not true. We choose, therefore, the smallest value, $\Delta = 3$, which can satisfy all input-output specifications.

The output of a memoryless robot at any instant is thus determined by some purely spatial input pattern and the reaction time to all such spatial input patterns is the same.

If we were going to construct one of these robots, we would have to decide on what kind of a memoryless robot we wanted. We would have to resolve on such things as the number of receptors, audioreceptors, tactile receptors, chemoreceptors, etc. Also, what accessory sensory apparatus we wanted — e.g., what sort of antenna joints and sockets, or eye structures and lenses, or external ear, or nostrils guiding air to the olfactory receptors, etc. Likewise, we would have to settle on the number of effectors and power units and their kinds — whether we wanted muscle-type power units, for example, or electric-motor type and just how they would be connected with the rest of the body. Also we would have to decide about a whole lot of mechanical specifications such as strength of materials, and various other non-neurological specifications concerning the sizes and shapes of the body parts and appendages and how they fit together. Also, finally, even such details as what kind of paint or cosmetics to use, if any.

But in all this robotics only the nerve net theory, or "theoretical neurology" as we might call it, is of interest to us here. We are concerned only with the robot's brain design, and this only in regard to how to connect its neurons together, not what to embed them in or how to maintain their power supplies and other metabolism, or any thing of that sort.

We wish to remark here that for our robot's brain-design we will use a completely standard, easy, fool-proof method in which there is always a layer of 2^n central cells where n is the number of receptors. We will use this completely standard, perfectly straightforward and obvious circuit design even though this many central cells is not necessary. We choose the simplest rather than the most economical design since, as explained above, we don't care just now how many neurons we use — neuro-economy, or minimizing the number of central cells, can be considered later.

So far as neurology is concerned then, we may design any memoryless robot if in regard to the proposed robot, we are given:

(1) The number of receptors. What kind each is does not matter.

(2) The number of effectors. What kind each is does not matter.

(3) Its input-output specifications (explained below).

If the robot's sensory requirements call for n receptors $r_1, r_2, \ldots r_n$ then there are 2^n classes or sets of these receptor cells. For convenience, the first thing we do is to order or label these sets in any way at all. The following is a very convenient way of ordering them.

1st receptor set	The null set	...00000
2nd " "	r_1	...00001
3rd " "	r_2	...00010
4th " "	r_1+r_2	...00011
5th " "	r_3	...00100
6th " "	r_1+r_3	...00101
7th " "	r_2+r_3	...00110
8th " "	$r_1+r_2+r_3$	...00111
9th " "	r_4	...01000
10th " "	r_1+r_4	...01001
11th " "	r_2+r_4	...01010
.	.	.
.	.	.
.	.	.
jth " "		
.		.
.		.
.		.
2^nth " "	All the receptor cells	...11111

Note that each set is indicated by a number consisting of n digits, each a 0 or a 1. The digit representing r_1 is at the extreme <u>right</u> in this number. The digit representing r_2 is second from the right, etc. In this way the jth set is indicated by the jth binary number.

Since we have 2^n sets of receptors, we put 2^n central neurons into the robot (one for each set of receptors), and we label them in any way — 1st central cell, 2nd central cell, 3rd central cell, ... jth central cell, ... 2^nth central cell, as shown in Figure 1.

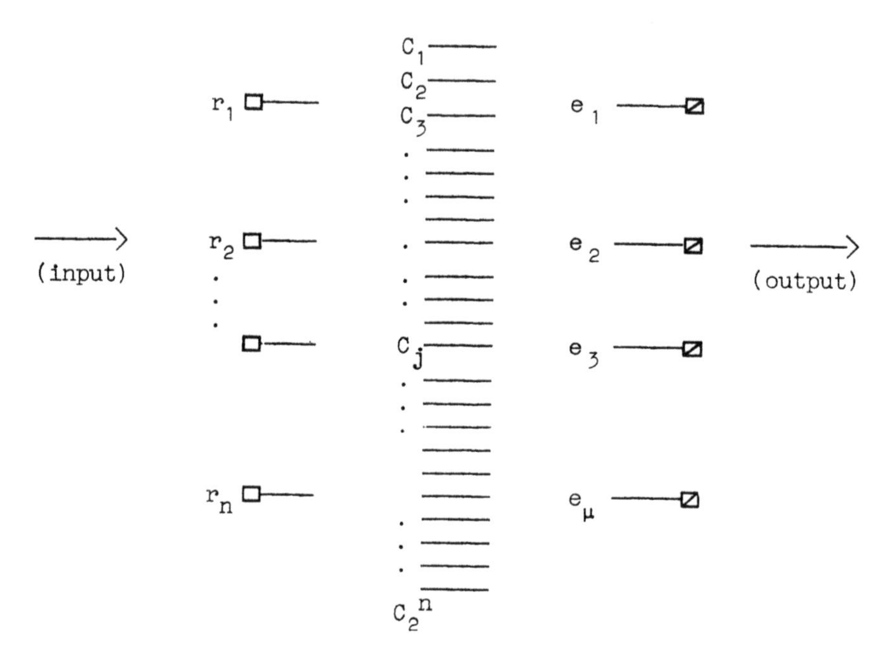

FIGURE 1. Memoryless Robot. Contacts between cells are not shown in this figure.

Now we can connect the receptor neurons to the central neurons as follows[1].

Receptor-to-Central Connections

Each receptor cell in the jth set has an excitatory bulb on the jth central cell, while each receptor cell which is not in the jth set has an inhibitory bulb on the jth central cell. Also we make the threshold on any central cell equal to the number of excitatory bulbs contacting it.

---

[1] In this simple type of design, these receptor-to-central connections are not influenced in the slightest by items (2) and (3) on the preceding page. In other words, the receptor-to-central connections depend only on the number of receptors, and not at all on the number of effectors or how we want the automaton to behave.

For example (listing the receptor sets 1st, 2nd, 3rd, etc., in the above manner), each receptor has an inhibitory bulb on $C_1(\theta=0)$. Note then that C_1 fires if and only if no receptor fires, i.e., if the 1st set, "the nullset", fires.

Receptor neuron r_1 constituting the 2nd set has an excitatory bulb on $C_2(\theta=1)$. Hence C_2 fires at t if and only if just receptor r_1 is fired at $t - 1$.

To take an arbitrarily chosen example, suppose $n > 5$. Then one of the receptor sets is the set $r_2 + r_3 + r_6$. This is the 39th set according to the above method of numbering the sets. There is a corresponding central neuron, C_{39}. Receptor neurons r_2, r_3, and r_6 each have one excitatory bulb on $C_{39}(\theta=3)$. The other receptor neurons have inhibitory bulbs on C_{39}. We note that C_{39} responds just to the firing of this one set $r_2 + r_3 + r_6$, namely the 39th set.

Each different input causes a different central cell to fire in this robot. Obviously, only one central cell can fire at any time.

This much of the brain we can construct merely by knowing how many receptors the robot is to have. We cannot continue assembling the brain, however, until we know how many effectors are desired and how the robot is to behave in every conceivable situation, i.e., what the desired input-output specifications are.

Input-Output Specifications

If the robot is to have μ effectors $e_1, e_2, \cdots e_\mu$ then there are 2^μ sets of these and we may order these effector sets in the same way as we ordered the receptor sets.

1st effector set	The null set	...0000
2nd " "	e_1	...0001
3rd " "	e_2	...0010
4th " "	e_1+e_2	...0011
5th " "	e_3	...0100
6th " "	e_1+e_3	...0101
7th " "	e_2+e_3	...0110
.		
.		
.		
ℓth " "		
.		
.		
.		
2^μth " "	All the receptor cells	...1111

To decide how we want the proposed robot to behave under any circumstances, is to specify for each receptor set an effector set such that if the former fires at t then the latter fires[2] at $t + 3$. There are $(2^{\mu})^{2^n}$ ways of deciding on these input-output specifications. The proposed input-output specifications, when chosen, can be indicated in a table. For example, the following table shows one of the 8^8 possible input-output specifications (proposed input-output relations) for a robot with 3 receptor cells and 3 effector cells.

Receptor sets	First 000	Second 001	Third 010	Fourth 011	Fifth 100	Sixth 101	Seventh 110	Eighth 111
Effector sets	100 Fifth	101 Sixth	110 Seventh	111 Eighth	000 First	001 Second	010 Third	001 Second

Central-to-Effector Connections

When we have decided on the input-output specifications for our memoryless robot then the neural connections of the robot may be completed as follows. If the specifications call for the jth receptor set to fire the ℓth effector set, then the jth central cell has one excitatory bulb on each member of the ℓth effector set. The threshold on each effector cell is made equal to 1.

For example, if the specifications are those given in the above table, then the fourth central cell has one endbulb on each effector cell. (Hence the fourth receptor set, 011, fires the eighth effector set, 111, as required by the specifications.) Actually the whole circuitry can be described succinctly, in summary, as follows:

(1) On each central cell, θ equals the number of excitatory bulbs contacting it.

(2) On each effector cell, $\theta = 1$.

(3) Each member of the jth receptor set has one excitatory bulb on the jth central cell, and each non-member of the jth receptor set has one inhibitory bulb on the jth central cell. (Hence the jth receptor set fires just the jth central cell.)

(4) If the specifications call for the jth receptor set to fire the ℓth effector set, then the jth central cell has one excitatory

[2] "Receptor set R fires at t" means "every receptor in R fires at t and no other receptors fire at t." Similarly, "effector set E fires at t" means "every effector in E fires at t and no other effectors fire at t." An effector, or exit of an effector cell, becomes active (here we say it "fires") one time unit after the effector cell itself fires.

bulb on each member of the ℓth effector set. (Hence the jth central cell fires just the ℓth effector set.)

It is of some interest that this method is completely general, since by it we can design memoryless robots to satisfy any proposed input-output specifications.

The method is very uneconomical except for small values of n, because 2^n increases very rapidly as n increases. Although a general method of reducing the number of central cells is not known at present, it is always possible to find some way of getting along with less than 2^n central cells and satisfying any input-output specifications. Whether or not these economies are trivial for some specifications when n is large, is not known. But, as explained at the start, we are not interested in neuroeconomy in the present discussion since we are merely illustrating certain simple general principles. It might be noted, however, that the input-output specifications of the preceding table can be satisfied, it so happens, without using any central cells at all, as shown in Figure 2. Incidentally, this reduces the reaction time to 2.

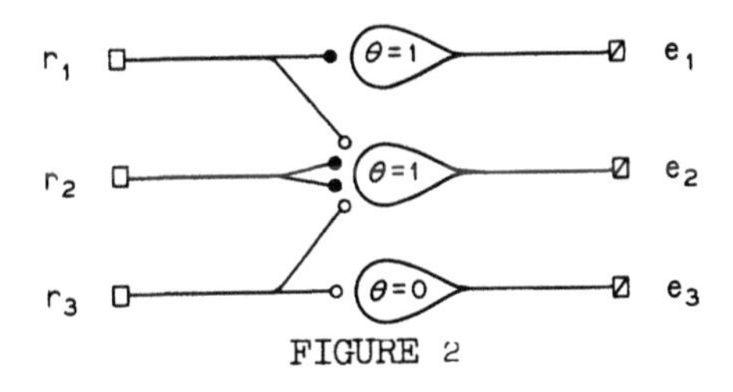

FIGURE 2

These memoryless robots will exhibit certain behavioral maladjustments due to the fact that having no memory they cannot react to events occurring over an interval of time. What reaction is suitable to a given input may depend on previous inputs, and such distinctions are a closed book to the memoryless robot. Thus adversaries with even rather short memories could soon take advantage of the purely spatial awareness of these robots. Also, these robots might easily get hung or trapped even in a completely inanimate environment. The following Rube Goldberg case is somewhat artificial but will illustrate this point.

Suppose a memoryless robot is constructed to destroy certain tall objects which it can distinguish from other objects by color or shape, etc. The robot proceeds directly forward except insofar as obstacles may deflect it. If an object-to-be-destroyed appears in its visual field it turns toward the object and then continues directly forward as before. If two or more such objects simultaneously appear, it turns toward the one producing the largest retinal image[3]. Being memoryless, the only way it can get back on course after being deflected by an obstacle is by the reappearance of the object (or the appearance of another one) in its visual field. When the

[3]We have already proved that a memoryless robot could be designed for any given input-output specifications.

robot contacts any one of these objects it destroys it and then proceeds on as before. Ordinarily all goes well and one object after another is destroyed. But there are possible features of the terrain that might repeatedly return the robot to the same position so that it would spend the rest of its natural life trudging in a circle. For instance, if the robot mounting the hill on the left in Figure 3 sights an object-to-be-destroyed at L, it proceeds toward L. This takes it to H where it slides down the steep incline HJ. The object at L is still in view so it continues upward to K where an obstacle deflects it toward the left so that it slides back over path KMHJ. Being in the same position at J as before, it retraces the path JKMHJ ad infinitum or at least until its power fails.

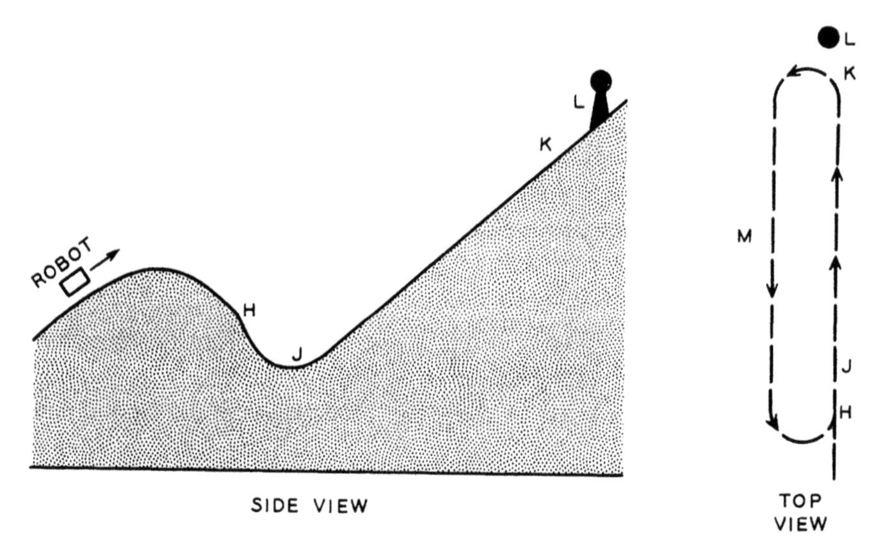

FIGURE 3. Indefinitely repeated behavior in a memoryless robot having completely fixed responses.

Memoryless robots can extricate themselves from traps of this kind if their behavior is probabilistic rather than completely deterministic. In the above completely deterministic robot, the probability, p, that the robot (if unobstructed) would move toward the object whenever it was in sight equaled 1. Suppose we replace some of the completely reliable neurons in the robot by slightly unreliable neurons, i.e., neurons which malfunction every now and then so that for each there is a small probability that it fail to fire when it should fire, or fire when it should not fire. Suppose in this way we make p somewhat less than 1, say $p = .89$, and let $p' = .1$ be the probability that the robot turn right through 90° when the object is in sight. Then sooner or later the robot climbing up from J will be far enough to the right to avoid the obstacle at K.

An outside observer (one unable to lift up the hood) could not predict the exact path of such a robot and might attribute to it some freedom of the will.

If some memoryless robots were constructed to destroy each other, it is clear that one built deterministically could be badly taken advantage of by those with properly weighted probabilistic responses.

A probabilistic behavior strategy can be simulated without inserting any unreliable elements if we use cycles. In Figure 4 the cells fire successively and each fires once every 4 time units. (The non-positive threshold on C_1 eliminates any problem about getting the cycle started and also allows it to start up again spontaneously after recovering from any accidental power failure or other stoppage.) It is clear that the period of each cell equals the number of cells in the cycle.

Let $Pr(x,y)$ be the probability that y fires at $t + 1$ if x fires at t. In general we may interrelate x, y and a cycle so as to simulate any given rational value of this probability as follows. If $Pr(x,y) = \alpha/\beta$ then

(1) let x have one excitatory bulb on y

(2) let θ on y equal 2

(3) let exactly α members in a cycle of β cells each have one excitatory bulb on y.

Since exactly α cycle members contact y the probability that any one of them fires at t equals α/β. Hence $Pr(x,y) = \alpha/\beta$. For example, in Figure 5 there are 7 cycle members (the inhibitory bulbs on C_1 are not shown) and 3 of them contact y_1. Hence $Pr(x,y) = 3/7$. For example, in Figure 5 there are 7 cycle members (the inhibitory bulbs on C_1 are not shown) and 3 of them contact y_1. Hence $Pr(x,y_1) = 3/7$; also, as shown, $Pr(x,y_2) = 2/7$ and $Pr(x,y_3) = 2/7$. The cycle is a kind of neurological roulette wheel, so to speak, determining whether or not the impulse is to pass through the synapse.

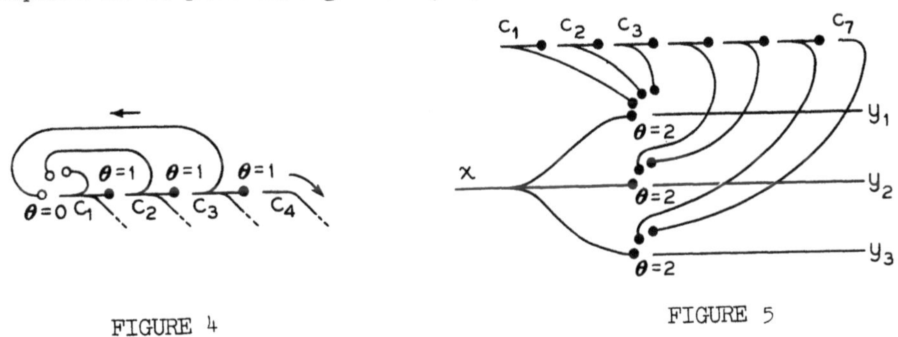

FIGURE 4

FIGURE 5

The possible input-output specifications for a probabilistic memoryless robot are not limited to $(2^{\mu})^{2^n}$ since for each receptor set and each effector set, we may assign any probability we want — the firing of the former cause the firing of the latter. Using these "roulette" cycles, however, it is easy to show in detail how to construct a memoryless robot satisfying any given probabilistic input-output specifications [2]. The more difficult but, no doubt, more practical reverse problem of getting deterministic behavior from unreliable elements has been worked out by von Neumann [5].

COMPLETE ROBOTS

Unlike deterministic memoryless robots, humans and animals do not react always the same way to a given spatial input. The way they respond may be influenced by inputs from the past — often the remote past. The way Professor Emeritus Jones behaves in the presence of ladies may depend on something that happened to him when he was four years old.

Human beings learn, they remember, and their behavior is modified by experience in various other ways. They ingeniously solve problems, compose symphonies, create works of art and literature and engineering, and pursue various goals. They seem to exhibit extremely complex behavior. The problem arises in constructing robots which would exhibit behavior of this kind.

In this section we will describe a general method for designing robots with any specified behavioral properties whatsoever. They can be designed to do <u>any</u> desired physically possible thing under any given circumstances and past experience, and certainly any naturally given "robot," such as Smith or Jones, can do no more. Remember, of course, that we are assuming, quite contrary to fact, that all the neurons we want are available, that they are small enough, etc., and that we have enough mechanical dexterity and enough time to assemble them.

We have examined robots with no memory at all; now let us go to the other extreme and examine robots with complete memory, i.e., a completely detailed memory, at any time t, of all their past receptor firings. Robots with complete memory do not resemble any actual organisms or any constructed hardware since these latter have incomplete (in the sense of <u>selective</u>) memory — they store only those parts of their past experience which are especially significant. Some appropriate definition of "significant" can be given in each case.

The robot with complete memory, although it is uneconomical, seems to be conceptually simpler than the robot with selective memory. We will examine the robot with complete memory first and the selective memory robot can be examined later as a modification of it. Hence, "memory" here means "complete memory" as just explained. Also we will consider only deterministic robots since it is clear how they could be made probabilistic.

These complete robots are just extensions of memoryless robots and their circuitry is essentially the same. It is merely that the receptor cells together with a large bank of other cells constitute the storage cells, and this necessitates a very large increase in the number of central cells $c_1, c_2, c_3, \ldots$. The complete robot may be constructed as shown in Figure 6.

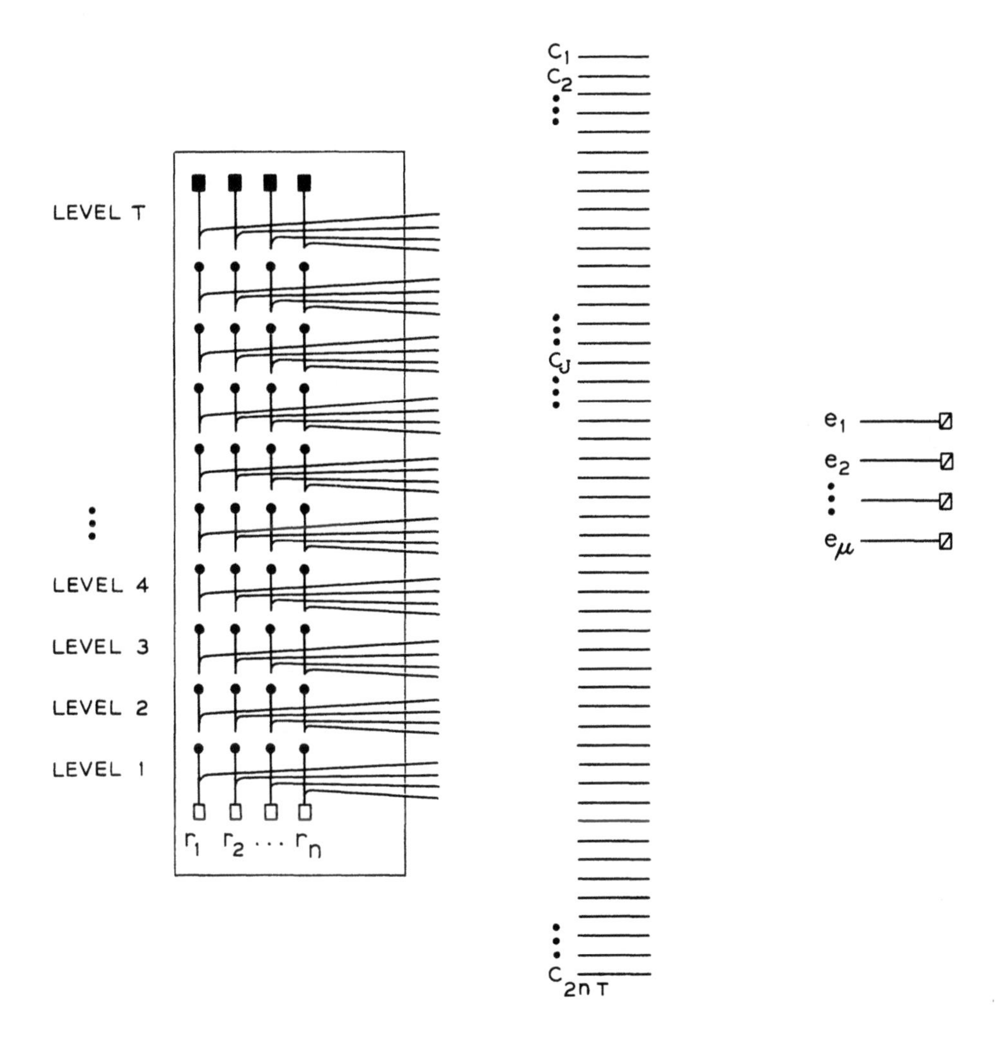

FIGURE 6. Complete Robot. Receptor cells r_1, r_2, . . . r_m. Storage cells enclosed by rectangle. Effector cells e_1, e_2, . . . e_μ.

Connections Between The Storage Cells

The storage consists of T levels, the receptor cells constituting level 1. Each level contains n cells. Each cell in levels 1 to $T - 1$ has an excitatory bulb on the cell immediately above it; that is, the ith cell in the kth level has one bulb on the ith cell in the $(k+1)$st level. There are no other connections between cells in the storage. The connections between the storage cells and the cells $c_1, c_2, \ldots c_{2^{nT}}$ will be described later.

If the robot is constructed, then there will be some time t_o which is the first time that any of its receptors have fired. Prior to t_o no receptors have fired, and at t_o one or more receptors fire. It will be convenient to call t_o the time of birth of the robot.

From the simple connections described above, we see that impulses move continually upward through the storage — if the ith receptor cell fires, then the cell immediately above that one fires, etc., until finally the ith cell in level T fires. It is clear that, however many receptors fire at any time, each input, however complex, is preserved intact, so to speak, as it moves upward through the storage and that at any time after birth and prior to $t_o + T$ there is in the storage a record of all inputs from birth up to that time.

The robot begins to function at birth, t_o, and ceases to function at $t_o + T$. We can arrange this by having each cell at level T connected directly to a charge of dynamite so that when any cell at this last level fires, then the total charge goes off immediately and the robot is destroyed. The time, $t_o + T$, is the time of death of the robot, and T is called the longevity of the robot.

Since in the present discussion we care nothing about neuroeconomy, we can make T as great as we please. Suppose we constructed a robot so that $T = 2.2 \times 10^{12}$ which gives it a life time of about seventy years. Then it would take about seventy years for the first, or birth, input to move upward all the way to the lethal cells at the last level. The robot would be destroyed T time units or about seventy years after its birth. At any time during its life there would be a completely detailed record, in the storage, of the robot's total past experience.

The nT storage cells we will designate $s_1, s_2, \ldots s_{nT}$. The subscripts may be assigned to them in any convenient fixed way.

There are then 2^{nT} classes or sets of these storage cells. In the usual way, already familiar to the reader, we may label these as follows:

1st storage set	The null set	...0000
2nd " "	s_1	...0001
3rd " "	s_2	...0010
4th " "	s_1+s_2	...0011
5th " "	s_3	...0100
6th " "	s_1+s_3	...0101
7th " "	s_2+s_3	...0110
.	.	
.	.	
.	.	
Jth " "		
.		
.		
.		
2^{nT}th " "	All the storage cells	...1111

Note that all 2^{nT} storage sets are designated, not just those corresponding to the rows or columns or anything of that kind.

Corresponding to these 2^{nT} sets of storage cells we have the 2^{nT} central cells $c_1, c_2, \ldots, c_J, \ldots, c_{2^{nT}}$ shown in Figure 6. Now we can connect the storage cells to these central cells as follows.

Connections From The Storage To The Central Cells

Each storage cell in the Jth set has an excitatory bulb on the Jth central cell, while each storage cell which is not in the Jth set has an inhibitory bulb on the Jth central cell. Also we make the threshold on any central cell equal to the number of excitatory bulbs contacting it. Thus the Jth storage set fires the Jth central cell.

Consider any time t $(t_0 < t < t_0 + T)$ in the life of the robot. At t some set of storage cells is firing. Which set of storage cells is firing is uniquely determined by the particular past experience of the robot — that is, by just which receptors fired at each instant from t_0 to t. For each possible past history of receptor firings there is a different set of storage cells. The actual past history up to any time t determines which storage set fires at t. Which storage set fires at t then determines which central cell fires at $t + 1$, as explained in the above paragraph. Thus, which central cell fires at time $t + 1$ is uniquely determined by the entire detailed past history of experience (receptor firings) of the robot. For a different past history a different central cell would fire. This involves an impossible number of central cells and endbulbs at each synapse, but we will remain calm even in the presence of such unnaturally large numbers, since certain issues become clearer when we do so.

Input-Output Specifications

If the robot is to have effectors $e_1, e_2, \ldots e_\mu$ then there are 2^μ sets of these and we may order these effector sets in the same way as we ordered them for the memoryless robot.

To decide on the input-output specifications of the robot is to specify for each storage set an effector set such that if the former fires at t then the latter fires at t + 3. There are $(2^\mu)^{2^{nT}}$ ways of deciding on these input-output specifications. Each storage set corresponds to a possible past history of experience up to some time t, and for each such possible past history we must specify what we want the robot to do. (Even if for some particular past history up to t we wanted the robot to perform some extended sequence of responses, it would still suffice merely to specify for each t what effectors are to fire at t + 3 since the past histories up to t + 1, t + 2, t + 3, t + 4, etc., can continue to produce the desired sequence of responses.)

Central-to-Effector Connections

We will suppose that drawing up the complete input-output specifications is practically possible, that is, we will suppose that we have enough time and enough perseverance. When we have decided on the input-output specifications for our complete robot, then the neural connections may be completed as follows. This is tantamount to saying that in the following way we may construct the robot so that it will behave any way we want it to, specifically, for each possible past history of experience up to each moment of its life.

The Jth storage set corresponds to one of the 2^{nT} past histories of experience. If the specifications call for the Jth storage set to fire the ℓth effector set, then the Jth central cell has one excitatory bulb on each member of the ℓth effector set. The threshold on each effector cell is made equal to 1.

Characteristics of Complete Robots

Since they can, in principle, satisfy any given input-output specifications, they can do any prescribed things under any prescribed circumstances — ingeniously solve problems, compose symphonies, create works of art and literature and engineering, and pursue any goals. They can be given any desired behavioral properties.

They have some peculiar features, or rather, they don't have some peculiar defects. For example, one could be constructed to behave in the same way as John Jones would under any circumstances, and hence, as a special case, it could simulate any of Jones's learning activity. On the other hand, it could be constructed so that learning processes were unnecessary. Jones has to learn each German word or phrase in learning German. It, however, according to the whim of its designer, might, for example, give no appropriate responses in German at all until it heard the word "schnell", and then right away after that speak perfect German. It is idle to give further examples since it could be designed to do any physically possible thing for any given circumstances and past history of experience.

Much less than a complete robot, then, can simulate Jones — in the sense that with Jones's actual past history it would do just what Jones does, and with any other past history it would do just what Jones himself would do with that past history.

It is easy, if we have some ingenuity (the more the better) to design all sorts of relatively simple and economical feedback devices like the RAND electromechanical animals or Grey-Walter's turtles, which delight the gadgeteer and are instructive and entertaining to all and seem so life-like if not pondered too closely.

It is also easy to indicate, as we have done, the totally uneconomical design of complete robots which can have any behavioral properties whatsoever, including human behavior as a special case.

Relatively easy then are the two extremes — the uneconomical but perfectly general method of designing robots with any behavioral properties whatsoever, however complex, and the economical and practical design of simple robots, each worked out almost as a special case. The intermediate field is the difficult one. How can we design automata with the intermediate, biologically given, behavioral complexity and yet do it with something like nature's economy? What sort of design can the brain have to accomplish human behavior strategy with only 10^{10} cells? It is certainly not any relay design ingeniously fitted into a neuroeconomical straight jacket. The neurologist rejects such a notion, even though interested in possible future electronic developments. Large portions of the brain can be removed, and the cells are not completely reliable although forming long chains. There is also plasticity in many other respects.

While the main attempt at present is to build simple devices to greater and greater complexity, it might be interesting for someone to examine the alternative approach of starting with the complete robot and then paring it down to human and animal proportions by biologically suggested economies.

BIBLIOGRAPHY

[1] CULBERTSON, J. T., Consciousness and Behavior, Dubuque, Iowa, Wm. C. Brown Co., 1950.

[2] CULBERTSON, J. T., "Hypothetical Robots," RAND Project P-296, April 1952.

[3] McCULLOCH, W. S. and PITTS, W., "A Logical Calculus of Ideas Immanent in Nervous Activity," Bulletin of Mathematical Biophysics, 5, 1943.

[4] RASHEVSKY, N., Mathematical Biophysics, 1948. Part Three.

[5] von NEUMANN, J., "Probabilistic Logics and the Synthesis of Reliable Organisms from Unreliable Elements," California Institute of Technology Lectures, January 1951, and these Automata Studies, pp. 43-98.

[6] KLEENE, S., "Representation of Nerve Nets and Finite Automata," RAND paper and these Automata Studies, pp. 3-41.

SOME UNIVERSAL ELEMENTS FOR FINITE AUTOMATA

M. L. Minsky

In recent years a number of theories have appeared all of which are concerned with the construction of complicated "machinery" from a small number of basic elements. The neurological models of Rashevsky, and of Pitts and McCulloch, the more general models of Kleene, and in logic the combinatorial constructions of Curry and several others, all display similar features in this regard. In the present paper, it is shown that a certain category of sets of elements are "universal" in the sense that one can assemble such elements into machines with which one can realize functions which are arbitrary to within certain reasonable restrictions. The discussion takes place within the framework of Kleene's theory of "finite automata" as generalized from the McCulloch-Pitts neurological model [1943].

1. Elements and Response Functions

An "element" J will be defined as an object with a finite number of "input fibres" S_i and a finite number of "output fibres" R_i. At any of a discrete set of "moments" each fibre may be in either of two states, "quiescent" or "firing." Each element will be assumed to act as though it were a "finite automaton" in the sense of Kleene (ref.) with the input fibres playing the role of "input cells" and the output fibres playing the role of some of the "inner cells" of the corresponding automaton. It will be further assumed that each element acts as though it were an "automaton without cycles," i.e., as though it has a finite set of internal states and that if no input fibre is fired for a sufficient interval (of duration $d(J)$ moments) the element reverts to a basic "quiescent" state and so remains at least until input activity is resumed.

A few examples of useful elements are listed below:

i. "Con-junctions." This is the McCulloch-Pitts element with two input fibres and "threshold 2" i.e., the element fires at time t if and only if both input fibres are fired at time $t - 1$. The element is represented both by the diagram and the equation below:

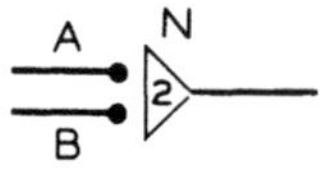

$$N(t) \equiv A(t-1).B(t-1)$$

ii. "Dis-junctions." This element has also two input fibres and fires at time t if and only if either or both input fibres fire at time $t - 1$.

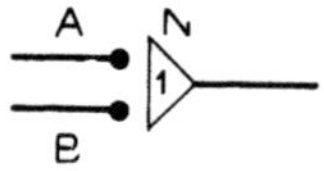

$$N(t) \equiv A(t-1) \vee B(t-1)$$

iii. "Binary inhibitory junction" or "b. i. j." This element has two input fibres E and I and one output fibre. The input fibres are different in that the element fires at time t if and only if fibre E is fired at $t - 1$ and fibre I is NOT fired at $t - 1$.

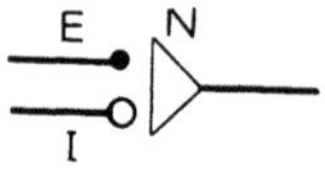

$$N(t) \equiv E(t-1).\overline{I(t-1)}$$

iv. Elements with a "unit refractory period." An element will be said to have a unit refractory period if, regardless of other firing conditions, it cannot fire on two successive moments. Thus if a con-junction N has a unit refractory period, its behavior can be described by the equation

$$N(t) \equiv A(t-1).B(t-1).\overline{N(t-1)}.$$

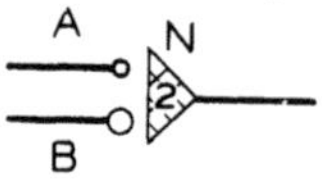

A "net" composed of copies of a set $J_1, \ldots, J_j$ of elements will be defined as an assembly of such elements in which output fibres of some elements will be connected to input fibres of others. It is permitted to split any output fibre into any number of fibres, but this is not allowed for input fibres. A certain collection of input fibres $S_1, \ldots, S_m$ will be distinguished as "input fibres of the net" and similarly for some distinguished collection $R_1, \ldots, R_n$ of output fibres.

A stimulus A_α is a finite set of firings of input fibres of the net; thus A_α is a union of a finite set of events $S_{\alpha_i}(t_{\alpha_i})$ $(i = 1, \ldots, n_\alpha)$ where $S_i(t_j)$ means that fibre S_i fires at time $t = t_j$.

A response of the net is defined similarly to be a union Z_β of events $R_{\beta_i}(t_{\beta_i})$ where $R_i(t_j)$ means that fibre R_i fires at time t_j.

(We will consider only nets which produce finite responses to finite stimuli.)

A <u>response function</u> f will be a finite pairing of stimuli with responses; $f: A_\alpha \longrightarrow Z_{f(\alpha)}$. (f may be regarded as operating either on the stimuli or their index sets.)

In order to eliminate complications related to the question of absolute time it is convenient to restrict the domain of definition of a response function to some <u>"admissible class of stimuli II,"</u> as follows: <u>A class II of stimuli is admissible if it has the property that if $A = \{S_{\alpha_i}(t_{\alpha_i})\}$ is in II then for no $k \neq 0$ is the stimulus $A^k = \{S_{\alpha_i}(t_{\alpha_i} + k)\}$ also in II.</u> Thus an admissible class contains no pair of stimuli which are equivalent under time translation. It follows that the null-stimulus cannot be a member of any admissible class. This restriction is very mild, since unless a net has some kind of internal absolute clock it cannot be expected to distinguish between time translated stimuli. If the net contains some kind of cyclic activity with period p (and if a finite net contains permanent activity this activity must be essentially periodic) then it might distinguish between time translations on a residue class basis; the definitions might be extended to cover this case but it seems hardly worth while, at this stage.

An example of an admissible class of stimuli which will be useful later is the class $II_a(b)$ containing <u>all</u> stimuli A for which

$$(\exists i) S_i(a) \in A . \left(S_j(t) \in A \implies a \leq t < a + b\right)$$

i.e., $II_a(b)$ is the class of all stimuli whose first firing occurs at time a and whose total duration is not longer than b moments.

2. Realization of Response Functions

In the general problem considered here we will be given a set of elements $J_1, \ldots, J_n$ and also a response function f defined over some admissible class of stimuli II. We are asked to construct a net out of copies of the J_i which will "realize" the given function. Now it may turn out that this cannot be done. Consider the following example. It is easy to show that if each of the elements of a net have a "monotonic"[1] response function then so will the net as a whole. Then no non-monotonic response function can ever be realized by any net composed of elements each of which is itself monotonic.

---

[1] A response function is <u>monotonic</u> if

$$A_1 \subset A_2 \implies f(A_1) \subset f(A_2).$$

Thus "monotonic" means "weakly monotone increasing, set-theoretically."

In this example the non-realizability is due to a kind of basic "functional" incompatibility between the given elements and the given function. It may happen, however, that the failure of a class of nets to realize a given function is due rather to trivial limitations on the "computation speed" or "information channel capacity" of the elements, than due directly to function-theoretic limitations. In order to maintain this distinctio and not to discriminate too harshly against nets which are simply "too slow," the notion of "realization" of a function will be coupled with notions of "computation delay" and of "expanded time scale."

DEFINITION: Given a response function f; $A_\alpha — Z_{f(\alpha)}$, a net will be said to "realize f with delay d" if the net in fact realizes the function

$$f^d: \quad A_\alpha \longrightarrow Z_{f^d(\alpha)} \qquad \text{all } A_\alpha \in II$$

where

$$R_k(t) \in Z_{f^d(\alpha)} \equiv R_k(t - d) \in Z_{f(\alpha)}.$$

DEFINITION: Under the same conditions a net will be said to realize f "in the p-expanded time scale" if the net in fact realizes the function

$${}_pf: \quad {}_pA_\alpha \longrightarrow {}_pZ_{f(\alpha)}, \quad \text{where}$$

$$S_i(pt) \in {}_pA \equiv S_i(t) \in A \quad \text{and}$$

$$R_j(pt) \in {}_pZ \equiv R_j(t) \in Z.$$

If f is defined over an admissible class, then so are f^d, ${}_pf$ and ${}_p(f^d)$ etc. In the p-expanded time scale, both the stimuli and the responses are stretched out uniformly in time. One can imagine a machine with an input channel capacity so limited that the stimuli cannot be absorbed at the given rate, but that if the signal is slowed down to match the input capacity the given problem can be handled. Our main result is the

<u>THEOREM: Given a response function f defined on an arbitrary admissible class II of stimuli, there are realizations of f^d (for some d) and of ${}_pf$ (for some $p > 0$) in nets which contain only copies of the following elements:</u>

i. "dis-junctions" and "con-junctions."

ii. Any element J whose response function is non-monotonic.

(Since it can be shown that any monotonic response function can be realized in the form f^d by a net composed only of con-junctions and dis-junctions, condition ii. can be replaced by the condition

ii'. Any element J which cannot be realized (with some delay) by the elements of i.)

The elements iii. and iv. of section 1. are examples of non-monotonic elements. In the following proof of the above theorem, no effort will be made to minimize d and p where simplicity of proof would suffer. The proof will be divided into several parts. Perhaps use of the "active loop" of C below should be permitted explicity as an hypothesis.

A. CONSTRUCTION OF THE ELEMENT Q^J. A net Q^J is constructed which isolates the non-monotonicity of the element J. See Figure 1.

Let J have m input fibres, n output fibres, and response function f_J. Since J is non-monotonic, there are two stimuli A_1 and A_2 for which A_1 A_2 but $f_J(A_1) \not\subset f_j(A_2)$. Then there must be an $R_{k*}(t*)$ for which $R_{k*}(t*) \in f_J(A_1)$ and $R_{k*}(t*) \notin f_J(A_2)$, where R_{k*} is some output fibre of J. Assume that $t = 1$ is the time of the first pulse of A_2 (otherwise time-translate A_1 and A_2 and f_J so that this will be the case). Supply now two input fibres N_1 and N_2 for the net, and for each $S_k(t)$ in A_1 run from N_1 a chain containing $t-1$ intermediate cells and ending on an input fibre of a dis-junction attached to the fibre S_k of J. Do the same for N_2 and each $S_k(t)$ in A_2. (NOTE: We shall not explicitly limit the number of input fibres to con-junctions or dis-junctions to 2, but will allow an arbitrary number. This can be reduced to the 2 - case by the construction of simple sub-nets; the additional delays introduced will lead to a higher value of "d" for the construction, but not alter in any essential way the main lines of the proof.)

The stimulus $N_1(0)$ will cause the stimulus A_1 to be presented to J, and likewise for N_2 and A_2. The stimulus $N_1(0).N_2(0)$ is equivalent to $N_2(0)$ as far as J is concerned. Now supply a con-junction R' with one input connected to R_{k*} and the other connected to the output of N_1 through a chain of $t*$ intermediate cells. Then the firing of R' will represent the response function $R'(t* + 1) = N_1(0).\overline{N_2(0)}$, assuming that pulses arrive at N_1 and/or N_2 only at $t = 0$. There is no telling what R' will do if sequential stimuli arrive at the input. However since J is a "finite network element", there is a certain time $d(J)$ such that if no S_i is fired in an interval of this duration then J will revert to its resting state. Let $T(J) = d(J) + (\text{time of latest pulse in } A_2)$. Supply a chain of length $T(J) - d(J)$ beginning with R' and ending on a final cell called R^J. Define the net Q^J to be the net so constructed and having N_1 and N_2 as input fibres and R^J as output fibre. This net will have the property that it will act as though it were a b. i. j. with delay $T(J)$ if it is used in such a way that not more than one input firing time occurs within any interval of duration $T(J)$.

B. THE "SEQUENTIAL RESOLVER" NET. The first step in construction of the realization net will be construction of a net N^{SR} which will convert the sequential patterns of stimuli into equivalent non-sequential patterns. In

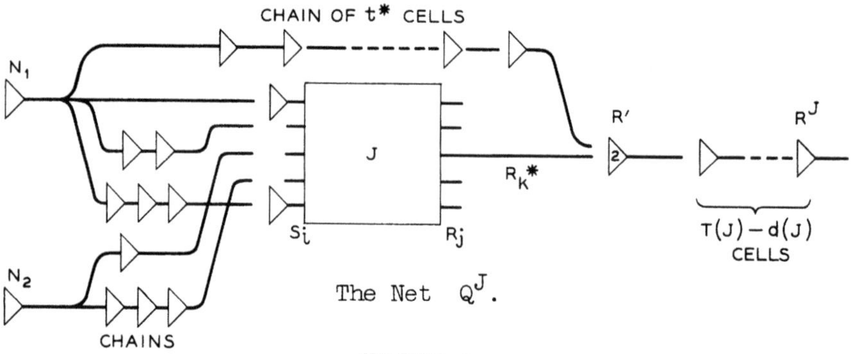

The Net Q^J.

FIGURE 1

computer language, the system is to be converted from serial to parallel operation, at the expense of duplication of machine components. Assume, for the present, that the stimuli are in $II_0(b)$; this will be generalized later.

The net N^{SR} makes use of identical copies of the following net N_i^{SR}, one for each input variable of the function to be realized.

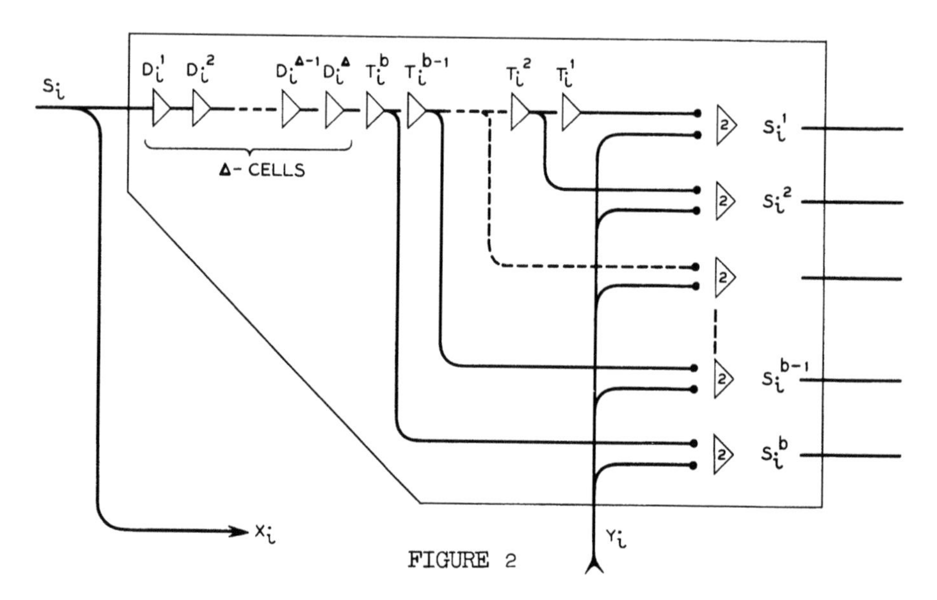

FIGURE 2

All cells in N_i^{SR} are disjunctions except for the con-junctions S_i^j. The net has the property that if $A \in II_0(b)$ and if <u>Y_i is fired only at time $\Delta + b$,</u> then $S_i^j(1 + \Delta + b) \equiv S_i(j) \in A$. (The value of Δ will be fixed later.) Hence the sequential stimulus occuring in the interval $(0, b - 1)$ is converted into a unit duration firing pattern on the S_i^j

The total net N^{SR} is composed of the N_i^{SR} by tying all the fibres Y_i to the output of a single cell Y and connecting fibres from each S_i to the input of a single dis-junction X. Between X and Y will go the "unit pulser", to be described in the next section.

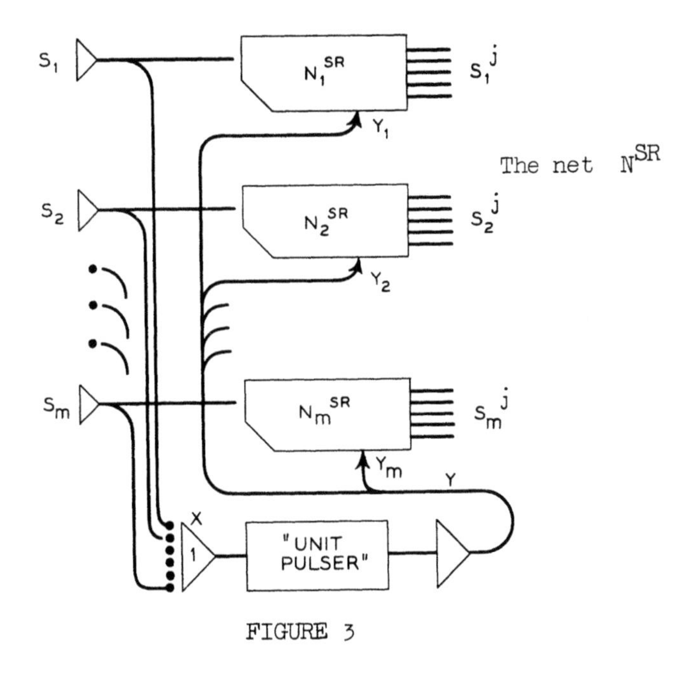

FIGURE 3

The latter is the device which is supposed to produce the unit pulse at $t = \underline{\Delta + b}$. Except for this the realization of the function f is easily completed. For each stimulus A_a of $II_0(b)$ we supply three elements; a dis-junction E_a, and con-junction F_a and a copy Q_a^J of Q^J. For each $S_i(j)$ in A_a connect a fibre from the output of S_i^j to an input of E_a. For each $S_i(j)$ NOT in A_a connect a fibre from S_i^j to an input of F_a. Connect the output of E_a to the input N_1 of I_a and the output of F_a to the input N_2 of I_a. Then the following may be verified: If A_a is the stimulus, the only activity of the output cells R_α^J of the nets Q_α^J will be the firing of the single cell R_a^J at time $(2 + \Delta + b + T(J))$, a time which is independent of $A_\alpha \in II_0(b)$. Now let the function to be realized be f: $A_\alpha \longrightarrow Z_{f(\alpha)}$. Supply a bank of dis-junctions R_k of "final output cells." For each $A_a \in II$ and each $R_k(t) \in Z_{f(a)}$ run a chain of $t - 1$ intermediate cells from the output of R_a^J to an input of R_k.[2] The resulting net, with the S_i taken as inputs and the R_k taken as outputs, can be seen to realize the function

$$f^{2 + \Delta + b + T(J)}.$$

---

[2] If there is an $R_k(t)$ with $t \le 0$ this could be realized here by inserting a uniform delay at this final level; it would be reasonable to rule out such an event as inconsistent with the notion of causality.

C. <u>CONSTRUCTION OF THE "UNIT PULSER."</u> We have to construct a net which produces a single pulse output in response to any stimulus of a given admissible class, and such that the time of the pulse is independent of the particular stimulus of the class. It will be shown that once a single pulse is available it is not difficult to get the pulse to be independent of the stimulus. But serious difficulties arise in constructing a device to get a single pulse even without regard to the time of its occurrence. For such a device is distinctly non-monotonic and must use copies of J. The non-monotonicity of J can be extracted by the construction of a device such as the net Q^J. But this device, and presumably any like it, must be protected from receiving closely spaced series of pulses (in general). Yet the "unit pulser" may be regarded as a device whose precise function is to remove pulses from arbitrary series of pulses until there is just one pulse left. Where can this reduction of the number of pulses occur? It is easy to see that no monotonic network can reduce the number of pulses in one given series without giving a null response to some other stimuli (e.g., to any single pulse input). Hence a net like Q^J will ultimately have to receive the series of pulses, and protection of its input by a non-monotonic net will just push the problem back to the protection of the Q^J's or equivalent in the "protecting" net. It seems some other technique must be invoked. First, one may take recourse to operating the whole net in the $T(J)$-expanded time scale. Then the Q^J's will not need protection, and the realization problem becomes simple for the function $T(J)^f$.

The problem can also be solved by the introduction of a permanently active "closed loop" into the network. While this is done with reluctance, it seems, in general, to be necessary. There are some important special cases in which this can be avoided and those will be mentioned after the general method is described.

The following net represents a complete "unit pulser." It has three parts: See Figure 4.

- I A device for converting arbitrary stimuli of $II_0(b)$ into the standard form of unbroken series of pulses (the length of which series may vary),
- II A closed loop which acts as a periodic pulse source, and
- III A network which makes the output of the assembly independent of the phasing of the loop activity with respect to the stimulus.

All elements are dis-junctions except for a bank M_i of conjunctions and a bank of copies of Q^J.

The loop of cells $(P_1, P_2, \ldots, P_{T(J)})$ is assumed to contain one circulating pulse so that P_1 fires exactly once in any interval of duration $T(J)$. Operation of the net is as follows: The cells K_i convert the input stimulus pattern into an unbroken firing sequence at $L_1 = K_b$. This train of pulses propagates out along the L_i and the <u>leading pulse</u> of the train remains in

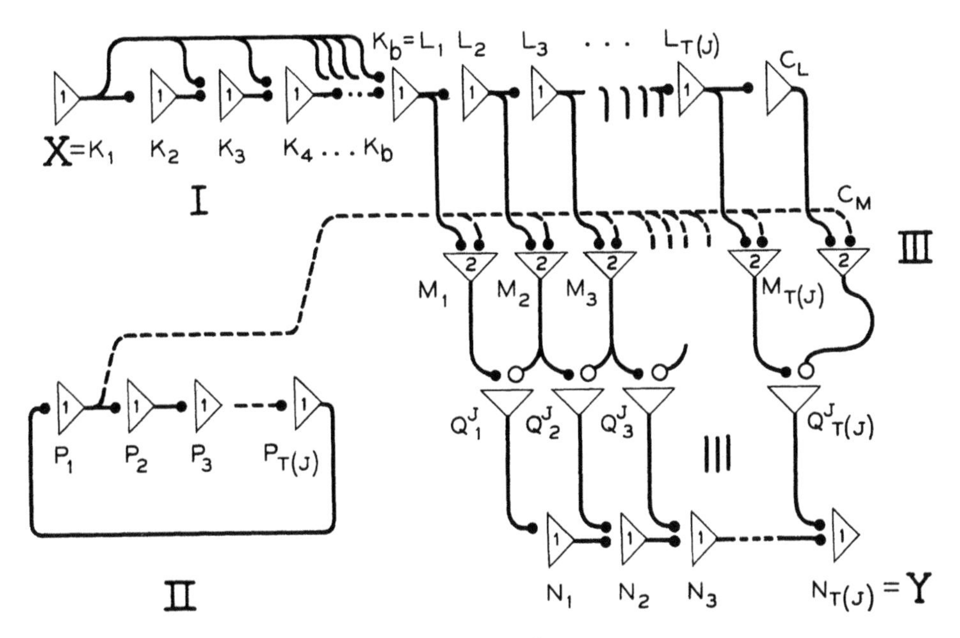

FIGURE 4

the L_i for exactly $T(J)$ moments. At exactly one moment in this interval the cell P_1 fires and an image of the activity in the L_i is copied at the M_i level. This pattern is presented at the next moment to the bank of Q^J_i units. Note that since no Q^J can have been excited in the previous $T(J) - 1$ moments, they are ready to act as b. i. j.'s. Exactly one Q^J_i will now fire; namely the one which is exposed to the leading pulse of the train. (At a time $T(J)$ moments later, there may still be pulses in the L_i, but the leading pulse will have left the bank of L-cells and no Q^J can fire at this time.[3] The cells marked C_L and C_M prevent the end of the L-bank from behaving like a leading pulse.) Firing of one Q^J_i allows one pulse to enter the bank N_i of cells. This bank adjusts the time of arrival of the single pulse at the final output cell Y so that the time of arrival is independent of the firing time of P_1! This time (for stimuli in $II_0(b)$) is exactly $1 + b + 2T(J)$. The "unit pulser" is complete. The value of Δ is fixed by the relation $\Delta + b = 1 + b + 2T(J)$ so that $\Delta = 1 + 2T(J)$. The entire system now realizes the function

$$f^{b+3+3T(J)}.$$

For operation in the $T(J)$-expanded time scale, first remove the loop of the P_i. Remove the bank of M_i cells and connect directly from the output fibres of the L_i to the corresponding inputs of the Q^J_i. Then between every element of the net and its output fibre insert a delay chain of length $T(J) - 1$. In the expanded time scale the Q^J's will not need

[3] This construction involving the L, M, and Q elements was discovered independently by Dr. L. S. Shapley.

the previous protection. After adjustment to a new value for Δ the net will realize the function ${}_{T(J)}(f^{d})$ for some d of the order of magnitude of $b + 3 + 3T(J)$. It is not difficult to show how to modify the net so that it realizes the function ${}_{p}f$ for a p somewhat greater than the above d. Smaller values of p and d could probably be obtained, but there is no reason to try to find minimal realizations at this point.

It remains to extend the result to the general admissible class II. Let b be the maximum duration of stimuli in II. Let t_a be the time of the first pulse in the stimulus A_a. Then if $f\colon\ A_a \rightarrow Z_{f(a)}$ is the given function, define $A_a{}^*$, $Z_b{}^*$ and f^* as follows:

$$S_k(t) \in A_a^* \equiv S_k(t + t_a) \in A_a$$

$$R_j(t) \in Z_{f(a)}^* \equiv R_j(t + t_a) \in Z_{f(a)},$$

$$f^*\colon\ A_a^* \rightarrow Z_{f(a)}^*.$$

Then A_a^* is in $II_0(b)$ for all A_a in II. And because II is an admissible class, $A_a \neq A_b \Rightarrow A_a^* \neq A_b{}^*$. The function f^* can then be realized as shown above. But the response function of the net constructed for the realization of f^* is invariant under time-translation of input and output. Hence the net which realizes f^* must also realize the given function f. This completes the proof.

3. Special Cases

For certain elements, realization of arbitrary functions can be achieved without recourse either to the expanded time scale or the loop device. A simple instance is that of the "binary inhibitory junction" of section 1. In this case $T(J) = 1$ so that realization in the $T(J)$-scale is realization in the ordinary time scale. A more interesting case is that of the cell with unit refractory period. It is actually possible to realize arbitrary functions in the ordinary time scale with nets composed entirely of con-junctions and dis-junctions <u>each of which</u> has a "unit refractory period" (or any longer period), provided only that we are supplied with banks of input and output cells which do not have this limitation, (an obviously necessary condition), and it is not necessary to use the loop device. The proof is much too specialized to present completely, but the method may be of interest. The net is similar to the previous constructions except in two respects; first the main net I is split into two identical halves and a device is introduced which shifts the stimulus pulses alternately between the halves so that each half can operate in the 2-expanded time scale. Instead of the loop device, the following net is introduced, for stimuli in $II_0(b)$: This net produces exactly one output pulse for any stimulus in $II_0(b)$; the reader is encouraged to trace through its operation to see how this is done. Remember that <u>no</u> element can fire on two successive moments. The time of

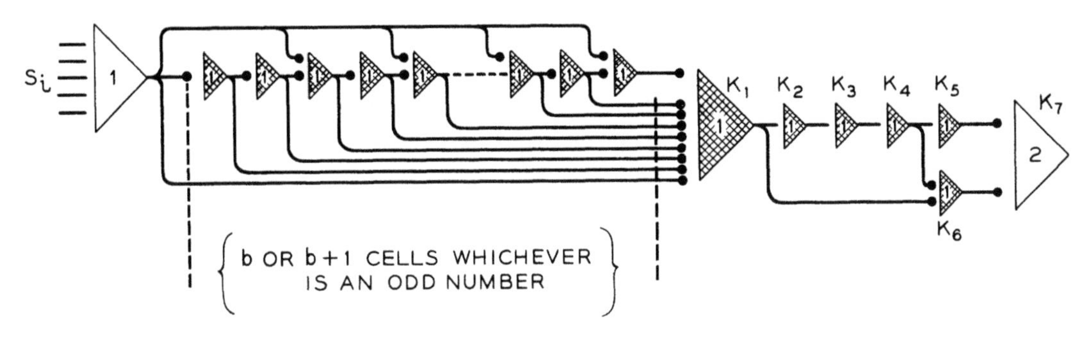

FIGURE 5

the pulse does depend on the stimulus, and a device similar to part III of the "unit pulser" (section C above) can be constructed to remedy this. In the "shifting" device above a similar method must be used.

Both con-junctions and dis-junctions are realizable with a single copy of the McCulloch-Pitts element

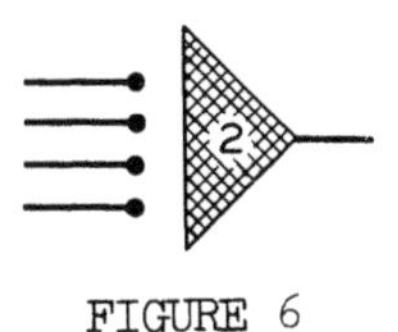

FIGURE 6

and if this element is provided with a refractory period, it becomes an essentially universal element, by itself. The same is true for the element since con-junction (with delay 2) can be obtained in the form

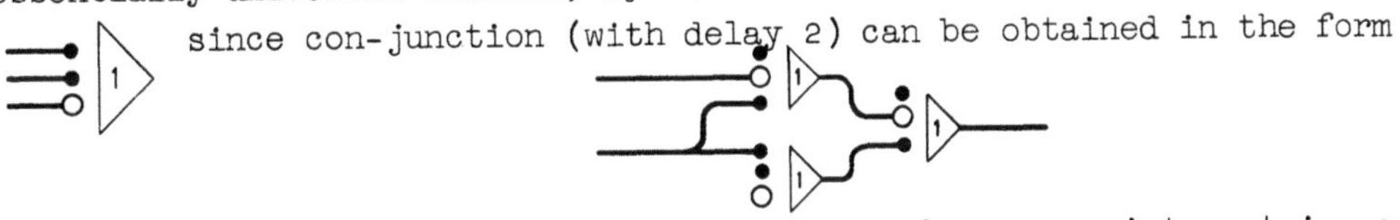

The former of these two universal elements may have some interest in connection with brain theory inasmuch as it shows that, generally speaking, the rather weak property of having a refractory period is sufficient, in principle, to account for physiological phenomena which might superficially appear to show a "more strongly non-monotonic" character such as "inhibition."

The hypothesis that "con-junctions" and "dis-junctions" are initially available is motivated by the prevailing opinion in neurophysiology that such elements are almost certainly represented among, and in fact are probably characteristic of, the cells of the central nervous system. On the other hand, the nature and distribution of the non-monotonic properties of the nervous system are not nearly so well understood, in particular the various forms of "inhibition". Thus the particular form of the theorem may be of some value in analysing those neural phenomena in which there appears to be an "inhibitory" quality but in which no specific inhibitory connection or mechanism has been isolated. The neuro-physiologist should be aware of the fact that the element of Figure 6, in particular, is universal, since it would not be rash to conjecture that a very large proportion of central nervous cells have some properties like those of this configuration, or differing only quantitatively. It is not my intention to suggest that the particular nets constructed for this proof bear any resemblance to mechanisms to be found in the central nervous system; their form reflects primarily the "reduction to canonical form" which is the major mathematical and logical tool used for the proof of the main theorem. To realize any particular response function, there are many other nets that could be used, not a few of which would be much more efficient in the number of elements used and perhaps much less orderly in appearance. It is perhaps worth adding that there would appear to be no reason why the recent results of von Neumann and Shannon on automata with "probabilistic" or "unreliable" elements could not be applied to the constructions of the present paper to satisfy the biological critic that the validity of the main theorem does not at all rest on what might be felt to be a perfection of hypothetical elements that could not mirror any biological situation.

Society of Fellows,
Harvard University.

BIBLIOGRAPHY

Most of this material is derived from Chapter II of my dissertation; the results there are in slightly more general form.

[1] CURRY, H. B., No specific reference is intended but see e. g., Am. J. Math., Vol. 52, 509 ff., 789 ff.

[2] KLEENE, S. C., See this volume.

[3] McCULLOCH, W. S. and PITTS, W., 1943, "A Logical Calculus of the Ideas Immanent in Nervous Activity," Bull. Math. Biophysics, 5, 115 ff.

[4] MINSKY, M. L., Neural-Analog Networks and the Brain-Model Problem, Princeton University, 1954. This is not in print, but microfilm copies are available. A revision is in preparation for publication.

[5] von NEUMANN, J., See this volume.

[6] RASHEVSKY, N., 1938. Mathematical Biophysics. Univ. of Chicago Press.

[7] SHANNON, C. E., See this volume.

GEDANKEN-EXPERIMENTS ON SEQUENTIAL MACHINES

Edward F. Moore

INTRODUCTION

This paper is concerned with finite automata[1] from the experimental point of view. This does not mean that it reports the results of any experimentation on actual physical models, but rather it is concerned with what kinds of conclusions about the internal conditions of a finite machine it is possible to draw from external experiments. To emphasize the conceptual nature of these experiments, the word "gedanken-experiments" has been borrowed from the physicists for the title.

The sequential machines considered have a finite number of states, a finite number of possible input symbols, and a finite number of possible output symbols. The behavior of these machines is strictly deterministic (i.e., no random elements are permitted in the machines) in that the present state of a machine depends only on its previous input and previous state, and the present output depends only on the present state.

The point of view of this paper might also be extended to probabilistic machines (such as the noisy discrete channel of communication theory[2]), but this will not be attempted here.

EXPERIMENTS

There will be two kinds of experiments considered in this paper. The first of these, called a simple experiment, is depicted in Figure 1.

---

[1]The term "finite" is used to distinguish these automata from Turing machines [considered in Turing's "On Computable Numbers, with an Application to the Entscheidungsproblem", Proc. Lond. Math. Soc., (1936) Vol. 24, pp. 230-265] which have an infinite tape, permitting them to have more complicated behavior than these automata.

[2]Defined in Shannon's "A Mathematical Theory of Communication", B.S.T.J. Vol. 27, p. 406.

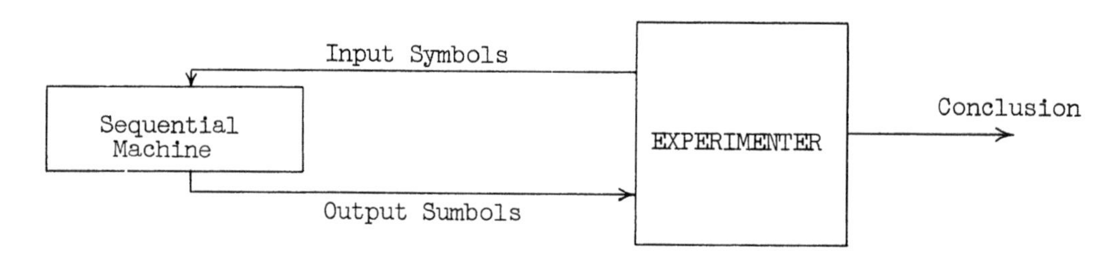

FIGURE 1. Schematic Diagram of a Simple Experiment

A copy of the sequential machine being observed experimentally will receive successively certain input symbols from the experimenter. The sequence of output symbols will depend on the sequence of input symbols (the fact that the correspondence is between sequences rather than individual symbols is responsible for the terminology "sequential machine") in a way that depends on which particular sequential machine is present and its initial state.

The experimenter will choose which finite sequence of input symbols to put into the machine, either a fixed sequence, or one in which each symbol depends on the previous output symbols. This sequence of input symbols, together with the sequence of output symbols, will be called the outcome of the experiment. In addition there can be a conclusion which the experimenter emits, the exact nature of which need not be specified. The conclusion might be thought of as a message typed out on a typewriter, such as "The machine being experimented on was in state q_1 at the beginning of the experiment". It is required that the conclusion depend only on which experiment is being performed and what the sequence of output symbols was.

The second kind of experiment considered in this paper is the multiple experiment, shown in Figure 2.

In this case the experimenter has access to several copies of the same machine, each of which is initially in the same state. The experimenter can send different sequences of inputs to each of these K copies, and receive from each the corresponding output sequence.

In each of these two kinds of experiments the experimenter may be thought of as a human being who is trying to learn the answer to some question about the nature of the machine or its initial state. This is not the only kind of experimenter we might imagine in application of this theory; in particular the experimenter might be another machine. One of the problems we consider is that of giving explicit instructions for performing the experiments, and in any case for which this problem is completely solved it is possible to build a machine which could perform the experiment.

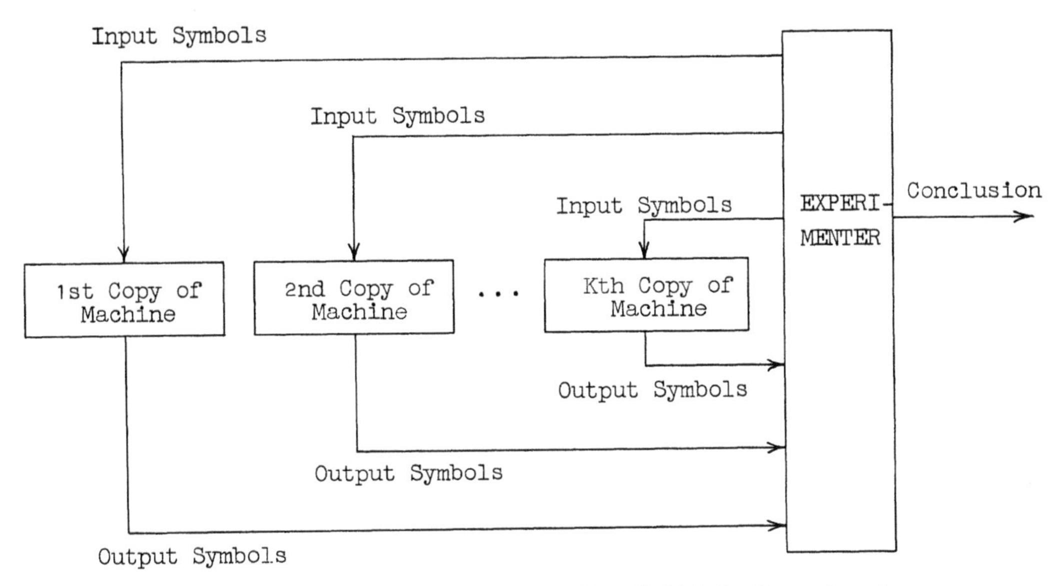

FIGURE 2. Schematic Diagram of a Multiple Experiment

EXAMPLES

It may be instructive to consider several situations for which this sort of theory might serve as a mathematical model.

The first example is one in which one or more copies of some secret device are captured or stolen from an enemy in wartime. The experimenter's job is to determine in detail what the device does and how it works. He may have partial information, e.g., that it is a bomb fuze or a cryptographic device, but its exact nature is to be determined. There is one special situation that can occur in such an experiment that is worthy of note. The device being experimented on may explode, particularly if it is a bomb, a mine, or some other infernal machine. Since the experimenter is presumably intelligent enough to have anticipated this possibility, he may be assumed to have conducted his experimentation by remote control from a safe distance. However, the bomb or mine is then destroyed, and nothing further can be learned from it by experimentation. It is interesting to note that this situation can be represented exactly by the theory. The machine will have some special state q_n, the exploded state. The transitions defining the machine will be such that there exists a sequence of inputs that can cause the machine to go into state q_n, but no input which will cause it to leave the state. Hence, if the experimenter happens to give the wrong sequence to the machine, he will be unable to learn anything further from this copy of the machine.

There is a somewhat artificial restriction that will be imposed on the action of the experimenter. He is not allowed to open up the machine and look at the parts to see what they are and how they are interconnected. In this military situation, such a restriction might correspond to the machine being booby trapped so as to destroy itself if tampered with. It might also correspond to an instance where the components are so unfamiliar that nothing can be gained by looking at them. At any rate, we will always impose this somewhat artificial restriction that the machines under consideration are always just what are sometimes called "black boxes", described in terms of their inputs and outputs, but no internal construction information can be gained.

Another application might occur during the course of the design of actual automata. Suppose an engineer has gone far enough in the design of some machine intended as a part of a digital computer, telephone central office, automatic elevator control, etc., to have described his machine in terms of the list of states and transitions between them, as used in this paper. He may then wish to perform some gedanken-experiments on his intended machine. If he can find, for instance, that there is no experimental way of distinguishing his design from some machine with fewer states, he might as well build the simpler machine.

It should be remarked that from this engineering point of view certain results closely paralleling parts of this paper (notably the reduction described in Theorem 4) have recently been independently found by D. A. Huffman in his Ph.D. thesis in Electrical Engineering (M.I.T.). His results are to appear in the Journal of the Franklin Institute.

Still another situation of which this theory is a mathematical model occurs in the case of the psychiatrist, who experiments on a patient. He gives the patient inputs (mainly verbal), and notes the outputs (again mainly verbal), using them to learn what is wrong with the patient. The black box restriction corresponds approximately to the distinction between the psychiatrist and the brain surgeon.

Finally, another situation of which this might conceivably be a mathematical model occurs when a scientist of any sort performs an experiment. In physics, chemistry, or almost any other science the inputs which an experimenter puts into his experiment and the outputs he gets from it do not correspond exactly to the things the experimenter wishes to learn by performing the experiment. The experimenter is frequently forced to ask his questions in indirect form, because of restrictions imposed by intractable laws of nature. These restrictions are somewhat similar in their effect on the organization of the experiment to the black box restriction.

The analogy between this theory and such scientific experimentation is not as good as in the previous situations, because actual experiments may be continuous and probabilistic (rather than finite and deterministic), and also because the experiment may not be completely isolated from the experimenter, i.e., the experimenter may be experimenting on a system of which he himself is a part. However, certain qualitative results of the theory may be of interest to those who like to speculate about the basic problems of experimental science.

CONVENTIONS

Each machine will have a finite number n of states, which will be called $q_1, q_2, \ldots, q_n$ a finite number m of possible input symbols which will be called $S_1, S_2, \ldots, S_m$, and a finite number p of possible output symbols, which will be called $S_{m+1}, S_{m+2}, \ldots, S_{m+p}$. In several examples used in this paper we will have $m = 2$, $p = 2$, $S_1 = S_3 = 0$, and $S_2 = S_4 = 1$.

Time is assumed to come in discrete steps, so the machine can be thought of as a synchronous device. Since many of the component parts of actual automata are variable in their speed, this assumption means the theory has not been stated in the most general terms. In practice, some digital computers and most telephone central offices have been designed asynchronously. However, by providing a central "clock" source of uniform time intervals it is possible to organize even asynchronous components so that they act in the discrete time steps of a synchronous machine. Digital computers and other electronic automata are usually built in this synchronous fashion. The synchronous convention is used in this paper since it permits simpler exposition, but the fact that these results can be translated with very little change into asynchronous terms should be obvious from the fact that Huffman wrote his paper in terms of the asychronous case.

The state that the machine will be in at a given time depends only on its state at the previous time and the previous input symbol. The output symbol at a given time depends only on the current state of the machine. A table used to give these transitions and outputs will be used as the definition of a machine. To illustrate these conventions, let us consider the following example of a machine:

Machine A

Previous State	Present State: Previous Input 0	Present State: Previous Input 1
q_1	q_4	q_3
q_2	q_1	q_3
q_3	q_4	q_4
q_4	q_2	q_2

Present State	Present Output
q_1	0
q_2	0
q_3	0
q_4	1

These two tables give the complete definition of a machine (labelled machine A, for future reference). In the left table, the present state of the machine is given as a function of the previous state and the previous input. In the right table, the present output of the machine is given as a function of the present state.

An alternate way of representing the description of a machine can also be used, which may be somewhat more convenient to follow. This other representation, called a transition diagram, consists of a graph whose vertices correspond to the states of the machine represented, and whose edges correspond to the possible transitions between those states. Each vertex of this transition diagram will be drawn as a small circle, in which is written the symbol for the corresponding state, a semicolon, and the output which the machine gives in that state.

Each pair of these circles will be joined by a line if there is a direct transition possible between the corresponding pair of states. An arrowhead will point in the direction of the transition. Beside each such line there will be written a list of the possible input symbols which can cause the transition. Below is given a transition diagram for machine A:

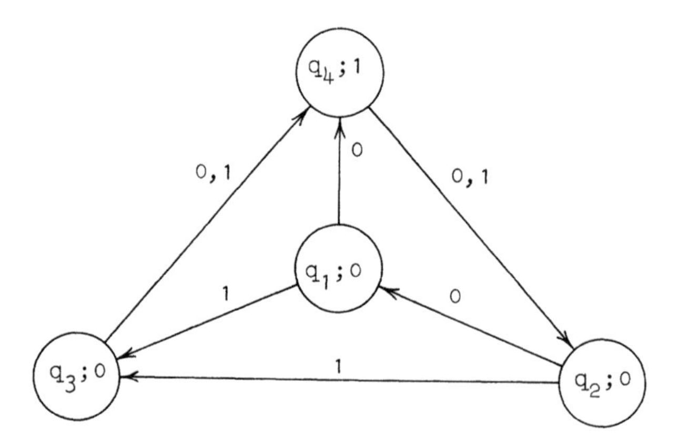

FIGURE 3. Transition Diagram of Machine A

An experiment can be performed on this machine by giving it some particular sequence of inputs. As an example, the sequence 000100010 might be used. If the machine is initially in the state q_1, the outcome of this experiment would be:

0	0	0	1	0	0	0	1	0
q_1	q_4	q_2	q_1	q_3	q_4	q_2	q_1	q_3
0	1	0	0	0	1	0	0	0

where the first line of the above is the sequence of inputs, the second line is the sequence of states, and the third line is the sequence of outputs. The last two lines can be obtained from the first by use of the tabular definition of machine A or its transition diagram. It should be emphasized that only the bottom line of the above is observable by the experimenter, and the sequence of states is hidden away, usable only in arriving at or explaining the observable results of the experiment.

Suppose that the same sequence of inputs mentioned above is presented to machine A, initially in some other state. The outcome of the experiment would be one of the following, according as the initial state is q_2, q_3, or q_4:

0	0	0	1	0	0	0	1	0
q_2	q_1	q_4	q_2	q_3	q_4	q_2	q_1	q_3
0	0	1	0	0	1	0	0	0

0	0	0	1	0	0	0	1	0
q_3	q_4	q_2	q_1	q_3	q_4	q_2	q_1	q_3
0	1	0	0	0	1	0	0	0

0	0	0	1	0	0	0	1	0
q_4	q_2	q_1	q_4	q_2	q_1	q_4	q_2	q_3
1	0	0	1	0	0	1	0	0

Even though this example of an experiment involved putting a predetermined sequence of input symbols into the machine, it should not be assumed that this is the only kind of experiment permitted. In general, the inputs to the machine can depend on its previous outputs, permitting the course of the experiment to branch.

There would be several ways of specifying such a branching experiment, but for the purposes of this paper, a loose verbal description of such experiments will be used. If it were desired to make these descriptions more formal, the experimenter could be described as another sequential machine, also specified in terms of its internal states, inputs, and outputs. The output of the machine being experimented on would serve as input to the experimenter and vice versa. The experimenter would also have another output in which it would summarize the results of the experiment, indicating what has been learned about the machine by the experiment.

In the simple sequence given above as an example of an experiment, it is natural to define the length of the experiment as 9, since this is the number of terms in the input sequence, and the number of discrete steps of time required to perform this experiment. But in the case of an experiment with possible branches during its performance, some of these branches may lead to a conclusion more quickly than others. In this case the length required for the longest possible alternative would be taken as the length of the experiment.

Although a branching experiment is the most general type of deterministic experiment, most of the experiments which will be required in the proofs of this paper can simply be sequences. For example, the shortest simple experiment which can be used to distinguish between two states (of the same or different machines) is merely a sequence. For if this is the shortest experiment, the result is not known until the last step, i.e., the output sequences coming to the experimenter are the same except for the last term. This term comes too late to affect any part of the experiment.

Two machines, S and T, will be said to be isomorphic if the table describing S can be obtained from the table describing T by substituting new names for the states wherever they occur as either the arguments or the entries of the table. Clearly, isomorphic machines will always have the same behavior, and will be indistinguishable from one another by any experiment.

Since distinguishability has already been referred to several times, and is vital to every proof in this paper, it will be explained in some detail.

A state q_i of a machine S will be said to be indistinguishable from a state q_j of S if and only if every experiment performed on S starting in state q_i produces the same outcome as it would starting in state q_j.

A pair of states will be said to be distinguishable if they are not indistinguishable. Hence, q_i is indistinguishable from q_j if and only if there exists some experiment of which the outcome depends on which of these two states S was in at the beginning of the experiment.

Similarly, we can say that a state q_i of a machine S is distinguishable (or indistinguishable) from a state q_j of a machine T if there exists an experiment (or there does not exist an experiment) of which the outcome starting with machine S in state q_i differs from the outcome starting with machine T in state q_j.

Finally, distinguishability and indistinguishability can be defined for pairs of machines. A machine S will be said to be distinguishable from a machine T if and only if at least one of the following two conditions hold: either

(1) there exists some state q_i of S, and some experiment the outcome of which beginning with S in state q_i differs from its outcome beginning with T in each of its states, or

(2) there exists some state q_j of T, and some experiment the outcome of which beginning with T in state q_j differs from its outcome beginning with S in each of its states.

S and T will be said to be indistinguishable if and only if they are not distinguishable, or, in other words, if both of the following two conditions hold:

(1) for every state q_i of S, and every experiment, there exists a state q_j of T such that the experiment beginning with machine S in state q_i produces the same outcome as the experiment beginning with machine T in state q_j, and

(2) for every state q_j of T, and every experiment, there exists a state q_i of S such that the experiment beginning with machine T in state q_j produces the same outcome as the experiment beginning with machine S in state q_i.

If S is indistinguishable from T, then the two machines are alike in their behavior (although they may differ in their structure), and may be thought of as being interchangeable. In any practical application of real machines, the manufacturer can take advantage of this equivalence, and produce whichever of the two machines is cheaper to build, easier to repair, or has some other desirable internal property.

Distinguishability and indistinguishability are defined here as binary relations. That is, they hold between a pair of machines or a pair of states. This does not mean that an experiment which distinguishes between them must be a multiple experiment. In many cases a simple experiment suffices. In any event, we perform the experiment on just one of the two machines or states we wish to distinguish, and its outcome depends on which of the two was present. In these cases we may think of the conclusion which the experimenter reaches as being of the form: "If the machine being examined was either S or T, then it is now known to be T."

This is certainly an extremely elementary kind of a conclusion, which makes a binary choice between two alternatives. Part of this paper will deal with methods of building up more complicated conclusions from such elementary ones.

An obvious modification of distinguishability is to state whether the machines which can be distinguished require multiple experiments to tell them apart or not. In the case of pairs of states, the two kinds of distinguishability can easily be seen to coincide.

In the course of the proofs given below, it will frequently be convenient to look at experiments in terms of what is actually happening inside the machines. Although the experimenters are not permitted to look inside the black boxes, we are under no such restriction. In fact, we will be able to learn more about the limitations imposed by the black box restriction if we have no such restriction on our observations, constructions, or proofs.

AN ANALOGUE OF THE UNCERTAINTY PRINCIPLE

The first theorem to be proved will be concerned with an interesting qualitative property of machines.

<u>Theorem</u> 1: There exists a machine such that any pair of its states are distinguishable, but there is no simple experiment which can determine what state the machine was in at the beginning of the experiment.

The machine A, already described on the previous pages, satisfies the conditions of the theorem. The previously described experiment will distinguish between any pair of states, except the pair (q_1, q_3). That is, given any other pair of states, if it is known that the machine is in one state of this pair at the beginning of the experiment, applying this experiment will give an output that depends on which state the machine was in. In order to distinguish between q_1 and q_3, the experiment should consist of applying the sequence 11. The outcome of this will be:

$$\begin{matrix} 1 & 1 \\ q_1 & q_3 \\ 0 & 0 \end{matrix} \qquad\qquad \begin{matrix} 1 & 1 \\ q_3 & q_4 \\ 0 & 1 \end{matrix}$$

Thus there exists a simple experiment which can distinguish between any pair of states. Furthermore, the multiple experiment which uses two copies of the machine, sending one of the two previously mentioned sequences to each, can obtain enough information to completely specify what state the machine was in at the beginning of the experiment.

To complete the proof, it need only be shown that given only one copy of machine A, there is no experiment which can determine whether it was in state q_1 at the beginning of the experiment.

It is clear that any experiment will distinguish between q_1 and q_4, since the first output symbol will be different. But any simple experiment that distinguishes q_1 from q_2, cannot distinguish q_1 from q_3. To see this, note that any experiment which begins with the input 1 does not permit q_1 to be distinguished from q_2 (since in either case the first output is 0 and the second state is q_3, so that no future inputs can produce different outputs). Similarly any experiment which begins with the input 0 does not permit q_1 to be distinguished from q_3.

This result can be thought of as being a discrete-valued analogue of the Heisenberg uncertainty principle. To point out the parallel, both the uncertainty principle and this theorem will be restated in similar language.

The state of an electron E will be considered specified if both its velocity and its position are known. Experiments can be performed which will answer either of the following:

(1) What was the position of E at the beginning of the experiment?

(2) What was the velocity of E at the beginning of the experiment?

In the case of machine A, experiments can be performed which will answer either of the following:

(1) Was A in state q_2 at the beginning of the experiment?

(2) Was A in state q_3 at the beginning of the experiment?

In either case, performing the experiment to answer question 1 changes the state of the system, so that the answer to question 2 cannot be obtained. In other words, it is only possible to gain partial information about the previous history of the system, since performing experiments causes the system to "forget" about its past.

By analogy with the uncertainty principle, could we also state that the future state of machine A cannot be predicted from past experimental results? Here the analogy ends. Even though we cannot learn by experiment what state machine A was in at the beginning of the experiment, we can learn what state it is in at the end of the experiment. In fact, at the end of the first experiment described, machine A will be in one particular predetermined state (independent of its initial state), namely the state q_3.

Despite the incompleteness of the analogy, it does seem interesting that there is an analogue of the uncertainty principle in this discrete, deterministic system. Any applications of this example to causality, free will, or other metaphysical problems will be left to the reader.

FURTHER THEOREMS ON DISTINGUISHABILITY

Theorem 2: Given any machine S and any multiple experiment performed on S, there exist other machines experimentally distinguishable from S for which the original experiment would have had the same outcome.

Let S have n states $q_1, \ldots q_n$, and let the experiment have length k. Then define a machine T having $n(k+1)$ states $q_1, q_2, \ldots, q_{n(k+1)}$ as follows:

If the machine S goes from state q_i to q_j when it receives the input symbol a, then let T go from q_{i+tn} to $q_{j+(t+1)n}$ under the same input, for all t such that $0 \leq t < k$, but let T go from q_{i+kn} to q_{j+kn}.

If the machine S has output symbol b in state q_i, let T have output symbol b in state q_{i+tn}, for $0 \leq t < k$, but let T have some output symbol different from b in state q_{i+kn}.

Then at the step t+1 of any simple experiment, the machine T will be in state q_{i+tn} whenever machine S is in state q_i and $0 \leq t < k$. But at any step later than the kth, machine T will be in state q_{i+kn}. Thus it can be seen that for the first k steps of any simple experiment, the outputs of S and T will be alike. But after the kth step, the outputs of S and T will always be different. The extension to multiple experiments is immediate.

This result means that it will never be possible to perform experiments on a completely unknown machine which will suffice to identify it from among the class of all sequential machines. If, however, we restrict the class to be a smaller one, it may be possible. In particular, much of the rest of this paper will be concerned with the case where the class consists of all machines with n states or fewer, m input symbols or fewer, and p output symbols or fewer. Such a machine will be called an (n, m, p) machine.

Definition: A machine S will be said to be strongly connected if for any ordered pair (q_i, q_j) of states of S, there exists a sequence of inputs which will take the machine from state q_i to state q_j.

The term "strongly connected" is used since any such machine will have a transition diagram which is a connected graph, but the converse is not true. A counter-example to the converse is given by the following machine:

Machine B

Present State		
Previous State	Previous Input	
	0	1
q_1	q_2	q_3
q_2	q_1	q_3
q_3	q_1	q_2
q_4	q_2	q_2

Present State	Present Output
q_1	0
q_2	1
q_3	0
q_4	0

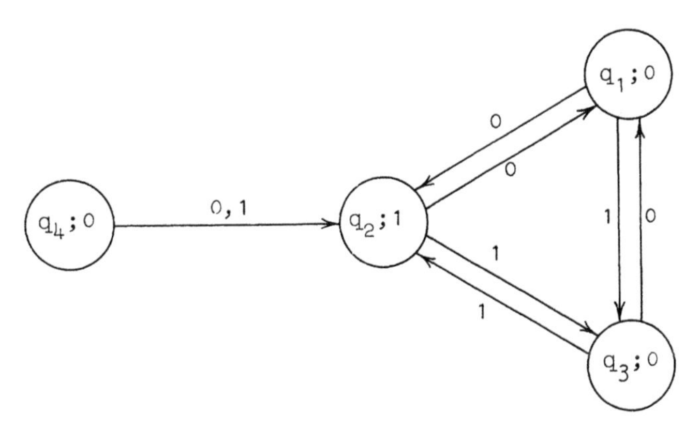

FIGURE 4. Transition Diagram of Machine B

Theorem 3: If S is a strongly connected machine, and T is indistinguishable from S by any simple experiment, then for every state q_i of S there exists a state q_j of T which is indistinguishable from q_i by any simple experiment.

Since T is indistinguishable from S by any simple experiment, we have, as one of the two conditions implied by indistinguishability, that given any state q_i of S, and any simple experiment on S beginning in state q_i, there exists a corresponding state q_j of T such that the same experiment, starting with a copy of T in state q_j, will produce the same sequence of output symbols. This theorem states that if S is strongly connected, q_j can be chosen independently of the experiment. That is, q_j corresponds to q_i for all experiments, rather than just this particular experiment.

To prove the theorem first note that if we consider an experiment consisting of any sequence of input symbols applied to machine S in state q_1, there must have been states of T which would have given the same sequence of outputs. With each such sequence of input symbols, we associate the set of states that machine T could be in at the end of this sequence after having produced the same sequence of outputs that S would produce starting in state q_1. Then if we lengthen the sequence of input symbols, the number of elements in the associated set can only decrease, but it can never become zero (or else this would give an experiment which distinguishes S from T).

Hence, we can choose a particular sequence and extend it until the number of elements in the associated set of states of T can no longer be decreased by any further extension. Then we add to this sequence a further sequence which will cause machine S to go successively into every one of its states at least once, if the entire sequence is applied starting in state q_1.

Then for each state q_i of S, consider the set Y of states which are associated with the subsequence obtained by truncating the original sequence at the last time it causes S to go into state q_i. Then Y is non-empty, and every member is indistinguishable from q_i. This follows from the fact that if q_j is a member of Y and is distinguishable from q_i, the experiment that distinguishes them defines a sequence, which when added to the truncated sequence above, would give a further reduction of the number of elements in its associated set. But this contradicts the definition of the original sequence.

Note the words "strongly connected" cannot be removed from the statement of Theorem 3. A counter-example is given by machine B, defined just before Theorem 3, which is indistinguishable by any simple experiment from the machine B', defined by removing the bottom row from each of the two tables that define machine B. However, the state q_4 of machine B is distinguishable from every state of B'.

Theorem 4: The class of all machines which are indistinguishable from a given strongly connected machine S by any simple experiment has a unique (up to an isomorphism) member with a minimal number of states. This unique machine, called the reduced form of S, is strongly connected, and also has the property that any two of its states are distinguishable.

Given any machine T, indistinguishable from S, define the relation R to hold between states of S and states of T if they are indistinguishable by a simple experiment. That is, the state q_1 of S will have the relation R to the state q_j of T if and only if there is no simple experiment which can distinguish them.

Then by Theorem 3 the domain of the relation R is the set of all states of S. And, after verifying the transitivity of indistinguishability it can be seen that any two states of S are indistinguishable from each other if and only if they are indistinguishable from the same state of T. Hence, the number of equivalence classes into which the states of S are partitioned by the equivalence relation of indistinguishability is the smallest number of states which T can have.

Let us define a machine T* with exactly this many states, associating each state with one such equivalence class. We can define the output symbol for each state of T* to be the output symbol for any state in its equivalence class, since if the states are indistinguishable, they must give the same output symbols. We define the transitions by letting state q_i of T* go into state q_j of T* upon receiving the input symbol a, if and only if some member of the equivalence class associated with q_i goes into some member of the equivalence class associated with q_j upon receiving the input symbol a. There is never any ambiguity in this definition, since indistinguishable states cannot have transitions which take them into distinguishable ones (or else this would give a way of distinguishing the original indistinguishable states).

Next, T* can be seen to be indistinguishable from S as an immediate consequence of its definition. Also T* is strongly connected, since to go between states q_i and q_j of T*, use the sequence which goes from any state in the equivalence class associated with q_i to any state in the equivalence class associated with q_j.

Then to show that T* is unique up to an isomorphism, consider any other machine T, having the same number of states, and also indistinguishable from S. Then since T will also be indistinguishable from T*, and T* is strongly connected we can apply Theorem 3. Then defining another relation R as done earlier in the proof, note that it can be seen to be a 1:1 correspondence between the states of the two machines, and in fact, it is the desired isomorphism.

<u>Definition</u>: A machine S will be said to be in reduced form, if and only if S is the reduced form of S.

<u>Theorem</u> 5: If S is a strongly connected machine, then S is in reduced form, if and only if any pair of its states are distinguishable. To prove the converse, consider the relation of indistinguishability as in the proof of Theorem 4: it partitions the states of S into equivalence classes, each having just one member. Hence, the reduced form of S as constructed above has exactly as many states as S, and the uniqueness of the reduced form of S completes the proof.

The following is an example of a machine which this theorem shows to be not in reduced form. This particular example has just one pair of states which are indistinguishable:

Machine C

Present State		
Previous State	Previous Input	
	0	1
q_1	q_2	q_1
q_2	q_3	q_1
q_3	q_2	q_1

Present State	Present Output
q_1	0
q_2	1
q_3	0

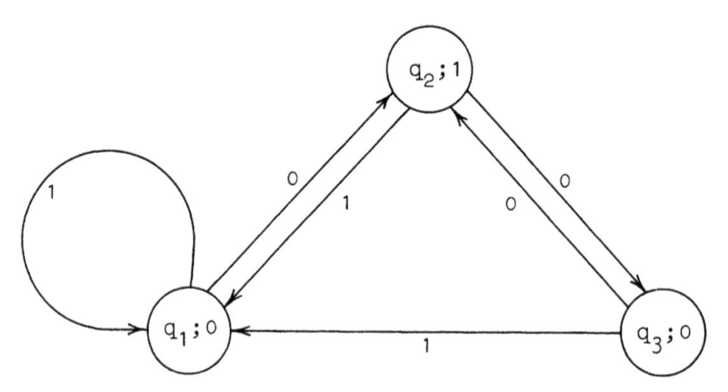

FIGURE 5. Transition Diagram of Machine C

In connection with these theorems, it might be mentioned that not every machine indistinguishable from a strongly connected machine is strongly connected. The machines B and B', previously described, also serve as an example of this.

However, since the reduced form of a machine is unique and has no indistinguishable states, it may be thought of as a simplified version of the machine, with all unessential parts of its description removed. The reduction of a machine to its reduced form is closely related to one of the steps proposed by D. A. Huffman as a step in the design of sequential machines.

The reduced form will be considered the natural form in which to describe a strongly connected machine, and the remaining theorems of this paper will be written in a form so as to apply directly to machines in reduced form. The indirect application of these results to other strongly connected machines is also sometimes possible.

THEOREMS CONCERNING LENGTHS OF EXPERIMENTS

The theorems proved heretofore have mainly been concerned with qualitative questions, i.e., whether or not it is possible to perform experiments which answer questions about the current state of a machine or its internal structure. The remaining theorems will be concerned with how many steps these experiments require, and their proofs will include methods for designing the experiments.

Theorem 6: If S is an (n, m, p) machine such that any two of its states are distinguishable, then they are distinguishable by a simple experiment of length $n-1$.

For each positive integer k, we define the relation R_k to hold between any two states q_i and q_j of S if and only if q_i is indistinguishable from q_j by any experiment of length k. Since each R_k can be seen to be an equivalence relation, it defines a partition P_k of the set Z of states of S into equivalence classes.

Then P_{k+1} is a refinement of P_k; that is, if two states are indistinguishable by any experiment of length $k+1$, they are indistinguishable by an experiment of length k. Further, if P_k does not subdivide Z into subsets having just one member, then P_{k+1} is a proper refinement of P_k. To show this, choose any two states q_i and q_j which are indistinguishable by an experiment of length k. Since by hypothesis they are distinguishable, consider the shortest sequence of inputs which will serve as an experiment to distinguish them. If this sequence of inputs is of length r, consider the pair of states which q_i and q_j are transformed into by the first $r-k-1$ inputs of this sequence. This pair of states is distinguishable by an experiment of length $k+1$ (namely, the rest of the above sequence) but not by any experiment of length k (for such an experiment would contradict the minimal length of the above sequence).

Since P_1 partitions Z into at least two subsets (for otherwise every state would have the same output associated, and hence no pairs of states are distinguishable) we can prove by induction from above that if $k \leq n - 1$, P_k partitions Z into at least $k+1$ subsets, which for the case $k = n - 1$ completes the proof of the theorem.

The above proof suggests a method for finding the shortest experiments for distinguishing between any two states. First construct P_1, by subdividing Z into sets of states giving the same output symbol. Then, proceeding by recursion, P_{k+1} can be constructed from P_k. If any two states q_i and q_j undergo transitions into states which belong to different classes of P_k upon receiving the same input symbol a, then q_i and q_j should be put into different classes of P_{k+1}, and a is the first symbol of an experiment for distinguishing between q_i and q_j in $k+1$ steps. If, however, under all input symbols q_i and q_j remain together in the same classes of P_k, they are indistinguishable by any experiment of length $k+1$, and hence belong in the same class of P_{k+1}. By continuing the recursion until any desired pair of states can be distinguished, this method constructs an experiment. It proceeds backwards; that is, the last step of the experiment is found first, and at the end of the construction the first step of the experiment is determined.

The following examples will show that the $n-1$ bound obtained in the theorem cannot be lowered. For each $n \geq 3$, define the machine D_n in accordance with the following table:

Machine D_n

Present State		
Previous State	Previous Input	
	0	1
q_1	q_2	q_2
q_2	q_3	q_1
...	...	...
q_i	q_{i+1}	q_{i-1}
...	...	...
q_{n-1}	q_n	q_{n-2}
q_n	q_n	q_{n-1}

Present State	Present Output
q_1	1
q_2	0
...	...
q_i	0
...	...
q_{n-1}	0
q_n	0

Then D_n is an $(n, 2, 2)$ machine such that any two of its states are distinguishable, but the shortest experiment which can distinguish q_n from q_{n-1} has length $n-1$.

For the case $n = 4$, D_n is represented by the following transition diagram:

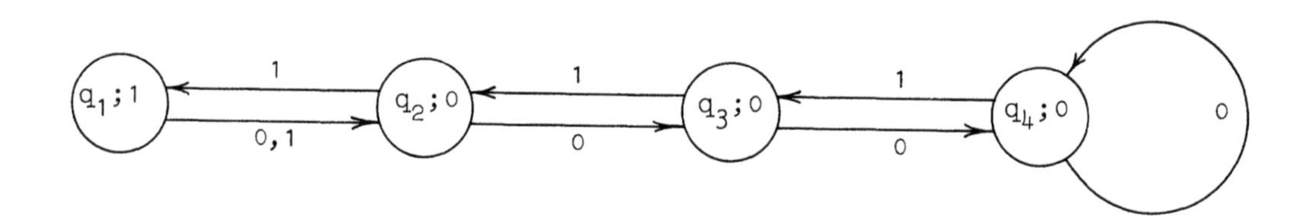

FIGURE 6. Transition Diagram of Machine D_4

Theorem 7: If S and T are (n, m, p) machines, such that some state q_i of S can be distinguished from state q_j of T, then this experiment can be of length $2n-1$.

First define the machine $S + T$, the direct sum of S and T. The table defining it will contain all of the entries and arguments of the table for S, plus entries and arguments obtained from those of the table for T by replacing q_i by q_{i+h}, for all i. This direct sum $S + T$ contains as submachines an isomorphic copy of S and one of T, but it is of course not strongly connected. Its transition diagram consists of the combined (but not connected) diagrams for S and T, with the names of the states of T changed to avoid ambiguity. Physically, the direct sum $S + T$ can be interpreted as a black box which has either the behavior of S or that of T, with no way of changing it between the two kinds of behavior. $S + T$ is a $(2n, m, p)$ machine such that certain pairs of its states are distinguishable, and hence by the methods used in proving Theorem 6, they can be shown to be distinguishable by an experiment of length $2n - 1$. The experiment distinguishing any two states of $S + T$ also obviously distinguishes between the corresponding states of S and T.

The following examples will show that the $2n-1$ bound obtained in this theorem cannot be lowered. For each $n \geq 3$, define the machine E_n in accordance with the following table:

Machine E_n

Present State		
Previous State	Previous Input 0	Previous Input 1
q_1	q_2	q_2
q_2	q_3	q_1
...	...	...
q_i	q_{i+1}	q_{i-1}
...	...	...
q_{n-1}	q_n	q_{n-2}
q_n	q_{n-1}	q_n

Present State	Present Output
q_1	1
q_2	0
...	...
q_i	0
...	...
q_{n-1}	0
q_n	0

It can easily be verified that the shortest experiment which distinguishes q_1 of D_n from q_1 of E_n has length $2n-1$. For the case $n = 4$, the transition diagram of E_n is shown below:

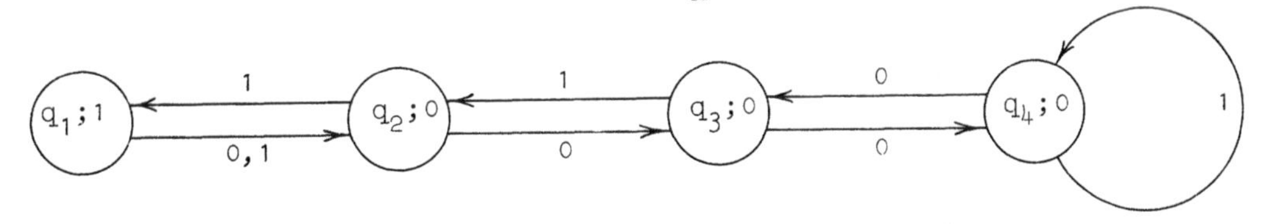

FIGURE 7. Transition Diagram of Machine E_4

Theorem 8: Given any (n, m, p) machine S such that any two of its states can be distinguished, there exists an experiment of length $n(n-1)/2$ which can determine the state of S at the end of the experiment.

This experiment will be constructed as it is being performed (since this will in general be a branching experiment, a complete formalization of this construction would involve defining a specific machine which could perform this experiment). As in the proof of Theorem 3, after each part of the experiment is performed there is a corresponding set of states which the machine could be in at the end of this experiment, i.e., which are compatible with all the outputs the machine has given during the experiment. Giving any one of certain sequences to the machine will reduce the number of elements in this set of states. Choose one of the shortest sequences having this property, and perform it as the next part of the experiment. Repeat this process until the set of possible states has just one element, i.e., the state of the machine is known.

It will be proved by induction on k that when the set of possible states of S has been reduced until it has $n-k$ members, at most $k(k+1)/2$ units of time will have elapsed. This is obvious for $k = 1$. For any $k < n$, let G_{k-1} be this set having at most $n-k+1$ members. Also the partition P_k, as constructed in the proof of Theorem 6, partitions the set of states of S into at least $k+1$ classes. Then G_{k-1} must have members belonging to at least two different classes of P_k (otherwise one class of P_k has at least $n-k+1$ members, and the other k have at least k members, so their union, the set of states of S, must have at least $n+1$ members). Consider such a pair of states belonging to different classes of P_k. An experiment distinguishing them has length k, and performing this experiment at this point will eliminate one or the other of the pair of states these will be transformed into by this experiment from the set of possible states of S. Hence by the fact that the shortest sequence having this property will be used in the construction, at most k more steps are required to reduce the set until it has $n-k$ members. Since by inductive hypothesis only at most $(k-1)k/2$ units of time had been used before this reduction, at most k more brings the total to at most $k(k+1)/2$. To complete the proof, let $k = n - 1$.

The following examples will show that the $n(n-1)/2$ bound obtained in this theorem is within a multiplicative constant of the best possible bound. For each $j > 3$, define the machine F_j in accordance with the following table:

Machine F_j

Present State				
Previous State	Previous Input 0	Previous Input 1	Present State	Present Output
q_1	q_{j+2}	q_{j+1}	q_1	1
q_2	q_3	q_2	q_2	0
...	...	...	...	...
q_j	q_{j+1}	q_j	q_j	0
q_{j+1}	q_{j+2}	q_{j+1}	q_{j+1}	0
q_{j+2}	q_{j+3}	q_2	q_{j+2}	0
...	...	...	...	...
q_{2j-1}	q_{2j}	q_{j-1}	q_{2j-1}	0
q_{2j}	q_{2j+1}	q_j	q_{2j}	0
q_{2j+1}	q_2	q_1	q_{2j+1}	0

Then F_j has $n = 2j+1$ states, is strongly connected, and any two of its states are distinguishable. It can be shown that the shortest experiment which can determine the final state of S consists of the sequence of length j^2+j-2 having a "0" in all positions except the first and those positions divisible by $j+2$, in which it has a "1". For the case $j = 3$, the transition diagram of F_j is shown below:

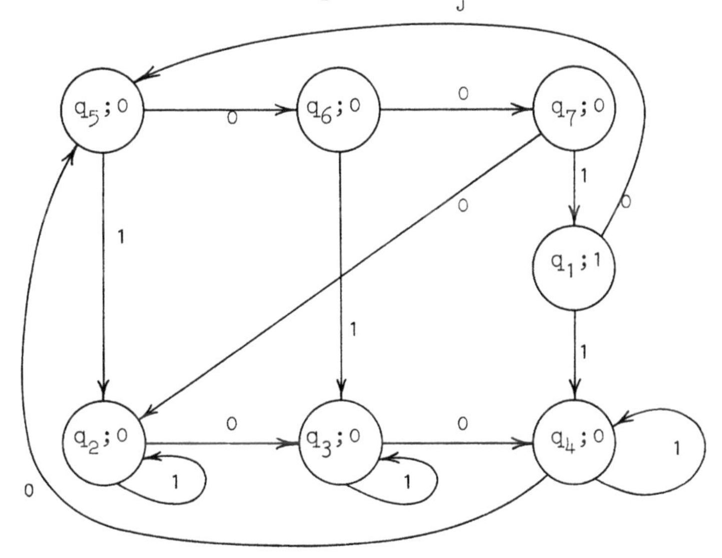

FIGURE 8. Transition Diagram of Machine F_3

Theorem 9: If $R_{n,m,p}$ is the class of all strongly connected (n, m, p) machines in reduced form, then there exists a simple experiment of length at most $n^{nm+2}p^n/n!$ which, when performed on a copy of any member S of $R_{n,m,p}$ will suffice to distinguish S from all other members of $R_{n,m,p}$

If $n = 1$, the result is obvious, so the proof will be concerned with the case $n \geq 2$. $R_{n,m,p}$ will be considered to have no two of its members isomorphic; that is, it will consist of just one of every essentially different strongly connected (n, m, p) machine in reduced form. Then define the machine Σ, the direct sum (as in the proof of Theorem 7) of all of the members of $R_{n,m,p}$. Apply to Σ the sort of experiment defined in the proof of Theorem 8, reducing the set of possible states it could be in until it has only one member. Then this identifies the machine S, up to an isomorphism.

To determine the length of this experiment, the first step is to note how many members $R_{n,m,p}$ has. Since there are exactly $n^{nm}p^n$ different (n,m,p) machines, the following correspondence between every member of $R_{n,m,p}$ and n! different (n, m, p) machines will show $R_{n,m,p}$ has at most $n^{nm}p^n/n!$ members. In the case of any member T of $R_{n,m,p}$ having exactly n states, the correspondence can be direct with all (n, m, p) machines obtained from T by all n! permutations of the names of the n states, since any two machines obtained from distinct permutations must be distinct. But if T is a member of $R_{n,m,p}$ having k states, with $k < n$, define the (n, m, p) machine T* whose transitions and outputs agree with T for all q_i with $i \leq k$, but for all q_i with $k < i < n$, let the output be 0 and let all inputs cause a transition into state q_{i+1}, and for $i = n$ let the output be 0 and all inputs cause a transition into state q_1. Then the correspondence can be defined between T and the n! different (n, m, p) machines obtained by permuting the names of the n states of T*. Then since no two (n, m, p) machines have been made to correspond to the same member of $R_{n,m,p}$, $R_{n,m,p}$ has at most $n^{nm}p^n/n!$ members.

Then, proceeding as in the proof of Theorem 8, we can estimate how many steps must be necessary to cut down the number of possible states of Σ. It will be convenient to consider the subsets of states of Σ obtained from each of the original machines. By Theorem 8, at most $n(n-1)/2$ steps are required to eliminate all but one of the members of any such set. But by Theorem 6, this last state can be eliminated (unless Σ actually is in this state) in at most $2n-1$ steps. But $n(n-1)/2 + 2n-1 \leq n^2$ for $n \geq 2$, so each of the $n^{nm}p^n/n!$ subsets require at most n^2 steps.

It seems probable that the $n^{nm+2}p^n/n!$ estimate of this theorem could be improved considerably, since in the early parts of the experiment many states of Σ can be eliminated simultaneously. But it can be seen that the bound cannot be lowered below m^{n-1} by considering the following abstract model of a combination lock. For each $n,m \geq 2$, define a basic machine $G_{n,m}$ as follows:

Machine $G_{n,m}$

Present State				
Previous State	Previous Input			
	S_1	S_2	...	S_n
q_1	q_1	q_1	...	q_1
q_2	q_1	q_1	...	q_1
...	...	...	...	...
q_n	q_1	q_1	...	q_1

Present State	Present Output
q_1	0
q_2	0
...	...
q_{n-1}	0
q_n	1

Then a combination lock will be defined as an (n, m, 2) machine whose tables are obtained from those of $G_{n,m}$ by replacing, for each i with $1 \leq i \leq n-1$, exactly one of the q_1 entries in the ith row of the left-hand table above with a q_{i+1} entry.

The only way to make this give a 1 output is putting it into state q_n and this will be said to be unlocking the combination lock. If the combination lock is originally in state q_1, it can be unlocked only by giving it exactly the proper input sequence for the last $n-1$ steps before unlocking it. This input sequence is, of course, called the combination of the lock. The machine H is an example of a combination lock having the combination 0,1,0:

Machine H

Present State		
Previous State	Previous Input	
	0	1
q_1	q_2	q_1
q_2	q_1	q_3
q_3	q_4	q_1
q_4	q_1	q_1

Present State	Present Input
q_1	0
q_2	0
q_3	0
q_4	1

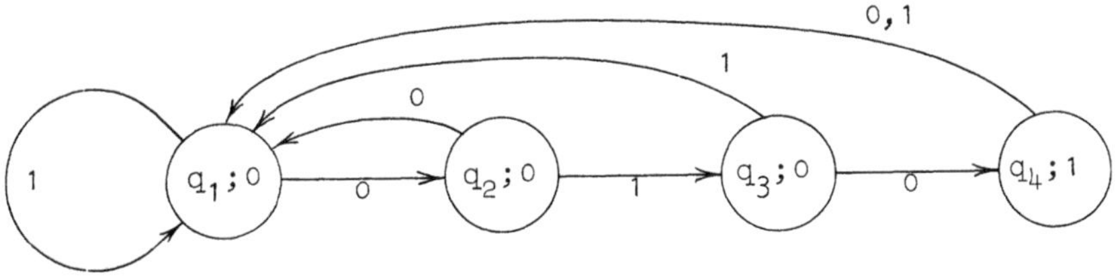

FIGURE 9. Transition Diagram of Machine H

For each n, m there are exactly m^{n-1} different combination locks. Suppose you are given some unknown combination lock, initially in state q_1, and are required to identify which one is present by an experiment which is as short as possible. Since in the first few steps in the experiment the lock cannot open, and at any later step in the experiment at most one combination lock can open, this experiment requires more than m^{n-1} steps.

FURTHER PROBLEMS

There are many further problems connected with this theory of sequential machines from the experimental point of view, which the author has not yet been able to solve.

One problem would be to find classes of machines more general than the strongly connected machines about which reasonable theorems can be proved. It should be pointed out that it was convenient to use direct sum machines (which are certainly not strongly connected) in two of the proofs. Infernal machines and the ordinary household electrical fuse provide important examples of machines which are not strongly connected.

Other problems which immediately suggest themselves are to improve the bounds given by Theorems 8 and 9. The author would like to conjecture, in this connection, that the best bound in Theorem 9 will be independent of p.

Still another problem of interest is the length of an experiment required to tell whether a given copy of an unknown strongly connected (n, m, p) machine S is indistinguishable from a known (n, m, p) machine T. This problem is akin to that faced by a maintenance man in checking whether a given machine is out of order. He knows what the machine is supposed to do, and he wishes to find out whether or not it does do this. If not, it is assumed that the machine is still a finite-state machine, differing in some subtle way from the supposed machine. A bound n on the number of states of the machine is helpful in view of Theorem 2, and is presumably derivable from the known number of relays or other components of which the machine is made. Theorem 9 does give a bound on the length of the experiment, although it seems fantastically large. A more reasonable experiment might be one which required the machine to undergo every transition only a few times.

Still other problems are suggested by permitting the inputs and the outputs of the machines to be k-tuples of symbols rather than single ones. The experimenters allowed in multiple experiments (see Figure 2) are already of this type, and many devices built out of relays or vacuum tubes have k-tuples of binary digits as their inputs or outputs. Such machines can be combined more freely than single input and output machines to make larger machines. Certain inputs of each machine are connected to the outputs of others, and other inputs and outputs of the individual machines are used as the inputs and outputs of the composite machine. If the k components of such a composite machine have $n_1, n_2, \ldots$ and n_k states each, the composite machine has

$$\prod_{i=1}^{k} n_i$$

states, namely the k-tuples of states of the component machines. Such composite machines are of particular interest if all or most of these states are distinguishable. Many problems exist in relation to the inverse question of decomposition into such components. Given a machine with n states, under what conditions can it be represented as a combination of two machines having n_1 and n_2 states, such that $n_1 n_2 = n$? Under what conditions is the decomposition unique?

One way of describing what engineers do in designing actual automata is to say that they start with an overall description of a machine and break it down successively into smaller and smaller machines, until the individual relays or vacuum tubes are ultimately reached. The efficiency of such a method might be determined by a theoretical investigation on such decompositions. This might also throw light on the validity with which the psychiatrists can hope to subdivide the mind into ego, superego, id, etc.

Acknowledgements The somewhat overlapping work of Dr. D. A. Huffman has already been mentioned, and should be acknowledged as an entirely independent contribution. The writer would particularly like to acknowledge his indebtedness to Dr. C. E. Shannon for several suggestions which led to this work.

TURING MACHINES

A UNIVERSAL TURING MACHINE WITH TWO INTERNAL STATES

Claude E. Shannon

INTRODUCTION

In a well-known paper[1], A. M. Turing defined a class of computing machines now known as Turing machines. We may think of a Turing machine as composed of three parts — a control element, a reading and writing head, and an infinite tape. The tape is divided into a sequence of squares, each of which can carry any symbol from a finite alphabet. The reading head will at a given time scan one square of the tape. It can read the symbol written there and, under directions from the control element, can write a new symbol and also move one square to the right or left. The control element is a device with a finite number of internal "states." At a given time, the next operation of the machine is determined by the current state of the control element and the symbol that is being read by the reading head. This operation will consist of three parts; first the printing of a new symbol in the present square (which may, of course, be the same as the symbol just read); second, the passage of the control element to a new state (which may also be the same as the previous state); and third, movement of the reading head one square to the right or left.

In operation, some finite portion of the tape is prepared with a starting sequence of symbols, the remainder of the tape being left blank (i.e., registering a particular "blank" symbol). The reading head is placed at a particular starting square and the machine proceeds to compute in accordance with its rules of operation. In Turing's original formulation

---

1 Turing, A. M., "On Computable Numbers, with an Application to the Entscheidungsproblem," Proc. of the London Math. Soc. 2 - 42 (1936), pp. 230 - 265.

alternate squares were reserved for the final answer, the others being used for intermediate calculations. This and other details of the original definition have been varied in later formulations of the theory.

Turing showed that it is possible to design a universal machine which will be able to act like any particular Turing machine when supplied with a description of that machine. The description is placed on the tape of the universal machine in accordance with a certain code, as is also the starting sequence of the particular machine. The universal machine then imitates the operation of the particular machine.

Our main result is to show that a universal Turing machine can be constructed using one tape and having only two internal states. It will also be shown that it is impossible to do this with one internal state. Finally a construction is given for a universal Turing machine with only two tape symbols.

THE TWO-STATE UNIVERSAL TURING MACHINE

The method of construction is roughly as follows. Given an arbitrary Turing machine A with an alphabet of m letters (symbols used on the tape, including the blank) and n internal states, we design a machine B with two internal states and an alphabet of at most $4mn + m$ symbols. Machine B will act essentially like machine A. At all points of the tape, except in the position opposite the reading head and one adjacent position, the tape of B will read the same as the tape of A at corresponding times in the calculation of the two machines. If A is chosen to be a universal Turing machine, then B will be a universal Turing machine.

Machine B models the behavior of machine A, but carries the information of the internal state of A via the symbols printed on the tape under the reading head and in the cell of the tape that the reading head of A will next visit. The main problem is that of keeping this state information up to date and under the reading head. When the reading head moves, the state information must be transferred to the next cell of the tape to be visited using only two internal states in machine B. If the next state in machine A is to be (say) state 17 (according to some arbitrary numbering system) this is transferred in machine B by "bouncing" the reading head back and forth between the old cell and the new one 17 times (actually 18 trips to the new cell and 17 back to the old one). During this process the symbol printed in the new cell works through a kind of counting sequence ending on a symbol corresponding to state 17, but also retaining information as to the symbol that was printed previously in this cell. The bouncing process also returns the old cell back to one of the elementary symbols (which correspond one-to-one with the symbols used by machine A), and in fact returns it to the particular elementary symbol that should be printed in that cell when the operation is complete.

The formal construction of machine B is as follows: Let the symbol alphabet of machine A be $A_1, A_2, \ldots, A_m$, and let the states be $S_1, S_2, \ldots, S_n$. In machine B we have m elementary symbols corresponding to the alphabet of the A machine, $B_1, B_2, \ldots, B_m$. We further define 4mn new symbols corresponding to state symbol pairs of machine A together with two new two-valued indices. These symbols we denote by $B_{i,j,x,y}$ where $i = 1, 2, \ldots, m$ (corresponding to the symbols), $j = 1, 2, \ldots, n$ (corresponding to the states), $x = +$ or $-$ (relating to whether the cell of the tape is transmitting or receiving information in the bouncing operation) and $y = R$ or L (relating to whether the cell bounces the control to the right or left).

The two states of machine B will be called α and β. These two states are used for two purposes: First, on the initial step of the bouncing operation they carry information to the next cell being visited as to whether the old cell is to the right (α) or left (β) of the new one. This is necessary for the new cell to bounce the control back in the proper direction. After the initial step this information is retained in the new cell by the symbol printed there (the last index y). Second, the states α and β are used to signal from the old cell to the new one as to when the bouncing operation is complete. Except for the initial step of bouncing, state β will be carried to the new cell until the end of the bouncing operation when an α is carried over. This signifies the end of this operation and the new cell then starts acting as a transmitter and controlling the next step of the calculation.

Machine B is described by telling what it does when it reads an arbitrary symbol and is in an arbitrary state. What it does consists of three parts: printing a new symbol, changing to a new state, and moving the reading head to right or left. This operation table for machine B is as follows.

symbol;	state ⟶	symbol;	state;	direction		
B_i;	α ⟶	$B_{i,1,-,R}$;	α;	R	$(i = 1, 2, \ldots, m)$	(1)
B_i;	β ⟶	$B_{i,1,-,L}$;	α;	L	$(i = 1, 2, \ldots, m)$	(2)
$B_{i,j,-,x}$;	β ⟶	$B_{i,(j+1),-,x}$;	α;	x	$(i = 1, 2, \ldots, m)$ $(j = 1, 2, \ldots, n-1)$ $(x = R, L)$	(3)
$B_{i,j,+,x}$;	α or β ⟶	$B_{i,(j-1),+,x}$;	β;	x	$(i = 1, 2, \ldots, m)$ $(j = 2, \ldots, n)$ $(x = R, L)$	(4)
$B_{i,1,+,x}$;	α or β ⟶	B_i;	α;	x	$(i = 1, 2, \ldots, m)$ $(x = R, L)$	(5)

So far, these operations do not depend (except for the number of symbols involved) on the operation table for machine A. The next and last type of operation is formulated in terms of the operation table of the machine being modeled. Suppose that machine A has the operation formula

(6) $$A_i;\ S_j \longrightarrow A_k;\ S_\ell;\ {R \atop L}\ .$$

Then machine B is defined to have

(7) $$B_{i,j,-,x};\ \alpha \longrightarrow B_{k,\ell,+,{R \atop L}};\ {\beta \atop \alpha}\ ;\ {R \atop L}$$

where if the upper letter (R) occurs in (6) the upper letters are used in (7) and conversely.

To see how this system works, let us go through a cycle consisting of one operation of machine A and the corresponding series of operations of machine B.

Suppose that machine A is reading symbol A_3 and is in state S_7, and suppose its operation table requires that it print A_8, go into state S_4 and move to the right. Machine B will be reading (by inductive assumption) symbol $B_{3,7,-,x}$ (whether x is R or L depends on preceding operations and is irrelevant to those which follow). Machine B will be in state α. By relation (7), machine B will print $B_{8,4,+,R}$, go into state β, and move to the right. Suppose the cell on the right contains A_{13} in machine A; in machine B the corresponding cell will contain B_{13}. On entering this cell in state β, by relation (2) it prints $B_{13,1,-,L}$, goes into state α, and moves back to the left. This is the beginning of the transfer of state information by the bouncing process. On entering the left cell, it reads $B_{8,4,+,R}$ and by relation (4) prints $B_{8,3,+,R}$, goes to state β and moves back to the right. There, by relation (3), it prints $B_{13,2,-,L}$, goes into state α and returns to the left. Continuing in this manner, the process is summarized in Table I.

The operations indicated complete the transfer of state information to the right cell and execution of the order started in the left cell. The left cell has symbol B_8 registered (corresponding to A_8 in machine A) and the right cell has symbol $B_{13,4,-,L}$ registered, with the reading head coming into that cell with internal state α. This brings us back to a situation similar to that assumed at the start, and arguing by induction we see that machine B models the behavior of machine A.

To get machine B started in a manner corresponding to machine A, its initial tape is set up corresponding to the initial tape of A (with A_i replaced by B_i) except for the cell initially occupied by the reading head. If the initial state of machine A is S_j and the initial symbol in this cell is A_i, the corresponding cell of the B tape has $B_{i,j,-,R(\text{or } L)}$ registered and its internal state is set at α.

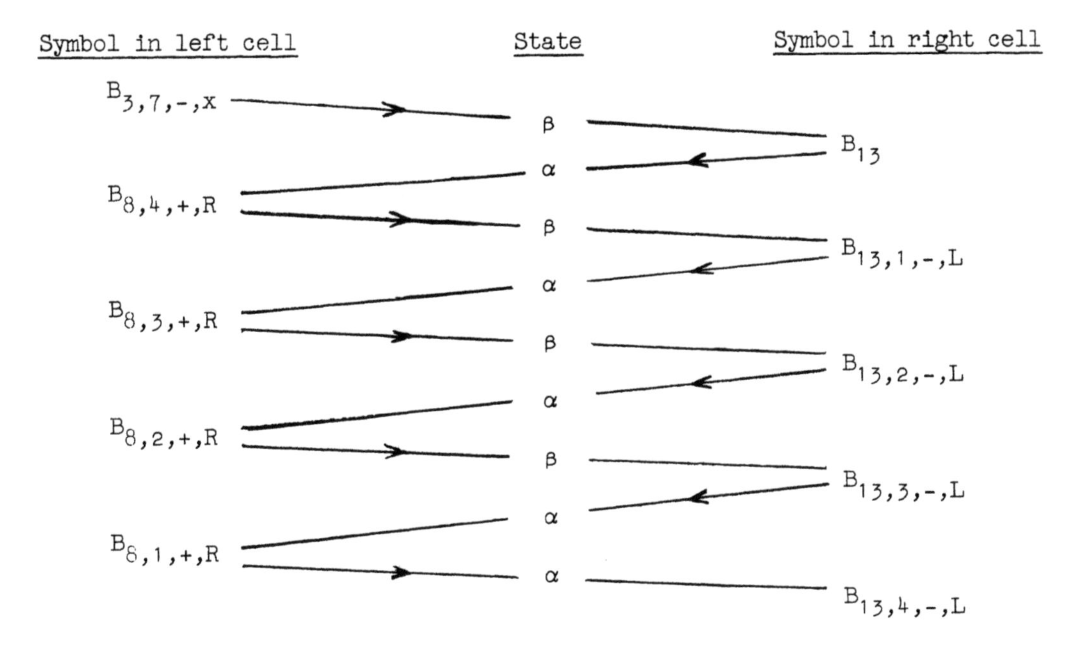

Table I

IMPOSSIBILITY OF A ONE-STATE UNIVERSAL TURING MACHINE

It will now be shown that it is impossible to construct a universal Turing machine using one tape and only one internal state.

Suppose we have a machine satisfying these conditions. By registering a suitable "description number" of finite length on part of the tape (leaving the rest of the tape blank), and starting the reading head at a suitable point, the machine should compute any computable number, in particular the computable irrational numbers, e.g., $\sqrt{2}$. We will show that this is impossible.

According to Turing's original conception, $\sqrt{2}$ would be computed in a machine by the machine printing the successive digits of $\sqrt{2}$ (say, in binary notation) on a specified sequence of cells of the tape (say, on alternate cells, leaving the others for intermediate calculations). The following proof assumes $\sqrt{2}$ to be calculated in such a form as this, although it will be evident that modifications would take care of other reasonable interpretations of "calculating $\sqrt{2}$."

Since $\sqrt{2}$ is irrational, its binary digits do not, after any finite point, become periodic. Hence if we can show that with a one-state machine either (1) all but a finite number of the cells eventually have the same symbol registered, or (2) all but a finite number of the cells change indefinitely, we will have proved the desired result.

Assume first a doubly infinite tape — an infinite number of blanks each side of the description number for $\sqrt{2}$. When the reading head enters a blank cell it must either stay there indefinitely or eventually move out either to right or left. Since there is only one state, this behavior does not depend on previous history of the computation. In the first case, the reading head will never get more than one removed from the description number and all the tape except for a finite segment will be constant at the blank symbol. If it moves out of a blank symbol to the left, either the left singly infinite section of blank tape is not entered in the calculation and therefore need not be considered, or if it is entered, the reading head from that time onward continues moving to the left leaving all these previously blank cells registering the same symbol. Thus the tape becomes constant to the left of a finite segment and blank to the right of this segment and could not carry $\sqrt{2}$. A similar situation arises if it emerges to the right from an originally blank cell. Hence the doubly infinite tape is no better than the singly infinite tape and we may assume from symmetry a singly infinite tape to the right of the description number.

Now consider the following operation. Place the reading head on the first cell of this infinite blank strip. The machine will then compute for a time and perhaps the reading head will be transferred back out of this strip toward the description number. If so, replace it on the first cell of the now somewhat processed blank tape. If it returns again off the tape, again replace it on the first cell, etc. The number of times it can be placed on the first cell in this fashion will be called the reflection number of the machine and denoted by R. This will be either an integer 1, 2, 3, ..., or ∞.

Now consider placing the reading head at its appropriate start for the description number to compute $\sqrt{2}$. After a certain amount of computation the reading head will perhaps emerge from the description number part of the tape. Replace it on the last cell of the description number. Again after a time it will possibly emerge. Continue this process as long as possible. The number of times it emerges will either be an integer 0, 1, 2, 3, ..., or ∞. This number, S, we call the reflection number for the $\sqrt{2}$ description.

If S is finite and R (possibly ∞) $> S$, the reading head after a finite time will be trapped in the part of the tape that originally contained the description number. Only a finite amount of the blank tape will have been changed and the machine will not have calculated $\sqrt{2}$.

If both R and S are infinite, the reading head will return indefinitely to the description number part of the tape. The excursions into the originally blank parts will either be bounded or not. If they are bounded, only a finite amount of the blank tape will have been changed as in the preceding case. If the excursions are unbounded, all but a finite segment of tape will be operated on by the reading head an unlimited number of times.

Since there is only one state and a finite alphabet of tape symbols, the symbol registered in a cell visited an unlimited number of times must either come to a constant (the same for all these cells) or else change cyclically an infinite number of times. In the first case, all the originally blank tape becomes constant and cannot represent $\sqrt{2}$. In the second case all the blank tape is continually changing and cannot be the computation of anything.

If $R \leq S$, the reading head eventually moves into the original blank part of the tape and stays there. In this case it can be shown that the symbols in the originally blank part become constant. For either it moves to the right out of the first blank cell into the second blank cell at least R times, or not. If not the reading head is trapped in what was the first blank cell after a finite time, and all but a finite amount of tape remains constant at the blank symbol. If it does move out R times it will not return to the first originally blank cell since R is the reflection number for blank tape. This first cell will then have registered the result of operating on a blank $2R$ times (R coming in from the left and R from the right). The second originally blank cell will eventually register the same constant symbol, since the same argument applies to it as to the first. In each case the machine works into the same tape (an infinite series of blanks) and enters the same number of times (R). This exhausts the cases and completes the proof.

MODELING A TURING MACHINE WITH ONLY TWO TAPE SYMBOLS

It is also possible, as we will now show, to construct a machine, C, which will act like any given Turing machine A and use only two symbols 1 and 0 on its tape, one of which, 0 say, is the symbol for a blank square. Suppose, as before, a given machine A has m tape symbols and n internal states. Let ℓ be the smallest integer such that m is less than or equal to 2^{ℓ}. Then we may set up an arbitrary association of the m symbols used by machine A with binary sequences of length ℓ, letting however the blank symbol of machine A correspond to the sequence of ℓ zeroes. Basically, the machine C will operate with binary sequences; an elementary operation in machine A will correspond in machine C to stepping the reading head to the right $\ell - 1$ squares (storing the read information in its internal state) then stepping back to the left $\ell - 1$ squares, writing the proper new symbol as it goes, and finally moving either to the right or to the left ℓ squares to correspond to the motion of the reading head of machine A. During this process, the state of machine A is also, of course, carried in machine C. The change from the old state to the new state occurs at the end of the reading operation.

The formal construction of machine C is as follows. Corresponding to states $S_1, S_2, \ldots, S_n$ of machine A we define states $T_1, T_2, \ldots, T_n$ in machine C (these will occur when machine C is at the beginning of an operation, reading the first symbol in a binary sequence of length ℓ). For each of these T_i we define two states T_{i0} and T_{i1}. If machine C is in state T_i and reads symbol 0, it moves to the right and goes into state T_{i0}. If it reads a 1, it moves to the right and goes into state T_{i1}. Thus, after reading the first symbol of a binary sequence, these two states remember what that symbol was. For each of these there are again two states T_{i00}, T_{i01} and T_{i10} and T_{i11}. If the machine is in the state T_{i0} for example and reads the symbol 0 it goes to the state T_{i00} and similarly for the other cases. Thus these states remember the initial state and the first two symbols read in the reading process. This process of constructing states is continued for $\ell - 1$ stages, giving a total of $(2^{\ell} - 1)n$ states. These states may be symbolized by

$$T_{i,x_1,x_2,\ldots,x_s} \quad i = 1, 2, \ldots, n; \; x_j = 0, 1; \; s = 0, 1, \ldots, \ell - 1.$$

If the machine is in one of these states $(s < \ell - 1)$ and reads 0 or 1, the machine moves to the right and the 0 or 1 appears as a further index on the state. When $s = \ell - 1$, however, it is reading the last binary symbol in the group of ℓ. The rules of operation now depend on the specific rules of machine A. Two new sets of states somewhat similar to the T states above are defined, which correspond to writing rather than reading:

$$R_{i,x_1,x_2,\ldots,x_s} \text{ and } L_{i,x_1,x_2,\ldots,x_s}.$$

A sequence $x_1, x_2, \ldots, x_{\ell-1}, x_\ell$ corresponds to a symbol of machine A. Suppose that when machine A is reading this corresponding symbol and is in state i it prints the symbol corresponding to the binary sequence $y_1, y_2, \ldots, y_{\ell-1}, y_\ell$, goes to state j and moves (say) right. Then we define machine C such that when in state $T_{i,x_1,x_2,\ldots,x_{\ell-1}}$, and reading symbol x_ℓ, it goes into state $R_{j,y_1,y_2,\ldots,y_{\ell-1}}$, prints y_ℓ and moves to the left. In any of the states $R_{i,y_1,y_2,\ldots,y_s}$ (or $L_{i,y_1,y_2,\ldots,y_s}$), machine C writes y_s, moving to the left and changes to state $R_{i,y_1,y_2,\ldots,y_{s-1}}$ (or $L_{i,y_1,y_2,\ldots,y_{s-1}}$). By this process the binary sequence corresponding to the new symbol is written in place of the old binary sequence. For the case $s = 1$, the writing of y_1 completes the writing operation of the binary sequence. The remaining steps are concerned with moving the reading head ℓ steps to the right or left according as the machine is in an R state or an L state. This is carried out by means of a set of U_{is} and V_{is} $(i = 1, 2, \ldots, n; \; s = 1, 2, \ldots, \ell - 1)$. In state R_{ix_1} the machine writes x_1, moves to the right, and goes into state U_{i1}. In each of the U states it continues to the right, printing nothing and going into the next higher indexed U state until the last one is reached.

Thus U_{is} produces motion to the right and state U_{is+1} $(s < \ell - 1)$. Finally $U_{i\ell-1}$ leads, after motion to the right, to T_i, completing the cycle. In a similar fashion, L_{ix_i} leads to motion to the left and state V_{i1}; V_{is} gives motion to the left and V_{is+1} $(s < \ell - 1)$; finally, $V_{i\ell-1}$ gives motion to the left and T_i.

The initial tape for machine C is, of course, that for machine A with each symbol replaced by its corresponding binary sequence. If machine A is started on a particular symbol, machine C will be started on the left-most binary symbol of the corresponding group; if machine A is started in state S_i, C will be started in state T_i.

Machine C has at most $n(1 + 2 + 4 \ldots + 2^{\ell-1}) = n(2^\ell - 1)$ T states, similarly at most $n(2^\ell - 2)$ R states and $n(2^\ell - 2)$ L states, and finally $2n(\ell - 1)$ U and V states. Thus altogether not more than $3n2^\ell + n(2\ell - 7)$ states are required. Since $2^\ell < 2m$, this upper bound on the number of states is less than $6mn + n(2\ell - 7)$, which in turn is certainly less than $8mn$.

The results we have obtained, together with other intuitive considerations, suggest that it is possible to exchange symbols for states and vice versa (within certain limits) without much change in the product. In going to two states, the product in the model given was increased by a factor of about 8. In going to two symbols, the product was increased by a factor of about 6, not more than 8. These "loss" factors of 6 and 8 are probably in part due to our method of microscopic modeling — i.e., each elementary operation of machine A is modeled into machine B. If machine B were designed merely to have the same calculating ability as A in the large, its state-symbol product might be much more nearly the same. At any rate the number of logical elements such as relays required for physical realization will be a small constant (about 2 for relays) times the base two logarithm of the state-symbol product, and the factor of 6 or 8 therefore implies only a few more relays in such a realization.

An interesting unsolved problem is to find the minimum possible state-symbol product for a universal Turing machine.

A NOTE ON UNIVERSAL TURING MACHINES

M. D. Davis

I. INTRODUCTION

Turing[1] has shown that it is possible to construct a definite computing machine U which is universal in the sense that any computation whatever, can be performed on U. Of course, this result presumes knowledge of what constitutes a computing machine and a computation. We assume familiarity with Turing's notion of computing machine (hereafter referred to as Turing machines), and with the theory of recursive functions.[2]

The universality of a Turing machine is manifested by its ability, *via a suitable encoding*, to perform any computation which could be performed by any given Turing machine. However, the condition must surely be added that the encoding itself be, in some suitable sense, simple. For there would not be much point in claiming universality for a Turing machine for which the encoding would require, in essence, another universal machine to carry it out. This raises the problem of explicitly defining *universal Turing machine*, which problem is the subject of the present note.[3]

II. THE DEFINITION

We begin with

Definition II.1. A set C of (non-negative) integers is called *complete* if it is recursively enumerable and if

---

1 Cf. A. M. Turing; *Computable Numbers with an Application to Entscheidungsproblem*. Proceedings, London Math. Soc. (1936-37).

2 Cf. M. Davis; *Computability and Unsolvability*, McGraw Hill. (Forthcoming.) R. Péter, *Rekursive Funktionia*, Budapest, 1951; S. C. Kleene, *Introduction to Metamathematics*, New York, 1952.

3 This problem was suggested at different times by Dr. John McCarthy and Dr. C. E. Shannon.

for every recursively enumerable set R, there is a recursive function $\sigma(x)$ such that

$$R = (x \mid \sigma(x) \in C).$$

Definition II.2.

$$U_n = \left(x \mid \bigvee_y T(n, x, y)\right)$$

Thus, the sequence $U_0, U_1, U_2, \ldots,$ consists of all[4] recursively enumerable sets (with repetitions).

Corollary II.3. If C is complete, there exists a recursive function $f(n, x)$ such that

$$U_n = (x \mid f(n, x) \in C).$$

PROOF. Let,

$$R = \left(x \mid \bigvee_y T(K(x), L(x), y)\right)^{5}.$$

Then, for some recursive $\sigma(x)$,

$$R = (x \mid \sigma(x) \in C).$$

But,

$$\begin{aligned} U_n &= \left(x \mid \bigvee_y T(n, x, y)\right) \\ &= \left(x \mid \bigvee_y T(K(J(n, x)), L(J(n, x)), y)\right) \\ &= (x \mid J(n, x) \in R) \\ &= (x \mid \sigma(J(n, x)) \in C). \end{aligned}$$

This proves the result with $f(n, x) = \sigma(J(n, x))$.

Definition II.4. With each Turing Machine Z we associate a set of D_Z of instantaneous

4 Cf. Davis loc. cit. or Kleene loc cit. Note that "recursively enumerable set" as used here is equivalent to "recursively enumerable set or the empty set" in Kleene's terminology.

5 $K(x)$, $L(x)$ are the Cantor pairing functions. Cf. Davis loc. cit. or Péter loc cit. I.e., in a suitable (effective) enumeration of the ordered pairs of integers, $(K(x), L(x))$ is the x-th ordered pair, and (x, y) is the $J(x, y)$-th ordered pair.

descriptions[6] as follows:

> α belongs to D_Z if and only if there exists a sequence α_i, $1 \leqq i \leqq n$, of instantaneous descriptions such that $\alpha = \alpha_1$, α_n is terminal, and $\alpha_i \rightarrow \alpha_{i+1}$ (Z), $i = 1, 2, \ldots, n + 1$.

Thus, D_Z is the set of all instantaneous descriptions α which have the property that Z, started at α, eventually halts, enabling an answer to be read off.

> Definition II.5. We define δ_Z to be the set of Gődel numbers of all elements of D_Z.

> Definition II.6. Z is universal if δ_Z is complete.

In order to make it plausible that this is a reasonable definition, we shall have to indicate:

1) how an arbitrary computation can be encoded, so as to be capable of being performed by a given universal machine.
2) that any Turing machine which would ordinarily be regarded as universal satisfies our condition, and
3) that the encoding involved in 1) may be entirely accomplished by Turing machines which are not themselves universal.

III. THE ENCODING

As is well known, it suffices to consider as computations, the computing of the values of partial recursive functions of a single variable.[7]

Thus, let M be a Turing machine which is universal in our sense. By $\psi_Z(x)$, we mean the function whose value for given x is obtained by placing Turing machine Z in internal configuration q_1, scanning the first of a consecutive block of $x + 1$, 1's in an otherwise blank tape, allowing the machine to compute, and, should it halt, counting the number of 1's

6 Recall that an instantaneous description is an expression which symbolizes the "complete present state" of a Turing machine. That is, it consists of an expression on a machine's tape, into which has been inserted a symbol, q_i, which represents the machine's present internal configuration, immediately preceding the scanned symbol. Cf. Davis, loc. cit.

7 Cf. Davis, loc. cit. or Kleene loc. cit.

which then occur on the tape; should the machine not halt, $\psi_Z(x)$ is undefined. Our problem is to encode the computation of $\psi_Z(x)$ in terms of M. By Corollary 2.3 there exists a recursive function $f(n, x)$ such that

$$U_n = \left(x \mid f(n, x) \in \delta_M\right). \tag{1}$$

Next, let us define the recursively enumerable sets R^n_m, S^n_m, by[8]

$$R^n_m = \left(x \mid \bigvee_y [T(n, x, y) \wedge U(y) < m]\right),$$

$$S^n_m = (x \mid \bigvee_y [T(n, x, y) \wedge U(y) \geqq m]).$$

As is easily shown, suitable (primitive) recursive functions $\sigma(m, n)$, $\theta(m, n)$ may be determined such that

$$\begin{aligned} R^n_m &= \left(x \mid \bigvee_y T(\sigma(m, n), x, y)\right) \\ S^n_m &= \left(x \mid \bigvee_y T(\theta(m, n), x, y)\right). \end{aligned} \tag{2}$$

If n is a Gödel number of a Turing machine Z, then R^n_m, S^n_m are the sets of all numbers x for which $\psi_Z(x)$ is defined and $< m$ or $\geqq m$, respectively.

By (2)

$$R^n_m = U_{\sigma(m, n)}$$

$$S^n_m = U_{\theta(m, n)} \cdot$$

Hence, by (1)

$$R^n_m = \left(x \mid f(\sigma(m, n), x) \in \delta_M\right)$$

$$S^n_m = \left(x \mid f(\theta(m, n), x) \in \delta_M\right).$$

Now, finally, let Z_o be any Turing machine, n_o a Gődel number of Z, and x_o some integer. M is employed as follows. First, the instantaneous description β_o whose Gődel number is $f\left(\theta(0, n_o), x_o\right)$ is

---

8 If y is the Gödel number of the sequence $\alpha_1, \ldots, \alpha_p$ of expressions, $U(y)$ is the number of 1's occurring in α_p.

obtained.[9] Then, we permit M to compute, beginning at β_0. Then, M halts, if and only if $x_0 \in S_0^{n_0}$, i.e., if and only if $\psi_{Z_0}(x_0)$ is defined. Thus, if M computes forever, it duplicates, in this case, the behavior of Z_0. If M does halt, it remains to determine the value of $\psi_{Z_0}(x_0)$. We define α_0, β_0, $M \geqq 1$, to be the instantaneous descriptions[9] whose Gődel numbers are $f\big(\sigma(m, n_0), x_0\big)$, $f(\theta(m, n_0), x_0\big)$, respectively. Then, we set M at α_1, and simultaneously set (a duplicate of) M at β_1 and permit the pair of computations to proceed. Precisely one of these computations will halt; which one depends on whether $\psi_{Z_0}(x_0) < 1$ (i.e., $= 0$), or $\psi_{Z_0}(x_0) \geqq 1$. In the former case, the value of $\psi_{Z_0}(x_0)$ is determined; in the latter case we repeat the process with α_m, β_m, $m \geqq 1$. Clearly, this procedure eventually provides the desired value.

IV. THE UNIVERSALITY OF CERTAIN MACHINES

One obvious way of obtaining a machine which must be regarded as universal on intuitive grounds is to choose a machine M, such that

$$\psi_M^{(2)}(m, x) = U\Big(\min_y T(m, x, y)\Big) \tag{3}$$

We prove:

THEOREM IV.1. Any Turing machine M which satisfies (3) is universal.

PROOF. Clearly δ_M is recursively enumerable. Let R be any recursively enumerable set. Then,

$$R = \Big(x \mid \bigvee_y T(m_0, x, y)\Big)$$

for suitable m_0. That is,

$$R = \text{Domain}\ \Big(\psi_M^2(m_0, x)\Big).$$

Let $w(x)$ be the Gődel number of the instantaneous description $q_1 1^{m_0+1} B 1^{x+1}$. Then, $w(x)$ is recursive and,

$$R = \Big(x \mid w(x) \in \delta_M\Big).$$

This proves the theorem.

9 Obviously, f may be so defined that its values are all Gődel numbers of instantaneous descriptions.

Any Turing machine which can be employed (via a uniform recursive encoding) to compute all partially computable functions will admit a similar result. For, every recursively enumerable set is the domain of definition of some partially computable function.

V. THE SIMPLICITY OF THE ENCODING

We begin with

THEOREM V.1. If δ_Z is recursive, Z is not universal.

PROOF. If δ_Z were complete, and if K is a non-recursive set which is recursively enumerable, we should have, for suitable recursive $\sigma(x)$,

$$K = \left(x \mid \sigma(x) \in \delta_Z\right),$$

which would render K recursive, and so, is impossible.

Now, our encoding in III employed only recursive functions. We wish to show that this encoding may be accomplished entirely by non-universal Turing machines. Actually, we prove the stronger result that the encoding may be performed by machines which satisfy the hypothesis of Theorem V.1. Namely, we shall prove

THEOREM V.2. If $f(x)$ is a recursive function, then there exists a Turing machine Z such that $\psi_Z(x) = f(x)$ and δ_Z is recursive.

This result is a consequence of the following:

THEOREM V.3. If $f(x)$ is a recursive function, then there exists a Turing machine Z such that $\psi_Z(x) = f(x)$, and such that if α is any instantaneous description, $\alpha \in D_Z$.[10]

We first introduce three operations by which new functions may be obtained from given functions. These are <u>substitution</u>, by which from $f(x)$ and $g(x)$, we obtain

10 That is, Z eventually halts, no matter "where" it is placed initially.

$$h(x) = f(g(x));$$

addition by which from $f(x)$ and $g(x)$, we obtain

$$h(x) = f(x) + g(x);$$

and inversion by which from a function $f(x)$ which takes on all values, we obtain

$$h(x) = f^{-1}(x) = \min_y (x = f(y)).$$

Then, we have the following well-known[11] result:

Lemma 1. All recursive functions of one variable are obtainable by a finite number of applications of substitution, addition, and inversion beginning with the functions $S(x) = x + 1$ and $E(x) = x - ([\sqrt{x}])^2$.

Definition V.4. We shall say that a Turing machine Z strongly computes the recursive function $f(x)$ if:

1) For every instantaneous description α, $\alpha \in D_Z$, and
2) For every x, there exists a sequence of instantaneous descriptions,

$$q_1 1^{x+1} = \alpha_1 < \alpha_2 < \cdots < \alpha_p$$
$$= B^{r_1} q_{r_2} 1^{f(x)} B^{r_3},$$

where α_p is terminal, for suitable r_1, r_2, r_3.
$f(x)$ is then said to be strongly computable.
Clearly, it follows that $\Psi_Z(x) = f(x)$.

We shall prove Theorem V.3 in the following slightly strengthened form:

Every recursive function is strongly computable.

Lemma 2. $S(x) = x + 1$ is strongly computable.

---

11 Cf. Julia Robinson, General Recursive Functions, Proc. of Amer. Math. Soc., vol. 1 (1950), pp. 703-718, Theorem 3.

PROOF. Take Z to be the empty set; or to consist of the sole quadruple

$$q_1 1 R q_1$$

Beginning anywhere such a Z will eventually halt; in the second case, at worst, when a B is encountered.

<u>Lemma 3</u>. $E(x) = x - [\sqrt{x}]^2$ is strongly computable.

PROOF. We do not give the detailed construction of a Turing machine Z, in this case, but content ourselves with a brief indication of how this could be accomplished.

Namely, we subtract successive odd numbers from the argument since

$$E(x) = x - \Big(1 + 3 + \ldots + (n + 1)\Big)$$

where n is chosen as large as is compatible with the non-negativeness of $E(x)$. It is easily seen that this may be accomplished as a strong computation.

<u>Lemma 4</u>. If $f(x)$ and $g(x)$ are strongly computable, then so is $f\big(g(x)\big)$.

PROOF. Let f be strongly computed by Z_1, g by Z_2. Let N be the largest integer such that q_N is an internal configuration of Z_2. Let Z_3 be obtained from Z_1 by adding N to all of the subscripts υ which occur in q_i's of Z_1. Let Z_4 consist of all quadruples $q_i S_j S_j q_{N+1}$ where $i \leqq N$, and no quadruple of Z_2 begins with $q_i S_j$. Finally, let $Z = Z_2 \cup Z_3 \cup Z_4$: Then Z strongly computes $f\big(g(x)\big)$.

<u>Lemma 5</u>. If $f(x)$ and $g(x)$ are strongly computable, then so is $f(x) + g(x)$.

PROOF. Here one of the functions must be computed, then enclosed in special signs, and then the other handled. In order to preserve strong computability, we must make certain that the operations of hunting for the special signs cannot be extended indefinitely. This is accomplished by interpreting every double blank as a sign-off.

Lemma 6. If $f(x)$ is strongly computable and takes on all values, then $f^{-1}(x)$ is strongly computable.

PROOF. Here the natural method will not work. Namely, this natural method would be as follows:

Beginning with x, compute $f(0)$, $f(1)$, $f(2)$, etc., comparing each with x. When one equal to x is determined, terminate the computation.

To see that this will not work suppose e.g. that $f(7) = 6$, $f(n) \neq 6$ for $n \neq 7$. Then, if we begin with an instantaneous description which corresponds to a stage in this computation when $f(8)$ is being determined, this will continue indefinitely.

We get around this difficulty by adding a loop after each comparison of $f(n)$ with x, by which $f(0)$, $f(1)$, ..., $f(n)$ are regenerated and compared with x. This loop itself will be terminated by a comparison with n. But now, a greater than or equal to criterion may be employed, precluding a non-halting computation.

Our result now follows from Lemmas 1 through 6.

THE INVERSION OF FUNCTIONS DEFINED BY TURING MACHINES

John McCarthy

Consider the problem of designing a machine to solve well-defined intellectual problems. We call a problem well-defined if there is a test which can be applied to a proposed solution. In case the proposed solution is a solution, the test must confirm this in a finite number of steps. If the proposed solution is not correct, we may either require that the test indicate this in a finite number of steps or else allow it to go on indefinitely. Since any test may be regarded as being performed by a Turing machine, this means that well-defined intellectual problems may be regarded as those of inverting functions and partial functions defined by Turing machines.

Let $f_m(n)$ be the partial function computed by the mth Turing machine. It is not defined for a given value of n if the computation does not come to an end. This paper deals with the problem of designing a Turing machine which, when confronted by the number pair (m, r), computes as efficiently as possible a function $g(m, r)$ such that $f_m(g(m, r)) = r$. Again, for particular values of m and r no $g(m, r)$ need exist. In fact, it has been shown that the existence of $g(m, r)$ is an undecidable question in that there does not exist a Turing machine which will eventually come to a stop and print a 1 if $g(m, r)$ does not exist.

In spite of this, it is easy to show that a Turing machine exists which will compute a $g(m, r)$ if such exists. Essentially, it substitutes integers in $f_m(n)$ until it comes to one such that $f_m(n) = r$. It will therefore find $g(m, r)$ if it exists, but will never know enough to give up if $g(m, r)$ does not exist. Since the computation of $f_m(n)$ may not terminate for some n, it is necessary to avoid getting stuck on such n's. Hence the machine calculates the numbers $f_m^k(n)$ in some order where $f_m^k(n)$ is $f_m(n)$ if the computation of $f_m(n)$ ends after k steps and is otherwise undefined.

Our problem does not end once we have found this procedure for computing $g(m, r)$ because this procedure is extremely inefficient. It corresponds to looking for a proof of a conjecture by checking in some order all possible English essays.

In order to do better than such a purely enumerative method, it is necessary to use methods which take into account some of the structure of Turing machines. But before discussing this, we must attempt to be somewhat (only somewhat) more precise about what is meant by efficiency.

The most obvious idea is to say that if T_1 and T_2 are two Turing machines each computing a $g(m, r)$, then for a particular m and r the more efficient one is the one which carries out the computation in the fewest steps. However, this won't do since for any Turing machine there is another one which does k steps of the original machine in one step. However, the new machine has many more different kinds of symbol than the old. It is probably also possible to increase the speed by increasing the number of internal states, though this is not so easy to show. (Shannon shows elsewhere in these studies that it is possible to reduce the number of internal states to two at the cost of increasing the number of symbols and reducing the speed.)

Hence we offer the following revised definition of the length of a computation performed by a Turing machine. For any universal Turing machine there is a standard way of recoding on it the computation performed by the given machine. Let this computation be recoded on a fixed universal Turing machine and count the number of steps in the new computation. There are certain difficulties here connected with the fact that the rate of computation is limited if the tape of the universal machine is finite dimensional, and hence the rate should probably be defined with respect to a machine whose tape is infinite dimensional but each square of which has at most two states and which has only two internal states. This requires a mild generalization of the concept of Turing machines.

The tape space is not required to be either homogeneous or isotropic. We hope to make these considerations precise in a later paper. For now, we only remark that dimensionality is meant to be

$$\lim_{n \to \infty} \frac{\log V_n}{\log n}$$

where V_n is the number of squares accessible to a given one in n steps. It will be made at least plausible that a machine with Q internal states and S symbols should be considered as making about $\frac{1}{2}\log QS$ elementary steps per step of computation and hence the number of steps in a computation should be multiplied by this factor to get the length of the computation.

Having now an idea of what should be meant by the length $\ell(m, r, T)$ of a particular computation of $g(m, r)$ by the machine T, we can return to the question of comparing two Turing machines. Of course, if $\ell(m, r, T_1) \leq \ell(m, r, T_2)$ for all m and r for which these numbers are finite, we should certainly say that T_1 is more efficient than T_2. (We only consider machines that actually compute $g(m, r)$ whenever it exists. Any machine can be modified into such a machine by adding to it facilities for testing a conclusion and having it spend a small fraction of its time trying the integers in order.) However, it is not so easy to give a function which gives an over-all estimate of the efficiency of a machine at computing $g(m, r)$. The idea of assigning a weight function $p(m, r)$ and then calculating

$$\sum_{m, r} p(m, r)\ell(m, r, T)$$

does not work very well because $\ell(m, r, T)$ is not bounded by any recursive function of m and r. (Otherwise, a machine could be described for determining whether the computation of $f_m(n)$ terminates. It would simply carry out some $k(m, n)$ steps and conclude that if the computation had not terminated by this time it was not going to.) There cannot be any machine which is as fast as any other machine on any problem because there are rather simple machines whose only procedure is to guess a constant which are fast when $g(m, r)$ happens to equal that constant.

We now return to the question of how to design machines which make use of information concerning the structure of Turing machines. The straightforward approach would be to try to develop a formalized theory of Turing machines in which length of a computation is defined and then try to get a decision procedure for this formal theory. This is known to be a hopeless task. Systems much simpler than Turing machine theory have been shown to have unsolvable decision procedures. So, we look for a way of evading these difficulties.

Before discussing how this may be done, it is worthwhile to bring up some more enumerative procedures. First, let $f_k(x, y)$ be the function of two variables computed by the kth Turing machine. Our procedure consists in trying the numbers $f_k(m, r)$ in order (again diagonalizing to avoid computations which don't end.) This is based on the plausible idea that, in searching for the solution to a problem, the given data should be taken into account.

The next complication which suggests itself is to revise the order in which recursive functions are considered. One way is to consider $f_{f_k(\ell)}(m, r)$ diagonalized on k and ℓ. This is based on the idea that the best procedure is more likely to be recursively simple, rather than merely to have a low number in the ordering. More generally, the enumeration of partial recursive functions should give an early place to compositions of the functions which have already appeared.

This process of elaborating the schemes by which numbers are tested can be carried much further. Intuitively it seems that each successive complication improves the performance on difficult problems, but imposes a delay in getting started.

The difficulty with the afore-mentioned methods and their elaborations is that they have no concept of partial success. It is a great advantage in conducting a search to be able to know when one is close. This suggests first of all that a function $f_k(x, y)$ be tried out first on simpler problems known to have simple solutions before very many steps of the computation $f_k(m, r)$ are carried out.

At this point it becomes unclear how to proceed further. What has been done is only a semiformal description of a few of the processes common to scientific reasoning and we have no guarantee of being exhaustive. Of course, exhaustiveness in examining for usefulness can be attained for any effectively enumerable class of objects simply by going through the enumeration. However, enumerative methods are always inefficient. What is needed is a general procedure which ensures that all relevant concepts which can be computed with are examined and that the irrelevant are eliminated as rapidly as possible. Here we remark that a property is useful mainly when it permits a new enumeration of the objects possessing it and not merely a test which can be applied. With the enumeration the objects not possessing the property need not even be examined. Of course, it is then necessary to be able to express all other relevant properties as functions of the new enumeration.

In order to get around the fact that all formal systems which are anywhere near adequate for describing recursive function theory are incomplete, we avoid restriction to any one of them by introducing the notion of a formal theory (not for the first time, of course).

For our purposes a formal theory is a decidable predicate $V(p, t_1, \ldots, t_k \rightarrow t)$ which is to be regarded as meaning that p is a proof that the statements $t_1, \ldots, t_k$ imply the statement t in the theory. We do not require that statements have negatives or disjunctions, although theories with these and other properties of propositional calculi will presumably belong to the most useful theories.

An interpretation of a formal theory is a computable function $i(t)$ mapping a class of statements of the theory into statements of other theories or concrete propositions. The only kind of concrete proposition which we shall mention for now is $P(m, n, r, k)$ meaning $f_m(n) = r$ in a computation of length $\leq k$. Such a proposition is of course verifiable.

The theories and interpretations of them can be enumerated. A function of their Gödel numbers is called a status function. (To be regarded as a current estimate of their validity and relevance, etc.) An action scheme is a computation rule which computes from a status function (its Gödel number) a new status function, perhaps gives an estimate of $g(m, r)$ if it has determined this, and computes a new action scheme.

REMARK 1. A status function would consist primarily of estimates of the validity of the separate theories and would place theories which had not been examined in an "unknown" category. However, an action scheme might modify the status of a whole class of theories simultaneously in some systematic way.

REMARK 2. The reader should note that a formal equivalence could probably be established between formal theories and certain action schemes. Thus, knowledge can be interpreted as a predisposition to act in certain ways under given conditions, and an action scheme can be interpreted as a belief that certain action is appropriate under these conditions. This equivalence should not be made because a simple theory may have a very complicated interpretation as an action scheme and conversely, and in this inversion problem the simplicity of an object is one of its most important properties.

This suggests an Alexandrian solution to a knotty question. Perhaps, a machine should be regarded as thinking if and only if its behavior can most concisely be described by statements, including those of the form "It now believes Pythagoras' theorem." This would not be the case for a phonograph record of a proof of Pythagoras' theorem. Such statements will probably be especially applicable to predictions of the future behavior under a variety of stimuli.

REMARK 3. In determining a procedure which has action schemes, theories, etc., we are imposing our ideas of the structure of the problem on the machine. The point is to do this so that the procedures we offer are general (will eventually solve every solvable problem) and also are improvable by the methods built into the machine.

REMARK 4. Except for the concrete proposition none of the statements of a formal theory have necessary interpretations which are stateable in advance. However, if the machine were working well, an observer of its internal mechanism might come to the conclusion that a certain statement of a theory is being used as though it were, say $(x)\ f_m(x) = 0$, etc. An important merit of a particular theory might be that in it the verification of a proof at the correlate of a certain concrete proposition is much shorter than the computation of the concrete proposition, i.e., shorter than k in $P(m, n, r, k)$.

REMARK 5. The relation of certain action schemes to certain theories might warrant our regarding certain statements in them as being normative, i.e., of the form "the next axiom scheme should be no. m" or "theory k should be dropped from consideration."

REMARK 6. Not every worthwhile problem is well-defined in the sense of this paper. In particular, if there exist more or less satisfactory answers with no way of deciding whether an answer already obtained can be improved on a reasonable time, the problem is not well-defined.

COMPUTABILITY BY PROBABILISTIC MACHINES

K. de Leeuw, E. F. Moore, C. E. Shannon, and N. Shapiro

INTRODUCTION

The following question will be considered in this paper: Is there anything that can be done by a machine with a random element but not by a deterministic machine?

The question as it stands is, of course, too vague to be amenable to mathematical analysis. In what follows it must be delimited in two respects. A precise definition of the class of machines to be considered must be given and an equally precise definition must be given of the tasks which they are to perform. It is clear that the nature of our results will depend strongly on these two choices and therefore our answer is not to be interpreted as a complete solution of the originally posed informal question. The reader should be especially cautioned at this point that the results we obtain have no application outside the domain implied by these choices. In particular our results refer to the possibility of enumeration of <u>infinite</u> sets and the computation of <u>infinite</u> sequences. They yield no information of the type that would be wanted if <u>finite</u> tasks were being considered; for example, the relative complexity of probabilistic machines which can perform a given finite task and their relative speeds.

This difficulty is implicit in any situation where mathematics is to be applied to an informally stated question. The process of converting this question into a precise mathematical form of necessity will highlight certain aspects. Other aspects, perhaps of equal or greater importance in another situation may be completely ignored.

The main body of the paper consists of definitions, examples, and statements of results. The proofs are deferred to the appendix since they consist in the most part of more or less elaborate constructions which are not absolutely essential for an understanding of the results.

Readers acquainted with recursive function theory will readily see that the results of this paper are actually results in that theory and can easily be translated into its terminology. In the light of this, the proofs of the results take on a dual aspect. They can be considered to be complete proofs, assuming the intuitively given notion of an effective process;[1] or they can be considered to be indications of how the theorems, if they were stated formally in the language of recursive function theory, could be proved formally using the tools of that theory. This formalization is not carried out in the paper since it would detract from the conceptual content of the proofs. However, if should be clear to anyone familiar with recursive function theory that the formalization can be carried out.

PROBABILISTIC MACHINES VS. DETERMINISTIC MACHINES

In this section we will first develop a precise definition of a class of computing machines. These machines will have an input and an output. The input will be an infinite sequence of binary digits which may be supplied in two different ways: Either as a fixed sequence (physically we may think of an infinite prepared tape) or as the output of a binary random device (with probability p of producing a 1). In this latter case we have a probabilistic machine. We will next formulate a precise class of tasks that we wish such machines to perform, namely, the enumeration with positive probability of sets of output symbols. The key result of the paper, Theorem 1, is then applied to answer our question. The answer is given in Theorem 2. What it states is that if the random device has a probability p, where p is a computable real number (that is, a real number such that there is an effective process for finding any digit of its binary expansion), any set that can be enumerated by a probabilistic machine of the type considered can be enumerated by a deterministic machine. This does not occur if p is a non-computable real number. Similar results are obtained if the sequential order of the output symbols is taken into consideration. The situation is summarized in Theorem 3.

We shall think of our machines as objects that accept as input a tape printed with 0's and 1's and puts forth as output a tape on which it may print using a finite or countably infinite selection of output symbols $s_1, s_2, s_3 \ldots$ (These symbols may be configurations formed from some finite alphabet.) The machine shall be assumed to have some initial state in which it is placed before any operation is started; (as, for example, a desk calculator must be cleared before a computation). Another requirement that we can reasonably require a machine to satisfy is the following: There

[1] For discussion of effective processes, see [9] or [2]. The reader who is not familiar with the notion of an effective process is advised to consult one of these before proceeding further. Effective processes are also discussed in [1], [7], and [8].

is an effective procedure whereby one can determine what the sequence of output symbols $(s_{j_1}, \ldots, s_{j_r})$ on the output tape will be if the machine is presented with the input sequence of 0's and 1's $(a_1, \ldots, a_n)$ in its initial state. This output sequence will be denoted by $f(a_1, \ldots a_n)$. Since we shall be interested only in the relationship between input and output in machines we can abstract away their interior and take this requirement as a definition. (Even though an abstract definition is given at this point, we shall continue to speak of machines in concrete terms.)

DEFINITION: A <u>machine</u> is a function that, for every n, to each n-tuple of 0's and 1's $(a_1, \ldots, a_n)$, associates some finite sequence $f(a_1, \ldots, a_n) = (s_{j_1}, \ldots, s_{j_r})$ consisting of elements from some fixed set S such that the following two conditions are satisfied.

1. $f(a_1, \ldots, a_n)$ is an initial segment of $f(a_1, \ldots, a_n, \ldots, a_{n+m})$ if $(a_1, \ldots, a_n)$ is an initial segment of $(a_1, \ldots, a_n, \ldots, a_{n+m})$.
2. f is a computable function, that is, there is an effective process[2] such that one can determine $f(a_1, \ldots, a_n)$ if $(a_1, \ldots, a_n)$ is given.

This definition is extended to infinite sequences as follows: If $A = (a_1, a_2, a_3, \ldots)$, $f(A)$ is the sequence which has as initial segments the $f(a_1, \ldots, a_n)$.

The operation of the machine can be thought of informally as follows: If it is supplied with input $(a_1, \ldots, a_n)$, it looks at a_1 and prints the sequence of symbols $f(a_1)$ on its output, it looks at a_2 and prints symbols after $f(a_1)$ on the tape to obtain the sequence $f(a_1, a_2)$, ..., it looks at a_n and prints symbols after $f(a_1, \ldots, a_{n-1})$ on the tape to obtain the sequence $f(a_1, \ldots, a_{n-1}, a_n)$ and then stops.

At this point several concrete examples of objects that are machines will be given. These are to serve two purposes, to illustrate the concept of machine introduced and to be referred to later to illustrate new concepts that arise. The examples need not all be read at this point.

MACHINE NO. 1: The output symbols of this machine are to be ordered pairs of integers (a, r). For each input symbol a the machine prints the output symbol (a, r) if a was the rth input symbol on the input tape. In other words, $f(a_1, \ldots, a_n) = ((a_1, 1), (a_2, 2), \ldots, (a_n, n))$.

---

[2] For discussion of effective processes, see [9] or [2]. The reader who is not familiar with the notion of an effective process is advised to consult one of these before proceeding further. Effective processes are also discussed in [1], [7], and [8].

MACHINE NO. 2: Let g be an integral valued function of positive integers which is computable, that is, there is an effective process for finding $g(n)$ if n is given. Let the machine print the output symbol $(r, g(r))$ after it has scanned the rth input symbol. In this case $f(a_1, \ldots, a_n) = \big((1, g(1)), (2, g(2)), \ldots (n, g(n))\big)$ and the output is independent of the input. This machine and Machine No. 1 are extreme cases. No. 2 is oblivious of the input and No. 1 essentially copies the input onto the output tape. Note that No. 2 would not be a machine according to our definition if the function g were not computable, for in this case there would be no effective process for determining what the output would be if the input were given.

MACHINE NO. 3: Let the machine print the symbol 0 as soon as it comes to the first zero in the input sequence. Otherwise it is to do nothing. Then $f(a_1, \ldots, a_n) = (0)$ if one of the a_j is a zero. Otherwise the machine prints nothing. This eventuality will be denoted by the "empty sequence," $(.)$, and we have $f(1, \ldots, 1) = (.)$.

MACHINE NO. 4: Let $f(a_1, \ldots, a_n) = (a_1)$. The machine merely copies the first input symbol onto the output tape and then prints nothing.

MACHINE NO. 5: Let the machine print the output symbol r if the maximum length of a string of 1's that the machine has observed is r. For example, $f(1) = (1)$, $f(1, 0) = (1, 1)$, $f(1, 0, 1) = (1, 1, 1)$, $f(1, 0, 1, 1) = (1, 1, 1, 2)$.

MACHINE NO. 6: Let $h(r, s)$ be a computable integral valued function whose domain is the positive integers. The machine prints nothing until it comes to the first 1 in the input; if this 1 is the first input symbol it never prints anything; if this 1 has occurred after r zeroes, it prints the symbol $(1, h(r, 1))$. After the next input the machine prints $(2, h(r, 2))$, after the next $(3, h(r, 3))$ etc. Let the function h_r with r fixed be defined by $h_r(s) = h(r, s)$. Then what this machine actually does is to compute the function h_r if it is presented with an initial string of r zeroes followed by a 1.

MACHINE NO. 7: This machine is given by $f(a_1, \ldots, a_n) = ((1, r_1), (2, r_2), \ldots, (q, r_q))$ where the integer r_s is obtained as follows: Look at the first 100^s digits of $(a_1, \ldots, a_n)$. Let p_s be the proportion of digits that are 1's. Let r_s be the sth digit in the binary expansion of the number p_s. q is the greatest s such that $100^s \leqq n$. A construction similar to this is used in the proof of part of Theorem 1.

Although the action of a machine is defined only for finite input sequences, it is clear what one of our machines would do if it were fed an infinite input sequence $A = (a_1, a_2, a_3, \ldots)$ and had available an infinite output tape on which it could print. Its action is determined by the fact that we know what its output will be for all finite input sequences. Thus the machine associates to each infinite input sequence an output sequence which may or may not be infinite. For example, if Machine No. 1 is fed an infinite tape on which only 1's are printed, the output sequence will be ((1, 1), (1, 2), (1, 3), ...). Machine No. 3 gives no output (the empty sequence (.)) and Machine No. 4 the output sequence (1) when presented with the same input.

We wish to consider the set of symbols that occur on the output tape of a machine that is supplied with some infinite input sequence A. (For example, if the output sequence is (1, 1, 1, ...) the set of symbols consists of the symbol 1 alone while if the output sequence is (1, 3, 5, 7, ...) the set of symbols is the set of all odd integers.) Thus the machine associates to each input sequence A some set of output symbols which it enumerates. (We do not wish to consider here the order in which the set is enumerated or whether or not it is enumerated without repetitions. This will be done later when computation of sequences is considered.)

For example, Machine No. 1 associates to any infinite input sequence $A = (a_1, a_2, \dots)$ the set of output symbols consisting of all (a_n, n) as n runs over the positive integers. Machine No. 2 associates any input sequence to the set of all (n, g(n)) as n runs over the positive integers. Machine No. 3 associates to the input sequence consisting of all 1's the empty set. To the same input Machine No. 5 associates the set of all positive integers.

DEFINITION: A machine supplied with the infinite input sequence $A = (a_1, a_2, a_3, \dots)$ will be called an A-machine. It will be said to A-enumerate the set of output symbols that it prints. A set of symbols is called A-enumerable if there is an A-machine that A-enumerates the set.

The sets that are A-enumerable if A is the sequence consisting of all 1's are usually referred to as recursively enumerable sets. We shall call them 1-enumerable sets and an A-machine shall be called a 1-machine if A consists entirely of 1's.

One can associate to each infinite sequence $A = (a_1, a_2, \dots)$ the set of all pairs (n, a_n) where a_n is the nth element in the sequence A and n runs over all positive integers. This set will be called A'. A sequence A is called computable if there is an effective process by means of which one can determine the nth term of A for every n. The following three lemmas will be proved in the appendix, although their proofs are almost immediate.

LEMMA 1. The sequence A is computable if and only if the associated set A' is 1-enumerable.

LEMMA 2. If A is a computable sequence, any A-enumerable set is 1-enumerable (and conversely).

LEMMA 3. If A is not a computable sequence, there exist sets that are A-enumerable but which are not 1-enumerable.

Lemma 3 together with the fact that non-computable sequences exist will be important later.

We now wish to attach a random device to a machine in order to construct a probabilistic machine. If one has a device that prints 0's and 1's on a tape, 1's occurring with probability p and 0's occurring with probability $1 - p$ $(0 < p < 1)$ and each printing being independent of the preceeding printings, the output of the device can be used as the input tape of a machine. The combination of the random device and the machine will be called a p-machine.

The question asked in the Introduction of whether there are probabilistic machines that are more general than deterministic machines can now be narrowed down to the more specific question: Can everything done by a p-machine be done by a 1-machine? (In the next section, it will be shown that there is little loss of generality in considering the rather narrow class of p-machines within the larger class of all conceivable probabilistic machines. We shall see that a large class of probabilistic machines are equivalent (in a sense that will be defined) to 1/2-machines, and many of their properties can be deduced from the properties of 1/2-machines.)

We must further restrict our question since it is clear that no precise answer can be given till we have a precise defininition of the task that the machines are to perform.

The task that the machines in this paper shall be set to perform is that of enumerating sets. This job is not as specialized as it at first may seem. For example, a machine can compute a function by enumerating symbols "$f(n) = m$"; or can decide the truth or falsity of a statement about positive integers by enumerating symbols like "$P(n)$ is true" and "$P(n)$ is false"; or can compute the binary expansion of a real number by enumerating symbols like "the nth digit of r is a".

We already have a definition of what is to be meant by a set of symbols being 1-enumerable. Namely, a set is 1-enumerable if there is a machine that has that set as its total output if a tape with all 1's is fed in as input. It is necessary to have some definition of enumerability to apply to p-machines so that a comparison can be made between them and 1-machines.

Two definitions will be proposed, p-enumerable sets and strongly p-enumerable sets. Our original informal question will have been reduced to the two precise questions: Is every p-enumerable set 1-enumerable? Is every strongly p-enumerable set 1-enumerable?

It remains to give the two definitions: If one has a given p-machine, one can assign to each set of possible output symbols its probability of occurrence as total output of that machine. This is the probability that that set will be precisely the set of symbols that occur on the output tape of the machine if it is started in its initial position, the random device connected and allowed to run forever. This probability is, of course, zero for most sets. Several examples will be given, calling upon the machines that we have used before. We shall assume that they have been attached to a random device with $p = 1/2$ and thus have become $1/2$-machines. Machine No. 1 prints any set with zero probability. It should be pointed out that finite sets cannot occur and that is the reason they have zero probability. However, there are infinite sets that <u>can</u> occur as output but nevertheless do so with zero probability. One should not forget the distinction between occurrence with probability zero and impossibility of occurrence. Machine No. 2 prints the set of all $(n, a(n))$ with probability 1 (and actually with certainty). Machine No. 3 has as output the set consisting of the symbol 0 alone with probability 1 but not with certainty. Machine No. 5 has as output the set of all positive integers with probability 1, but not certainty; since an infinite sequence of 0's and 1's contains arbitrarily long runs of 1's with probability 1, but not certainty. Machine No. 6 puts out the set of all $(n, h(r, n))$ with r fixed and n running over all positive integers with probability 2^{-r}. In other words, the machine computes the function h_r (which was defined by $h_r(n) = h(r, n)$) with probability 2^{-r}.

DEFINITION: For some fixed p, a set S of symbols will be called <u>strongly</u> <u>p-enumerable</u> if there is a p-machine that produces that set of output symbols in any order with non-zero probability.

The previous examples showed that the following sets are strongly $1/2$-enumerable: The set of all $(n, g(n))$ the set consisting of 0 alone, the set consisting of all positive integers and the set of all $(n, h(r, n))$ with r fixed.

It is clear how a p-machine could be used to give information about some strongly p-enumerable set S that it produces with positive probability. One would operate the p-machine having confidence that the set it enumerates is S in accordance with the probability of S occurring as output. So long as S has some positive probability, the p-machine could be used with some measure of confidence. If the probability of S occurring as output were zero, one would have no confidence in the machine's ability to enumerate <u>precisely S.</u> However, it might have a high probability of being right for <u>any particular output symbol</u> and still enumerate S with zero probability. So that this situation can be considered, a weaker definition will be given below.

There exists a definite probability that a given p-machine M will eventually print some given output symbol. Let S_M be the set of all output symbols that M eventually will print with probability $> 1/2$. For example, let the machines that have been considered as examples be connected to a random device with $p = 3/4$, so that they become $3/4$ machines. S_M for Machine No. 1 is the set of all $(n, g(n))$, for Machine No. 3 is the set consisting of 0 alone, for Machine No. 4 is the set consisting of 1 alone, for Machine No. 5 is the set of positive integers and for Machine No. 6 is the empty set.

If a p-machine M is started, the probability is $> 1/2$ that it will eventually print any given element of S_M, while the probability is $\leqq 1/2$ that it will print any given element not in S_M. Thus it is clear how a machine M could be used to give information about S_M. One would operate the machine having a certain degree of confidence that any particular element of S_M will eventually occur as an output symbol and that any particular element not in S_M will not occur as output. However, the set S_M itself might occur with probability zero, in which case, one would have no confidence that the machine will enumerate precisely S_M. (It is clear that all this remains true if a number greater than $1/2$ is used in the definition of S_M. All of the following results remain valid in this case.)

DEFINITION: A set of symbols S will be called p-enumerable if there is a machine M such that S is precisely S_M.

The condition that a set be p-enumerable is quite weak and we have thought of no weaker definition that would still ensure that there be a p-machine that give some information about each element of the set. One should note that a p-machine p-enumerates exactly one set while it may strongly p-enumerate many or no sets.

Superficially strong p-enumerability seems a much stronger concept that p-enumerability, and indeed it will be shown that a strongly p-enumerable set is p-enumerable. A bit deeper and perhaps unexpected is the converse result which implies actual equivalence of the two concepts.

To formulate the key result of the paper, which will be used to answer the two basic questions that we have enunciated, one more definition is needed. Let p be a real number, $0 < p < 1$. Then A_p is to be the infinite sequence of 0's and 1's $(a_1, a_2, a_3, \ldots)$ where a_n is the nth digit of the binary expansion of the real number p. That is, $p = .a_1a_2a_3\ldots$. Since A_p is an infinite sequence of 0's and 1's, it can be used as input to machines and we shall speak of A_p-machines, A_p-enumerability, etc.

For any number p between 0 and 1, there are now three concepts, A_p-enumerability, p-enumerability and strong p-enumerability. The first is a strictly deterministic notion while the others are probabilistic.

The key result is:

THEOREM 1: Let S be a set of symbols and p a real number, $0 < p < 1$. The following three statements are equivalent:

1. S is A_p-enumerable
2. S is p-enumerable
3. S is strongly p-enumerable.

A proof of this theorem is given in the appendix but a sketch of the proof will be given after it is shown that this theorem gives immediate answers to our two questions.

First, let p be a non-computable real number between 0 and 1. It is known that such exist and in fact almost all (in Lebesgue measure) numbers are non-computable. The non-computability of p is equivalent to the non-computability of A_p. Then, according to Lemma 3, there exists a set S that is A_p-enumerable but which is not 1-enumerable. This set, because of Theorem 1, is a set which is both p-enumerable and strongly p-enumerable but is not 1-enumerable. Thus, if one admits random devices that will print the symbol 1 with a probability that is not a computable real number, one can construct p-machines that will "enumerate" sets that are not 1-enumerable.

The situation is entirely different if p is a computable real number (in particular if it is $1/2$). This is equivalent to A_p being a computable sequence. Theorem 1 shows that any set which is p-enumerable or strongly p-enumerable is A_p-enumerable. Since A_p is computable, Lemma 2 shows that the set must be 1-enumerable. Thus a p-enumerable or strongly p-enumerable set must be 1-enumerable. The conclusion is that if p is restricted to be a computable real number (or in particular is $1/2$), a machine with a random device cannot "enumerate" anything that a deterministic machine could not.

Both cases can be summarized in:

THEOREM 2: If p is a computable real number, any p-enumerable or strongly p-enumerable set is already 1-enumerable. If p is not a computable real number, there exist p-enumerable and strongly p-enumerable sets that are not 1-enumerable.

An indication of how Theorem 1 will be proved is given at this point. The proof proceeds by demonstrating statement 3 implies statement 2 implies statement 1 implies statement 3. This chain of implications gives the equivalence.

That 3 implies 2 is proved by showing that if one has a p-machine that has output S occurring with positive probability one can construct a new machine which has output S occurring with probability $> 1/2$. Then every element of S occurs with probability $> 1/2$ and any element not in S occurs with probability $< 1/2$ so that S is p-enumerable. Thus any strongly p-enumerable set is p-enumerable.

That 2 implies 1 is proven as follows: Let a p-machine M be given. We wish to construct an A_p-machine that A_p-enumerates the set S_M. Let us assume that we have an infinite tape on which A_p is printed. By the use of larger and larger initial segments of the infinite sequence A_p, one can compute lower bounds, which are arbitrarily good, for the probability of the truth of statements of the form "By the time n inputs from the random device have occurred M will have printed the symbol s." M actually does print s with probability $> 1/2$ if and only if there is an n such that the above statement is true with probability $> 1/2$. This in turn occurs only if some one of the arbitrarily good lower bounds (that we can compute using initial segments of A_p) for the probability of the above statement being true is $> 1/2$. One can compute successively (using larger and larger initial segments of A_p) all of the lower bounds for the probabilities for every n and every s. As soon as one comes to a lower bound that is $> 1/2$, the corresponding symbol s is written down. One sees that the set of symbols thus enumerated is S_M and the enumeration has been accomplished by a process which is actually an A_p-machine. Thus, any set which is p-enumerable is also A_p-enumerable.

1 implies 3 is proved as follows: A p-machine is constructed which computes with probability $> 1/2$ the binary expansion of the real number p. That is, the output of the machine is, with probability $> 1/2$, the set of all (n, a_n) where a_n is the nth digit in the binary expansion of p. The machine that does this is a modification of Machine No. 7. If the output of this p-machine is used as input of another machine M, the consequent output will be, with probability $> 1/2$, exactly what the output of M would be if it were supplied as input the sequence A_p. If S is any A_p-enumerable set and M a machine that A_p-enumerates it, the composite machine constructed above is a p-machine that strongly p-enumerates S. Thus, any A_p-enumerable set is strongly p-enumerable. This proves that 1 implies 3 and the sketch of the proof of Theorem 1 is complete.

The question that will be considered now is whether the same results are obtained if one considers the sequential order of the outputs of the machines instead of only the totality of output symbols. It will be shown that this situation can be reduced to a particular case of the results that we have already obtained.

DEFINITION: An infinite sequence of symbols $S = s_1, s_2, s_3 \ldots$ is called

<u>A-computable</u> if it occurs as the output, in the given order, of some A-machine. It is <u>1-computable</u> if A consists entirely of 1's. S is <u>strongly p-computable</u> if there is a p-machine that puts it out, in the given order, with positive probability. S is <u>p-computable</u> if there is a p-machine that has as its nth output symbol the nth symbol of S with probability $> 1/2$.

It will be shown that the exact analog of Theorem 2 holds for these new concepts.

Let S be a <u>sequence</u> of symbols $(s_1, s_2, s_3, \ldots)$. S' is to denote the <u>set</u> of all pairs (n, s_n) where n runs over all positive integers. The following lemmas will be proven in the appendix but are immediate consequences of the definitions.

LEMMA 4: S is a 1-computable sequence if and only if S' is a 1-enumerable set.

LEMMA 5: If S is strongly p-computable, S' will be strongly p-enumerable. If S is p-computable, S' will be p-enumerable.

Applying the preceding two lemmas and Theorem 2 one gets

LEMMA 6: Let p be a computable real number (in particular 1/2). Then any p-computable or strongly p-computable sequence is already 1-computable.

PROOF: Let the sequence S be p-computable or strongly p-computable. Then S' is either p-enumerable or strongly p-enumerable because of Lemma 5. By Theorem 2, S' is 1-enumerable. By Lemma 4, S must be 1-computable.

This settles the case where p is a computable real number. If p is not computable, the proof of Theorem 1, given in the appendix, shows that there is a sequence that is p-computable and strongly p-computable but not 1-computable. (This sequence is the binary expansion of p.) Combining the two cases, one has the following analog of Theorem 2 for the computability concepts.

THEOREM 3: If p is a computable real number, any p-computable or strongly p-computable sequence is already 1-computable. If p is not computable, there exists a p-computable and strongly p-computable sequence that is not 1-computable.

As a special case of this one has the result that a 1/2-machine cannot print out any one particular non-computable sequence with positive probability.

A MORE GENERAL CLASS OF MACHINES

In this section, a generalization of the result that Theorem 2 gives for p a computable number will be given. A wide class of probabilistic machines, the computable-stochastic-machines or c-s-machines, which are not covered by our previous definitions, will be defined. A concept of equivalence for probabilistic machines will be proposed. It will be shown that any one of the new class of machines defined is equivalent to a 1/2-machine and as a consequence of Theorem 2, one will have that any set enumerated by a machine of this class must already be 1-enumerable. Furthermore, there is an effective process such that if one is supplied with the description of a machine in the class one can find a description of a 1/2-machine equivalent to it and a description of a 1-machine that 1-enumerates the same set that it enumerates.

To simplify the presentation, in this section, only one definition of enumerability, that corresponding to p-enumerability, is considered while the stronger definition corresponding to strong p-enumerability is ignored. It can be shown that the same results hold for the strong definition.

First, the class of stochastic machines will be defined. A stochastic machine is to be an object having a countable set of states $X_1, X_2, X_3, \ldots$, a distinguished initial state X_0, and a countable collection of output symbols $s_1, s_2, s_3 \ldots$. It is to be supplied with a rule that gives the probability that the machine will be next in state X_p and print output symbol s_q if it has passed successively through states $X_0, X_{j_1}, \ldots, X_{j_n}$ so far. This probability will be denoted by $P\left(X_{j_1}, \ldots, X_{j_n}; X_p; s_q\right)$. (The rule need not give an effective procedure for computing the P's.)

To any p-machine M one can associate a stochastic machine in the above sense. The states of the associated stochastic machine are to be finite sequences of 0's and 1's $(a_1, \ldots, a_n)$ where the initial state is to be the "empty" sequence (.). (What we are actually going to do here is say the "state" a p-machine is in the input sequence that it has received. These may not correspond to the internal states it may have if it is a concrete machine but determine them completely. The probability of going from state $(a_1, \ldots, a_n)$ to $(a_1, \ldots, a_n, 1)$ is p and from $(a_1, \ldots a_n)$ to $(a_1, \ldots, a_n, 0)$ is $1 - p$. The associated stochastic machine prints on arrival in state $(a_1, \ldots, a_n, a_{n+1})$ the same thing that the p-machine prints on receiving the $(n + 1)^{st}$ input symbol a_{n+1} after having received $(a_1, \ldots, a_n)$. Thus the stochastic machine is completely specified. It shall be referred to as the _stochastic machine associated with the p-machine M_. Thus, p-machines can be identified with special cases of stochastic machines.

When p is a computable number, it is easily seen that the associated machine has the following special property: The function P is _computable_; that is, there is an _effective process_ such that if one is given a positive integer m, the states $X_{j_1}, \ldots, X_{j_n}$ and X_p, and the output symbol s_q, one can compute the first m digits of the binary expansion of the number $P\left(X_{j_1}, \ldots, X_{j_n} : X_p; s_q\right)$. We shall call stochastic machines having this special property _computable stochastic machines_ or c-s-machines. As a direct consequence of the definitions, one has

LEMMA 7: The stochastic machine associated with a p-machine is a c-s-machine if and only if p is a computable real number.

The remainder of this section is devoted to showing that the result stated in Theorem 2 for p-machines with p a computable number is also valid for the much wider class of c-s-machines. That is, any set that can be enumerated by a c-s-machine is already 1-enumerable.

With stochastic machines and p-machines, one can speak of the probability of the machine eventually printing the output symbols $s_{j_1}, \ldots, s_{j_n}$ (as before we spoke of the probability of a p-machine eventually printing a single symbol s). Since we are having our machines enumerate sets, any two machines that eventually do the same thing with the same probability can be considered to be identical from the point of view of eventual output. Thus, the following definition will be made:

DEFINITION: Two objects which are either stochastic machines or p-machines will be said to be _equivalent_ if for any finite set of symbols $s_{j_1}, \ldots, s_{j_n}$, they have the same probability of eventually having that set included in their output. If the machines are M_1 and M_2, this will be denoted by $M_1 \sim M_2$.

For example, a p-machine and its associated stochastic machine are equivalent.

One recalls that a p-machine is said to p-enumerate a set of symbols if every element in that set occurs as an output symbol with probability $> 1/2$ and any output symbol not in the set occurs with probability $\leqq 1/2$. The definition can be used directly for stochastic machines and c-s-machines. If M is a stochastic machine, S_M will be the set of all output symbols that occur with probability $> 1/2$, in agreement with the previous notation for p-machines. S_M is the set of symbols that M enumerates. As an immediate consequence of the definition of equivalence one has the fact that if $M_1 \sim M_2$, then $S_{M_1} = S_{M_2}$. That is, two equivalent machines enumerate the same set.

DEFINITION: A set S is c-s-enumerable if there is a c-s-machine M such that $S = S_M$.

The question that we want answered now is the following: Is every c-s-enumerable set 1-enumerable? The answer is a consequence of the following which is proved in the appendix.

> THEOREM 4: Every c-s-machine is equivalent to a 1/2 machine.

Thus, if M is a c-s-machine, the set S_M that M enumerates is also enumerated by some 1/2-machine. According to Theorem 2, S_M must be 1-enumerable. Thus

> THEOREM 5: Any c-s-enumerable set is already 1-enumerable.

This is the result that we set out to obtain that any set which can be enumerated by a c-s-machine can be enumerated by a 1-machine.

The proof of Theorem 4 actually yields more than has been stated. It might have occured that the proof only demonstrated the existence of a 1/2-machine equivalent to any c-s-machine but gave no effective means of finding one. This, however, is not true and it does yield an effective process.

> THEOREM 4: (Supplement) There is an effective process such that if one is supplied with a description of a c-s-machine, one can produce a description of a 1/2-machine equivalent to it.

Thus, if one is given a description of a c-s-machine, M_1, there is an effective process for finding a description of a 1/2-machine M_2 such that $S_{M_1} = S_{M_2}$. This does not yet give an effective extension of Theorem 5. What is needed in order to obtain that is the following effective extension of a part of Theorem 1.

> THEOREM 1: (Supplement) There is an effective process such that if one is supplied with a description of a

p-machine M (with p a computable number), one can produce a description of a 1-machine that 1-enumerates the same S_M that M p-enumerates.

Combining the supplements of Theorem 1 and Theorem 4, one has

THEOREM 5: (Supplement) There is an effective process such that if one is supplied with a description of a c-s-machine that c-s-enumerates a set S, one can produce a description of a 1-machine that 1-enumerates S.

The results at the end of the preceding section about computability of sequences remains valid for c-s-machines. One defines in the obvious way c-s-computable sequences. The proofs of Lemmas 4, 5, and 6 are valid in this case and it follows that any c-s-computable sequence is already 1-computable.

APPENDIX

LEMMA 1: The sequence A is computable if and only if the associated set A' is 1-enumerable.

PROOF: Assume that the sequence $A = (a_1, a_2, a_3, \ldots)$ is computable. Define a machine by means of $f(b_1, \ldots, b_n) = ((1, a_1), \ldots, (n, a_n))$. Since A is computable, this actually defines a machine. The machine 1-enumerates the set A'. To prove the converse, assume that the set A' is 1-enumerable. Then there exists a 1-machine whose total output is the set of all symbols (n, a_n), but perhaps not occurring in the proper order. To determine a_n, the output tape of the machine is observed until a pair occurs whose first entry is the number n. This is certain to eventually happen. The second entry of this pair will be a_n. Thus, there is an effective process for determining a_n and A as a computable sequence.

LEMMA 2: If A is a computable sequence, then any A-enumerable set is 1-enumerable (and conversely).

PROOF: Let the machine M enumerate the set S if the computable sequence $A = (a_1, a_2, a_3, \ldots)$ is used as input. Because of Lemma 1 there is a 1-machine that has as output the set of all (n, a_n). This output can be converted in an effective manner into the sequence $(a_1, a_2, a_3, \ldots)$ and used as input by M. The composite object is a 1-machine that 1-enumerates S. It is clear that the converse is true, that any 1-enumerable set is A-enumerable. This is true because a 1-machine that 1-enumerates a set S can be converted into a machine with input, that is oblivious to that input, and will enumerate S given any input.

> LEMMA 3: If A is not a computable sequence, there exists sets that are A-enumerable but which are not 1-enumerable.

PROOF: The set A' is such a set.

The concept of "random device" will now be formalized. Let D be the set of all infinite sequences, $A = (a_1, a_2, a_3, \ldots)$, of 0's and 1's. Let $C(a_1, \ldots, a_n)$ be the subset of D consisting of all sequences starting with $(a_1, \ldots, a_n)$. D shall be considered to be a measurable space [5] whose measurable sets are the σ-ring [5] generated by the sets $C(a_1, \ldots, a_n)$. Let M be a machine (a precise definition has already been given) and S the set of possible output symbols of M. Let Q_S be the set of all finite or infinite sequences, $T = (s_{j_1}, s_{j_2}, \ldots)$, whose elements are in S. Q_S will be considered to be a measurable space whose measurable sets are the σ-ring generated by the $C(s_{j_1}, \ldots, s_{j_n})$ and the finite sequences. Let R_S be the set of all subsets of S. Let $E(j_1, \ldots, j_n; k_1, \ldots, k_m)$ be the subset of R_S consisting of all subsets of S that contain the s_{j_r} and do not contain the s_{k_r}. R_S shall be considered to be a measurable space whose measurable sets are the σ-ring generated by the sets E.

The machine M associates to each infinite input sequence in D some output sequence in Q_S. This determines a function f_M: $D \rightarrow Q_S$. One also has the mapping q_S: $Q_S \rightarrow R_S$ that takes a sequence into the set of its entries. The composite map $q_S f_M$ will be denoted by h_M. If A is in D, $h_M(A)$ is the set that will be enumerated by M if A is used as input.

The following is an immediate consequence of the definitions.

> LEMMA: The maps f_M, q_S, and h_M are measurability preserving.

Let p be a real number, $0 < p < 1$. D can be considered to be the sample space of a random device which prints 1's with probability p and 0's with probability $1 - p$. That is, D is the space of all possible events that can occur if infinitely many independent operations of the random device are observed. There is associated to each p, in a natural manner, a measure m_p on D which assigns to each measurable subset of D the probability that the random device will print an infinite sequence in that subset.[3] Thus m_p induces in the usual way probability measures on the spaces Q_S and R_S. Every measurable subset E of Q_S is assigned probability $m_p(f_M^{-1}(E))$ and every measurable subset F of R_S is assigned probability $m_p(h_M^{-1}(F))$. These induced measures will be referred to as $m_{M,p}$ and sometimes as m_M. It shall always be clear from the context which measure is being used.

The significance of the measures $m_{M,p}$ is clear: If E is a measurable subset of Q_S, $m_{M,p}(E)$ is the probability of the output sequence of M being in E if M is operated as a p-machine; if F is a measurable subset of R_S, $m_{M,p}(F)$ is the probability that the set of output symbols that M will enumerate is an element of F if M is operated as a p-machine.

If U is some subset of S, then {U}, the set consisting of U alone, is a measurable subset of R_S. Thus, one can in all cases speak of the probability of a p-machine having some set U as output. $E(j_1, \ldots, j_n;)$ is the subset of R_S consisting of all sets that contain all the s_{j_r}. Since $E(j_1, \ldots, j_n;)$ is measurable, one can in all cases speak of the probability that a p-machine will eventually print some finite set of output symbols.

The proof of Theorem 1 will now be given. First 3 implies 2 will be proven: If a set S is strongly p-enumerable it must be p-enumerable.

Let M be a machine that strongly p-enumerates the set S, that is, has S as its total output with positive probability if used with a random device having probability p of printing 1's. It will be sufficient to find a new machine M' that has S as output with probability $> \frac{1}{2}$ if used with a random device having probability p. This machine M' will also p-enumerate the set S. For every element of S will occur with probability $> \frac{1}{2}$ and every element not in S will occur with probability $< \frac{1}{2}$.

Machines $M(b_1, \ldots, b_n)$ will be constructed which are described as follows: The output of $M(b_1, \ldots, b_n)$ in response to input $(a_1, \ldots, a_m)$ is the same as the output of M corresponding to input $(b_1, \ldots, b_n, a_1, \ldots, a_m)$. Thus, $M(b_1, \ldots, b_n)$ acts in its initial state as if it were M and had already received an input of $(b_1, \ldots, b_n)$. It is intuitively rather compelling that if the set S occurs as output of M with non-zero probability, that there will be machines $M(b_1, \ldots, b_n)$ whose output is S with probability arbitrarily close to 1. (Such an $M(b_1, \ldots, b_n)$ could be taken as M' and the proof of 3 implies 2 would be complete.) That this actually does occur is seen as follows: Let D_S be $h_M^{-1}((S))$, that is the subset of D consisting of all sequences that will give the set S as output when supplied as input to M. Since M produces S with non-zero probability $M_p(D_S) > 0$. Recall that $C(b_1, \ldots, b_n)$

3 For a discussion of this, see [3].

is the subset of D consisting of all sequences that have initial segment $(b_1, \ldots, b_n)$. The probability that S will occur as output of $M(b_1, \ldots, b_n)$ is $m_p(D_S C(b_1, \ldots, b_n))/m_p(C(b_1, \ldots, b_n))$. Thus, our task will be completed if sequences $(b_1, \ldots, b_n)$ can be found such that $m_p(D_S C(b_1, \ldots, b_n))/m_p(C(b_1, \ldots, b_n))$ is arbitrarily close to 1. Actually, more than this is true.

LEMMA: Let $B = (b_1, b_2, b_3, \ldots)$ be some sequence in D. Then

$$\lim_{n \to \infty} m_p(D_S C(b_1, \ldots, b_n))/m_p(C(b_1, \ldots, b_n))$$

exists and is equal to 1 for every point of D_S and is equal to 0 for every point not in D_S, for all B in D except for a set of measure 0.

This lemma will be proven by setting up a measure preserving correspondence between the space D, supplied with the measure m_p, and the unit interval I, supplied with Lebesgue measure; and applying the following result about Lebesgue measure m, which is the precise correlate of the above lemma.

METRIC DENSITY THEOREM:[4] Let F be a measurable subset of I, x a point of I and $I_n(x)$ a decreasing sequence of intervals whose intersection is x. Then

$$\lim_{n \to \infty} m(F \cap I_n(x))/m(I_n(x))$$

exists and is equal to 1 for x in F and 0 for x not in F, for all points x except for a set of measure zero.

The correspondence is set up as follows: Let $f\colon D \to I$ be the map given by

$$f(A) = \sum_{n=1}^{\infty} a_n 2^{-n}$$

if $A = (a_1, a_2, a_3, \ldots)$. Every infinite sequence is taken into that real number which has the sequence as a binary expansion. The map f is onto and is 1 - 1 except for a countable set of points. The map f and the measure m_p induce a measure m'_p on I. If p were $\frac{1}{2}$, m'_p would be ordinary Lebesgue measure and f would be the correspondence wanted. If

[4] For a proof see Vol. 1, page 190 of [6].

p is not $\frac{1}{2}$, another step must be performed. Let $g\colon I \longrightarrow I$ be given by $g(x) = m'_p(I_x)$ where I_x is the closed interval from 0 to x. It is clear that g has the following properties: It is $1 - 1$, onto, monotone and bicontinuous. Let $h : D \longrightarrow I$ be the composite map of f and g. h is the correspondence wanted. It has the following properties:

No. 1: If E is a measurable subset of D, $h(E)$ as a Lebesgue measurable subset of I and $m(h(E)) = m_p(E)$.

No. 2: Let $B = (b_1, b_2, b_3, \dots)$ be a sequence in D. Then $h(C(b_1, \dots, b_n))$ is an interval containing the point $h(B)$ of I.

It is clear that since the mapping h has these properties, a statement corresponding to the Metric Density Theorem is true for m_p on D, and that the lemma stated is included in this statement.

Thus 3 implies 2 is proven. Actually slightly more has been proven. What has been shown is that if there is a p-machine that has a set S as output with non-zero probability, there are p-machines that have S as output with probability arbitrarily close to 1. It should be noted that the proof is non-constructive and no effective process is given for finding such machines.

Next 2 implies 1 will be proven: If a set S is p-enumerable, it must be A_p-enumerable.

Let p be any real number, $0 < p < 1$. Let M be any p-machine. To prove that 2 implies 1, it will be sufficient to construct a machine that A_p-enumerates S_M, which is the set that M p-enumerates.

Let n be a positive integer. Let s be a possible output symbol of M. Let $q(n, s)$ be the probability that M has printed the symbol s by the time n input symbols have been scanned. Note that $q(n, s) \leqq q(n + 1, s)$ and

$$\lim_{n \to \infty} q(n, s)$$

is the probability of s being eventually printed. S_M is the set of all s such that

$$\lim_{n \to \infty} q(n, s) > \tfrac{1}{2}.$$

Thus, S_M is the set of all s such that there exists some n with $q(n, s) > \frac{1}{2}$. Let $B_{n,s}$ be the set of all input sequences $(b_1, \dots, b_n)$ of length n which will, when supplied as input to M, produce an output containing s. The probability of an element of $B_{n,s}$ occurring as the first n outputs of the random device is precisely $q(n, s)$. Let $q(b_1, \dots, b_n)$ be the probability of the sequence $(b_1, \dots, b_n)$ occurring as output of the random device. $q(b_1, \dots, b_n) = p^r(1 - p)^t$, where r is the number of b's that are 1 and t is the number of b's that are 0. $q(n, s) = \Sigma\, q(b_1, \dots, b_n)$ where the summation is over all sequences in $B_{n,s}$. Let the binary expansion of the real number p be $.\, a_1\, a_2\, a_3 \dots$.

Let $p_m = .a_1 a_2 \ldots a_m$ and $p'_m = .a_1 a_2 \ldots a_m + 2^{-m}$. $p_m \leqq p \leqq p'_m$. A lower bound for $q(b_1, \ldots, b_n)$ is $p_m^r (1 - p'_m)^t$. This will be denoted by $q_m(b_1, \ldots, b_n)$. Let $q_m(n, s) = \Sigma\, q_m(b_1, \ldots, b_n)$ where the summation is over all sequences in $B_{n,s}$; $q_m(n, s)$ is a lower bound for $q(n, s)$. Since $p_m \leqq p_{m+1}$ and $p'_m \geqq p'_{m+1}$, $q_m(b_1, \ldots, b_n) \leqq q_{m+1}(b_1, \ldots, b_n)$ for all sequences and $q_m(n, s) \leqq q_{m+1}(n, s)$. Since $p_m \to p$ and $p'_m \to p$, $q_m(b_1, \ldots, b_n) \to q(b_1, \ldots, b_n)$ for all sequences and $q_m(n, s) \to q(n, s)$. Thus S_M, which is the set of all s such that there exists some n with $q(n, s) > \frac{1}{2}$, is also the set of all s such that exist some m and some n with $q_m(n, s) > \frac{1}{2}$.

Now it is clear from the construction that if one were presented a tape on which the sequence A_p were printed, one could compute successively in an effective manner all of the rational numbers $q_m(n, s)$, for every positive integral n and m and every possible output symbol s. If one writes down the symbol s each time some $q_m(n, s)$ occurs that is $> \frac{1}{2}$, the set that is enumerated is precisely S_M and the process by means of which it is enumerated is an A_p-machine.

Next 1 implies 3 will be proven: If a set S is A_p-enumerable, it must be strongly p-enumerable.

The case of p having a terminating binary expansion must be treated separately. In this case A_p is computable. Then if a set S is A_p-enumerable, it is 1-enumerable because of Lemma 2. Thus, a machine, which is oblivious of input, and which enumerates S, given any input, can be constructed. If the output of a random device having probability p of printing 1 is used as input for the machine, it enumerates S. Thus, S is strongly p-enumerable.

A machine M will now be constructed which has the following property:[5] Let p be a real number, $0 < p < 1$, which does not have a terminating binary expansion. If the output of a random device that prints 1 with probability p is fed into M, the output tape of M will, with probability $> \frac{3}{4}$, be printed with the binary expansion of p. (The $\frac{3}{4}$ is not significant. It can be replaced, uniformly in all that follows, by any X with $\frac{1}{2} < X < 1$).

Let N be a machine that A_p-enumerates the set S. If the output tape of the machine M, supplied with a random device having probability p, is used as input to N, the composite object has as output, with probability $> \frac{3}{4}$, the set S. The composite object will be a p-machine and will strongly p-enumerate the set S. Thus, any A_p-enumerable set will be shown to be strongly p-enumerable if M can be constructed. If this can be done, the two cases, p terminating and p not terminating, will have been covered, so that it is sufficient to construct the machine M.

---

[5] We are indebted to Hale Trotter at this point for the suggestion that led to the following construction.

Let $(c_1, c_2, c_3, \ldots)$ be a computable sequence of rational numbers such that $0 < c_j < 1$ and

$$\prod_{j=1}^{\infty} c_j > \frac{3}{4} .$$

M will be constructed so that it operates as follows: It scans the input from the random device having probability p, and computes the proportion of 1's that one has at the end of each output. It waits till it is certain[6] with probability $> c_1$ what the first binary digit of p must be to have produced the proportion of 1's observed. Then it prints this digit on its output tape. It forgets the proportion it has observed so far and starts computing the proportions of 1's that it next receives from the random device. It waits till it is certain with probability $> c_2$ what the first two binary digits of p must have been to have produced the proportion of 1's observed. It then prints the second digit of the two that it has on the output tape and starts again. Proceeding in this manner, it prints on its output tape the correct binary expansion of p with probability

$$> \prod_{j=1}^{\infty} c_j > \frac{3}{4} .$$

The precise construction is as follows. Let X_n be the nth coordinate function on D. That is, if $A = (a_1, a_2, a_3, \ldots)$, $X_n(A) = a_n$. D is now being considered as the set of all possible output sequences of a random device.

$$\sum_{j=1}^{n} X_j$$

is the number of 1's that have occurred during the first n operations of the device and

$$Y_n = \frac{1}{n} \sum_{j=1}^{n} X_n$$

is the proportion of 1's that have occurred. Recall that m_p is the measure on D that is used if the device prints 1's with probability p.

Define the function f as follows:

$$f(m, q, n, r) = m_{\frac{q}{2^m}} \left\{A: \mid Y_n(A) - q/2^m \mid > r/2^n\right\}$$

for all positive integers m and n, all $q = 1, \ldots, 2^m - 1$ and all $r = 1, \ldots, 2^n - 1$. This is the probability of a random device, having

[6] Most of the complications of the rest of the proof will be to show that this certainty can be obtained, uniformly in p, even for those arbitrarily close to terminating binary numbers.

$p = \frac{q}{2^m}$, printing out a sequence of n symbols which has its proportion of 1's deviating from $\frac{q}{2^m}$ by more than $\frac{r}{2^n}$. One easily sees that $f(m, q, n, r)$ is a rational number and that the function f is a computable function. Note that $f(m, q, n, 2^n) = 0$ so that if the function $g(m, q, n)$ is defined to be $\frac{1}{2^n}\left[\text{the least } r \text{ such that } (f(m, q, n, r) < \frac{1 - c_m}{5.2^m.n^2})\right]$ the function is well defined. Since the sequence of c_m has been chosen to be a computable sequence of rational numbers, the function g is a computable function which takes on only rational values.

LEMMA: $\lim_{n \to \infty} g(m, q, n) = 0$

PROOF: The theorem on page 144 of [4] shows, if one takes $x = n^{1/3}/(pq)^{1/2}$, that

$$m_p\left\{A: |Y_n(A) - p| > \frac{1}{n^{1/6}}\right\} \sim \sqrt{\frac{2p(1-p)}{\pi}}\ \frac{e^{\frac{-n^{2/3}}{2pq}}}{n^{1/3}} = o\left(\frac{1}{n^2}\right)$$

this implies, taking $p = \frac{q}{2^m}$, that $g(m, q, n) = o\left(\frac{1}{n^{1/6}}\right)$ and in particular that

$$\lim_{n \to \infty} g(m, q, n) = 0$$

The operation of the machine can now be described as follows: It accepts an input sequence $A^1 = (a_1^1, a_2^1, a_3^1, \ldots)$ from a random device. It computes, in order, all of the numbers $Y_1(A^1), Y_2(A^1), Y_3(A^1), \ldots$. As soon as it comes to an n such that $|Y_n(A^1) - \frac{1}{2}| > g(1, 1, n)$ (we shall see that this will occur with probability 1) it prints on its output tape the first binary digit of the number $Y_n(A)$. We shall see that this digit will be with probability $> c_1$ the first digit of the number p. It then accepts a new input sequence $A^2 = (a_1^2, a_2^2, a_3^2, \ldots)$ from the random device. It computes in order the numbers $Y_n(A^2)$ and the numbers $g(2, 1, n)$, $g(2, 2, n)$ and $g(2, 3, n)$. As soon as it comes to an n such that $|Y_n(A) - \frac{q}{4}| > g(2, q, n)$ for $q = 1, 2, 3,$ it writes the second binary digit of $Y_n(A^2)$ on the output tape. It will be the second digit of p with probability $> c_2$. At the mth stage the machine works in a similar manner. It accepts an input sequence $A^m = (a_1^m, a_2^m, a_3^m, \ldots)$, computes the numbers $Y_n(A^m)$ and when it reaches an n such that $|Y_n(A^m) - \frac{q}{2^m}| > g(m, q, n)$ for all $q = 1, \ldots, 2^m - 1$, it prints the mth digit of $Y_n(A^m)$ on its output tape. This digit will be the mth digit of the number p with probability $> c_m$. Thus the machine prints all the digits of p with probability $> \Pi c_m > \frac{3}{4}$.

It remains to verify that the mth digit printed is the mth digit of p with probability $> c_m$. Define the function U_m on D as follows: If A is a sequence in D, let $U_m(A) = Y_r(A)$, where r is the least n such that $|Y_n(A) - \frac{q}{2^m}| > g(m, q, n)$ for all $q = 1, \ldots, 2^m - 1$, if such an integer n exists. Let $U_m(A)$ be undefined if such an n does not exist. If the sequence A occurs as output of the random device and is used by the machine, according to the rules of operation that have been given, to determine the mth digit of p, the result will be correct if and only if $U_m(A)$ is defined and $\frac{q}{2^m} \leqq U_m(A) < \frac{q+1}{2^m}$ where q is such that $\frac{q}{2^m} \leqq p < \frac{q+1}{2^m}$. (Note that there will be no result if $U_m(A)$ is undefined.) The probability that $U_m(A)$ is undefined will first be determined. $U_m(A)$ is undefined if and only if for each n there is some q such that $|Y_n(A) - \frac{q}{2^m}| \leqq g(m, q, n)$. According to the lemma proved,

$$\lim_{n \to \infty} g(m, q, n) = 0,$$

so that there is some single q_o such that for all n sufficiently large $|Y_n(A) - \frac{q_o}{2^m}| \leqq g(m, q_o, n)$. But then

$$\lim_{n \to \infty} Y_n(A) = \frac{q_o}{2^m}.$$

Because of the Strong Law of Large Numbers [4], for every A except a set of m_p-measure zero,

$$\lim_{n \to \infty} Y_n(A) = p.$$

Since p does not have a terminating binary expansion, it is not equal to any of the $\frac{q}{2^m}$ so that $m_p\{A : U_m(A) \text{ undefined}\} = 0$.

Since the probability that U_m is undefined is zero, the probability that $U_m(A)$ has its first m digits correct is 1 minus the probability that it has an incorrect digit among the first m. Thus, it is sufficient to compute the latter. It is $\Sigma m_p \left\{ A : U_m(A) \text{ defined } \& \frac{q}{2^m} \leqq U_m(A) < \frac{q+1}{2^m} \right\}$ where the sum is over all q such that it is not true that $\frac{q}{2^m} \leqq p < \frac{q+1}{2^m}$. Each of these terms will be estimated separately and shown to be $\leqq \frac{1}{2^m}(1 - c_m)$. It will then follow that the sum is $\leqq 1 - c_m$ and that the probability of the mth digit of $U_m(A)$ being correct is $> c_m$. Thus, all that remains to conclude the proof is the estimation. Take any one of the sets $\left\{A: U_m(A) \text{ defined } \& \frac{q}{2^m} \leqq U_m(A) < \frac{q+1}{2^m}\right\}$ such that p is not between $\frac{q}{2^m}$ and $\frac{q+1}{2^m}$. Call this set E. It will be assumed that $p < \frac{q}{2^m}$. A procedure similar to that which follows covers the case $p > \frac{q+1}{2^m}$. Because of the manner in which U_m is defined, the set E is seen to be a disjoint union of sets $C(b_1, \ldots, b_r)$ where the proportion of 1's in the b_j is between $\frac{q}{2^m}$ and $\frac{q+1}{2^m}$.

(One should recall that $C(b_1, \ldots, b_r)$ is the subset of D consisting of all sequences that have $(b_1, \ldots, b_r)$ as initial segment. For any $(b_1, \ldots, b_r)$, let B be an infinite sequence that has the r-tuple as initial segment. Then E is the disjoint union of the $C(b_1, \ldots, b_r)$ where the $(b_1, \ldots, b_r)$ as such that for each s less than r there is some t, $t = 1, \ldots, 2^m - 1$, such that $\left| Y_s(b) - \frac{t}{2^m} \right| \leqq g(m, t, s)$ while $\left| Y_r(B) - \frac{t}{2^m} \right| > g(m, t, r)$ for all t and $\frac{q}{2^m} \leqq U_n(B) < \frac{q+1}{2^m}$.) Since $p < \frac{q}{2^m}$, $m_p\left(C(b_1, \ldots, b_r)\right) < m_{\frac{q}{2^m}}\left(C(b_1, \ldots, b_r)\right)$ if the proportion of 1's in $(b_1, \ldots, b_r)$ is between $\frac{q}{2^m}$ and $\frac{q+1}{2^m}$. Thus, $m_p(E) < m_{\frac{q}{2^m}}(E)$. But $m_{\frac{q}{2^m}}(E) = m_{\frac{q}{2^m}}\left\{A : U_m(A) \text{ defined } \& \frac{q}{2^m} \leqq U_m(A) < \frac{q+1}{2^m}\right\} \leqq m_{\frac{q}{2^m}}$ $\{A : U_m(A) \text{ defined}\}$. $U_m(A)$ is defined only if there is some n such that $\left| Y_n(A) - \frac{q}{2^m} \right| > g(m, q, n)$. Thus, $m_{\frac{q}{2^n}}\{A : U_m(A) \text{ defined}\}$

$$\leqq \sum_{n=1}^{\infty} m_{\frac{q}{2^n}} \left\{A : \left| Y_n(A) - \frac{q}{2^m} \right| > g(m, q, n)\right\} \leqq \frac{1-c_m}{5.2^m} \sum \frac{1}{n^2} < \frac{1-c_m}{2^m}$$

because of the definition of $g(m, q, n)$. Thus, each term is shown to be $\leqq \frac{1-c_m}{2^m}$ and the proof is complete.

It should be pointed out again that the $\frac{3}{4}$ could be replaced by any X satisfying $\frac{1}{2} < X < 1$ and that the construction is the same for all p that do not have a terminating binary expansion. Thus, the following is true. For any ϵ there is a machine M_ϵ such that if M_ϵ is used as a p-machine, the output is the binary expansion of p, with probability greater than $1 - \epsilon$, for any p not a terminating binary.

The proof of Theorem 1 is completed.

LEMMA 4: S is a 1-computable sequence if and only if S' is a 1-enumerable set.

PROOF: Same as Lemma 1.

LEMMA 5: If S is strongly p-computable, S' will be strongly p-enumerable. If S is p-computable, S' will be p-enumerable.

PROOF: Let M be a machine. Modify the M to form the machine M' as follows: If the output of M in response to some input is $\left(s_{j_1}, \ldots, s_{j_n}\right)$, the response of M' to the same input is to be $\left((1, s_{j_1}), \ldots (n, s_{j_n})\right)$.

M' is a machine and if M strongly p-computes the sequence S, M' strongly p-enumerates the set S'. If M p-computes the sequence S, M' p-enumerates the set S'.

Let M be a stochastic machine. Let G be the set of all sequences $X = \left(X_{j_1}, X_{j_2}, X_{j_3}, \ldots\right)$ whose elements are names of states of M. Let $C\left(X_{j_1}, \ldots, X_{j_n}\right)$ be the set of all sequences that have $\left(X_{j_1}, \ldots, X_{j_n}\right)$ as initial segment. G is to be considered as a measurable space whose measurable subsets are to be the σ-ring generated by the $C\left(X_{j_1}, \ldots, X_{j_n}\right)$. The underlying stochastic process of the machine gives rise in a natural manner to a measure m on G which associates to each measurable subset the probability that the machine will pass through a sequence of states in that subset.[7] Let S be the set of possible output symbols of M. Q_S and R_S are as before the set of all sequences of elements of S and the set of all subsets of S, respectively. We have the map $f_M : G \rightarrow Q_S$ which associates to each infinite sequence of states the output sequence that will be printed if the machine passes through that sequence of states. This combined with the natural map $g_S : Q_S \rightarrow R_S$, which takes a sequence into the set of elements that occur in it, gives rise to the composite map $h_M : G \rightarrow R_S$. If X is in G, $h_M(X)$ is the set of symbols that will be enumerated if the machine passes through the sequence of states X. The map h_M and the measure m on G induce a measure on R_S which will be denoted by m_M. If E is a measurable subset of R_S, $m_M(E) = m(h_M^{-1}(E))$ is the probability that the machine will enumerate a set in E.

Let $E(j_1, \ldots, j_n; k_1, \ldots, k_m)$ be the subset of R_S consisting of all subsets of S that contain $s_{j_1}, \ldots, s_{j_n}$ and do not contain $s_{k_1}, \ldots, s_{k_m}$. (Either n or m may be zero.) Two stochastic machines M_1 and M_2 have been defined to be equivalent, $M_1 \sim M_2$, if they have the same set S of possible output symbols and $m_{M_1}(E(j_1, \ldots, j_n;)) = m_{M_2}(E(j_1, \ldots, j_n;))$.

LEMMA: If $M_1 \sim M_2$, m_{M_1} and m_{M_2} agree on all sets of the form $E(j_1, \ldots, j_n; k_1, \ldots, k_m)$.

PROOF: The proof proceeds by induction on m. It is true for $m = 1$ because $E(j_1, \ldots, j_n;) = E(j_1, \ldots, j_n, k_1) + E(j_1, \ldots, j_n; k_1)$ where the sum is disjoint. m_{M_1} and m_{M_2} agree on the first two sets and thus must agree on the third. The induction from $m-1$ to m proceeds in a similar manner since $E(j_1, \ldots, j_n; k_1, \ldots, k_{m-1}) =$
$E(j_1, \ldots, j_n, k_m; k_1, \ldots, k_{m-1}) + E(j_1, \ldots, j_n; k_1, \ldots, k_{m-1}, k_m)$, the sum being disjoint. m_{M_1} and m_{M_2} agree on the first two and therefore must agree on the third.

---

7 For a discussion of this, see [3].

The set of all subsets of R_S of the form $E(j_1, \ldots, j_n; k_1, \ldots, k_m)$ is a ring of sets and generates the σ-ring of sets. Thus, if m_{M_1} and m_{M_2} agree on the ring, they agree on all measurable subsets of R_S [5, p. 54]. The converse is trivially true. Thus, we have.

LEMMA: $M_1 \sim M_2$ if and only if $m_{M_1} = m_{M_2}$.

This shows that $m_{M_1} = m_{M_2}$ could have been taken as definition of equivalence instead of the weaker statement.

In order to prove Theorem 4, that every stochastic machine is equivalent to a $\frac{1}{2}$-machine, a reduction step must be made. The underlying stochastic process of a stochastic machine will be reduced in the usual way to a markoff process.

DEFINITION: A stochastic machine is <u>markoff</u> if:

No. 1: $P\left(X_{j_1}, \ldots, X_{j_{n-1}}, X_{j_n}; X_p; s_q\right)$ is independent of $X_{j_1}, \ldots, X_{j_{n-1}}$.

No. 2: For every X_p there is only one s_q such that $P\left(X_{j_1}, \ldots, X_{j_n}; X_p; s_q\right)$ is non-zero.

That is, not only is the underlying stochastic process a markoff process but the symbol printed on arrival in any state depends deterministically on that state, and depends only on that state.

LEMMA: Every stochastic machine M is equivalent to a stochastic machine M' which is markoff. If M is a c-s-machine, M' can be chosen to be a c-s-machine.

PROOF: The usual trick of converting a stochastic process into a markoff process by considering it to be in "state" $\left(X_{j_1}, \ldots, X_{j_n}\right)$ if it has passed successively through the $X_{j_1}, \ldots, X_{j_n}$ works in this case. The states of M' are to have names $\left(X_{j_1}, \ldots, X_{j_n}; s_r\right)$. The probability of M' going from state $\left(X_{j_1}, \ldots, X_{j_n}; s_r\right)$ to state $\left(X_{j_1}, \ldots, X_{j_n}, X_p; s_q\right)$ and printing the symbol s_q on arrival is $P\left(X_{j_1}, \ldots, X_{j_n}; X_p; s_q\right)$, and all other probabilities are zero. M' is a markoff machine and $M \sim M'$. If M is a c-s-machine, M' is also a c-s-machine. Also, it is clear that an effective process has been furnished which will supply the description of M' if a description of M is given.

Thus, our attention may be restricted to c-s-machines that are markoff. A simplification in the notation can be made. The states of a markoff c-s-machine can be assumed to be named by the positive integers and $P(n, m)$ will be used to denote the probability of moving from the nth state to the mth state. Note that

$$\sum_{m=1}^{\infty} P(n, m) = 1.$$

We shall now use for the first time the fact that the objects under consideration are c-s-machines. This means that there is an effective process such that if one is given positive integers, r, n and m, one can find the first r digits of the binary expansion of $P(n, m)$. Denote the rational number $.a_1a_2 \ldots a_r$ by $P(n, m; r)$ if $(a_1, a_2, \ldots, a_r)$ are the first r digits of the binary expansion of $P(n, m)$. $P(n, m; r)$ is a computable function. Let $Q(n, m; r)$ be the number $.b_1b_2 \ldots b_r$ if $(b_1, b_2, \ldots, b_r)$ are the first r binary digits of

$$\sum_{j=1}^{m} P(n, j).$$

Note that $Q(n, m; r) \leqq Q(n, m; r + 1)$ and that

$$\lim_{r \to \infty} Q(n, m; r) = \sum_{j=1}^{m} P(n, j).$$

We now wish to construct a $\frac{1}{2}$ - machine that is equivalent to a given markoff c-s-machine. If this can be done Theorem 4 will be proven since every c-s-machine is equivalent to one that is markoff. The main trick is the use of a random device printing 1's with probability $\frac{1}{2}$ in conjunction with the computable function $Q(n, m; r)$ to obtain events occuring with the probabilities $P(n, m)$. A simplified but relevant example will be given first that will demonstrate the idea involved.

Let us assume that we are given a random device that prints 1's with probability $\frac{1}{2}$ and wish to construct a device that prints 1's with probability p, where p is a computable number. First construct a 1-machine N_p that generates the binary expansion $.b_1b_2b_3 \ldots$ of p. Let the output of our random device be the sequence $a_1, a_2, a_3, \ldots$. Compare the successive approximations

$$.b_1, .b_1b_2, .b_1b_2b_3, \ldots, \sum_{r=1}^{n} b_r 2^{-r}, \ldots$$

that one obtains for the number p by observing the output tape of N_p, with the numbers

$$.a_1,\ .a_1a_2,\ .a_1a_2a_3,\ \ldots,\ \sum_{r=1}^{n} a_r\, 2^{-r},\ \ldots$$

Eventually, if

$$\sum_{r=1}^{\infty} a_r\, 2^{-r} \neq p$$

(and thus with probability 1), one will come to the first n such that

$$\sum_{r=1}^{n} a_r\, 2^{-r} \neq \sum_{r=1}^{n} b_r\, 2^{-r}.$$

At this point, write down a 1 if

$$\sum_{r=1}^{n} a_r\, 2^{-r} < \sum_{r=1}^{n} b_r\, 2^{-r}$$

and a 0 if the inequality is in the other direction. Thus, since the probability is precisely p that

$$\sum_{r=1}^{\infty} a_r\, 2^{-r} < p$$

and is $1 - p$ that

$$\sum_{r=1}^{\infty} a_r\, 2^{-r} > p,$$

the event "a 1 is written" occurs with probability p and the event "a 0 is written" occurs with probability $1 - p$. (Note that the event "nothing ever happens" occurs with probability 0 but may still occur.) If this process is repeated, the object that is described is a device that prints 1 with probability p.

Furthermore, this construction demonstrates that every p-machine (remember that p is a computable number) is equivalent to a $\frac{1}{2}$ - machine. For any machine M, supplied with the output of the device constructed above, becomes a $\frac{1}{2}$ - machine, and is equivalent to the p-machine that one obtains when one supplies M with the output of a random device that prints 1's with probability p. The construction of a $\frac{1}{2}$ - machine equivalent to a given c-s-machine will be seen to be similar to the above construction.

We shall now assume that we have a markoff c-s-machine M and shall proceed to construct a $\frac{1}{2}$ - machine M' equivalent to it. It can be assumed that the machine M starts in the state 0. Let the output sequence of a random device printing 1 with probability $\frac{1}{2}$ be $(a_1, a_2, a_3, \ldots)$. The probability that the number

$$\sum_{n=1}^{\infty} a_n 2^{-n}$$

lies between

$$\sum_{j=0}^{m-1} P(0, j)$$

and

$$\sum_{j=0}^{m} P(0, j)$$

is precisely $P(0, m)$. Let N be a 1-machine that computes in <u>succession</u> the values of the computable function $Q(n, m; r)$ for every n, m and r. The machine M' does the following: At the sth stage, it compares the number

$$\sum_{j=1}^{s} a_j 2^{-j}$$

with the first s of the values $Q(0, m; r)$ that N has computed. Using this process, with probability 1 (but not certainty), these comparisons will eventually find some number t_1 such that

$$\sum_{r=1}^{t_1-1} P(0, r) < \sum_{n=1}^{\infty} a_n 2^{-n} < \sum_{r=1}^{t_1} P(0, r).$$

At this point the machine M' prints what M would print on arrival in state t_1. Thus M', imitating the first transition of M, has printed with probability $P(0, m)$ what M would print on arrival in state m. M' must now imitate the second transition of M. It again accepts an input $(b_1, b_2, b_3, \ldots)$ from the random device and compares the numbers

$$\sum_{j=1}^{s} b_j 2^{-j},$$

this time to the first s values of $Q(t_1, m; r)$ that N produces. With probability 1, these comparisons will eventually find some integer t_2 such that

$$\sum_{r=1}^{t_2-1} P(t_1, r) < \sum_{n=1}^{\infty} b_n 2^{-n} < \sum_{r=1}^{t_2} P(t_1, r).$$

At this point M' imitates the second transition of M and prints what M would print on arrival in state t_2. Thus, M' has printed with probability $P(t_1, m)$ what M would print on arrival in state m. M' then proceeds in the same manner to imitate the next transitions of M. It is clear that M' is a $\frac{1}{2}$ - machine that is equivalent to M and also that the above construction yields an effective process for transforming a description of M into a description of M'.

The proof of Theorem 4 is complete.

BIBLIOGRAPHY

[1] CHURCH, Alonzo, "An Unsolvable Problem of Elementary Number Theory," American Journal of Mathematics, Vol. 58, pp. 345 - 363 (1936).

[2] DAVIS, Martin, Forthcoming Book on Recursive Functions and Computability, to be published by McGraw-Hill.

[3] DOOB, J. L., "Stochastic Processes," John Wiley & Sons, (1953).

[4] FELLER, William, "An Introduction to Probability Theory and its Applications," John Wiley & Sons, (1950).

[5] HALMOS, Paul R., "Measure Theory," Van Nostrand, (1950)

[6] HOBSON, E. W., "The Theory of Functions of a Real Variable and the Theory of Fourier's Series," (1926).

[7] KLEENE, S. C., "Introduction to Metamathematics," Van Nostrand, (1952).

[8] POST, Emil L., "Finite Combinatory Processes - Formulation I." Jour. Symbolic Logic, Vol. 1, pp. 103 - 105, (1936).

[9] TURING, A. M., "On Computable Numbers, with an Application to the Entscheidungsproblem," Proc. London Math. Society, Ser. 2, Vol. 42 (1936), pp. 230 - 265, and Vol. 43 (1937), pp. 544 - 546.

SYNTHESIS OF AUTOMATA

DESIGN FOR AN INTELLIGENCE-AMPLIFIER

W. Ross Ashby

SECTION I

1. Introduction

For over a century Man has been able to use, for his own advantage, physical powers that far transcend those produced by his own muscles. Is it impossible that he should develop machines with "synthetic" intellectual powers that will equally surpass those of his own brain? I hope to show that recent developments have made such machines possible — possible in the sense that their building can start today. Let us then consider the question of building a mechanistic system for the solution of problems that are beyond the human intellect. I hope to show that such a construction is by no means impossible, even though the constructors are themselves quite averagely human.

There is certainly no lack of difficult problems awaiting solution. Mathematics provides plenty, and so does almost every branch of science. It is perhaps in the social and economic world that such problems occur most noticeably, both in regard to their complexity and to the great issues that depend on them. Success in solving these problems is a matter of some urgency. We have built a civilisation beyond our understanding and we are finding that it is getting out of hand. Faced with such problems, what are we to do?

Our first instinctive action is to look for someone with corresponding intellectual powers: we think of a Napoleon or an Archimedes. But detailed study of the distribution of Man's intelligence shows that this method can give little. Figure 1, for instance, shows the distribution of the Intelligence Quotient in the normal adult population, as found by Wechsler [1]. What is important for us now is not the shape on the left but the absolute emptiness on the right. A variety of tests by other workers have always yielded about the same result: a scarcity of people

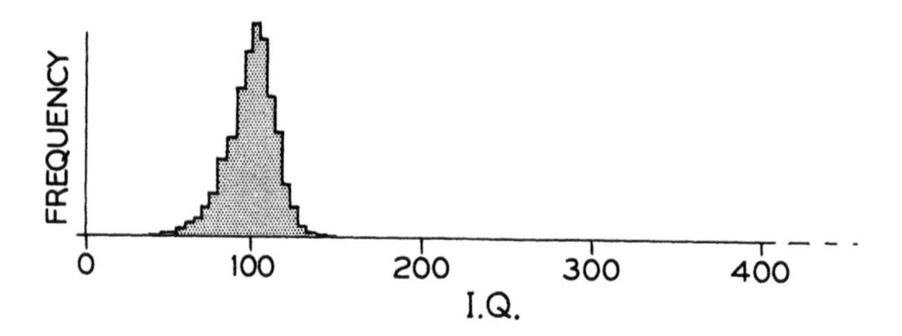

FIGURE 1: Distribution of the adult human Intelligence Quotient (after Wechsler, 1).

with I.Q.s over 150, and a total absence of I.Q.s over 200. Let us admit frankly that Man's intellectual powers are as bounded as are those of his muscles. What then are we to do?

We can see something of how to proceed by comparing our position today in respect of intellectual problems with the position of the Romans in respect of physical problems. The Romans were doubtless often confronted by engineering and mechanical problems that demanded extreme physical strength. Doubtless the exceptionally strong slave was most useful, and doubtless the Romans sometimes considered the possibility of breeding slaves of even greater strength. Nevertheless, such plans were misdirected: only when men turned from their own powers to the powers latent in Nature was the revolution inaugurated by Watt possible. Today, a workman comes to his task with a thousand horsepower available, though his own muscles will provide only about one-tenth. He gets this extra power by using a "power-amplifier". Had the present day brain-worker an "intelligence-amplifier" of the same ratio, he would be able to bring to his problems an I.Q. of a million.

If intellectual power is to be so developed, we must, somehow, construct amplifiers for intelligence – devices that, supplied with a little intelligence, will emit a lot. To see how this is to be done, let us look more closely at what is implied.

2. The Criterion of Intelligence

Let us first be clear about what we want. There is no intention here to enquire into the "real" nature of intelligence (whatever that may mean). The position is simple: we have problems and we want answers. We proceed then to ask, where are the answers to be found?

It has often been remarked that any random sequence, if long enough, will contain all the answers. Nothing prevents a child from doodling

$$\cos^2 x + \sin^2 x = 1,$$

or a dancing mote in the sunlight from emitting the same message in Morse or a similar code. Let us be more definite. If each of the above thirteen symbols might have been any one of fifty letters and elementary signs, then as 50^{13} is approximately 2^{73}, the equation can be given in coded form by 73 binary symbols. Now consider a cubic centimeter of air as a turmoil of colliding molecules. A particular molecule's turnings after collision, sometimes to the left and sometimes to the right, will provide a series of binary symbols, each 73 of which, on some given code, either will or will not represent the equation. A simple calculation from the known facts shows that the molecules in every cubic centimeter of air are emitting this sequence correctly over a hundred thousand times a second. The objection that "such things don't happen" cannot stand.

Doodling, then, or any other random activity, is capable of producing all that is required. What spoils the child's claim to be a mathematician is that he will doodle, with equal readiness, such forms as

$$\cos^2 x + \sin^2 x = 2 \qquad \text{or} \qquad \text{ci)xsi=nx1}$$

or any other variation. After the child has had some mathematical experience he will stop producing these other variations. He becomes not more but less productive: he becomes selective.

The close, indeed essential, relation between intelligence and selection is shown clearly if we examine the tests specially devised for its objective measurement. Take, for instance, those of the Terman and Merrill [2] series for Year IV. In the first Test the child is shown a picture of a common object and is asked to give its name. Out of all the words he knows he is asked to select one. In the second Test, three model objects — motor-car, dog, shoe — are placed in a row and seen by the child; then all are hidden from him and a cover is placed over the dog; he is then shown motor-car, cover, shoe, and asked what is under the cover. Again his response is correct if, out of all possible words, he can select the appropriate one. Similarly the other Tests, for all ages, evoke a response that is judged "correct" or "incorrect" simply by the subject's power of appropriate selection.

The same fact, that getting a solution implies selection, is shown with special clarity in the biological world. There the problems are all ultimately of how to achieve survival, and survival implies that the essential variables — the supply of food, water, etc. — are to be kept within physiological limits. The solutions to these problems are thus all selections from the totality of possibilities.

The same is true of the most important social and economic problems. What is wanted is often simple enough in aim — a way of ensuring food for all with an increasing population, or a way of keeping international frictions small in spite of provocations. In most of these problems the aim is the keeping of certain variables within assigned limits; and the problem is to find, amid the possibilities, some set of dynamic linkages that will keep the system both stable, and stable within those limits. Thus, finding the answer is again equivalent to achieving an appropriate selection.

The fact is that in admiring the productivity of genius our admiration has been misplaced. Nothing is easier than the generation of new ideas: with some suitable interpretation, a kaleidoscope, the entrails of a sheep, or a noisy vacuum tube will generate them in profusion. What is remarkable in the genius is the discrimination with which the possibilities are winnowed.

A possible method, then, is to use some random source for the generation of all the possibilities and to pass its output through some device that will select the answer. But before we proceed to make the device we must dispose of the critic who puts forward this well known argument: as the device will be made by some designer, it can select only what he has made it to select, so it can do no more than he can. Since this argument is clearly plausible, we must examine it with some care.

To see it in perspective, let us remember that the engineers of the middle ages, familiar with the principles of the lever and cog and pulley, must often have said that as no machine, worked by a man, could put out more work than he put in, therefore no machine could ever amplify a man's power. Yet today we see one man keeping all the wheels in a factory turning by shovelling coal into a furnace. It is instructive to notice just how it is that today's stoker defeats the mediaeval engineer's dictum, while being still subject to the law of the conservation of energy. A little thought shows that the process occurs in two stages. In Stage One the stoker lifts the coal into the furnace; and over this stage energy is conserved strictly. The arrival of the coal in the furnace is then the beginning of Stage Two, in which again energy is conserved, as the burning of the coal leads to the generation of steam and ultimately to the turning of the factory's wheels. By making the whole process, from stoker's muscles to factory wheel, take place in two stages, involving two lots of energy whose sizes can vary with some independence, the modern engineer can obtain an overall amplification. Can we copy this method in principle so as to get an amplification in selection?

3. The Selection-amplifier

The essence of the stoker's method is that he uses his (small) power to bring into action that which will provide the main power. The designer, therefore, should use his (small) selectivity to bring into action that which is going to do the main selecting. Examples of this happening are common-place once one knows what to look for. Thus a garden sieve selects stones from soil; so if a gardener has sieves of different mesh, his act of selecting a sieve means that he is selecting, not the stones from the soil, but that which will do the selecting. The end result is that the stones are selected from the soil, and this has occurred as a consequence of his primary act; but he has achieved the selection mediately, in two stages. Again, when the directors of a large firm appoint a Manager of Personnel, who will attend to the selection of the staff generally, they are selecting that which will do the main selecting. When the whole process of selection is thus broken into two stages the details need only a little care for there to occur an amplification in the degree of selection exerted.

In this connexion it must be appreciated that the _degree_ of selection exerted is not defined by what is selected: it depends also on what the object is selected from. Thus, suppose I want to telephone for a plumber, and hesitate for a moment between calling Brown or Green, who are the only two I know. If I decide to ring up Green's number I have made a one-bit selection. My secretary, who will get the number for me, is waiting with directory in hand; she also will select Green's number, but she will select it from 50,000 other numbers, a 15.6-bit selection. (Since a 1-bit selection has directly determined a 15.6-bit selection, some amplification has occurred.) Thus two different selectors can select the same thing and yet exert quite different degrees of selection.

The same distinction will occur in the machine we are going to build. Thus, suppose we are tackling a difficult social and economic problem; we first select what we want, which might be:

An organisation that will be stable at the conditions:

Unemployed	$<$	100,000 persons
Crimes of violence	$<$	10 per week
Minimal income per family	$>$	£500 per annum

This is _our_ selection, and its degree depends on what other conditions we might have named but did not. The solving-machine now has to make _its_ selection, finding this organisation among the multitudinous other possibilities in the society. We and the solving-machine are selecting the same entity, but we are selecting it from quite different sets, or contexts, and the degrees of selection exerted can vary with some independence. (The similarity of this relation with those occurring in information theory is unmistakable; for in the latter the information-content of a message depends not only on what is in the message but on what population of messages it came from [3,4].)

The building of a true selection-amplifier – one that selects over a greater range than that covered when it was designed – is thus possible. We can now proceed to build the system whose selectivity, and therefore whose intelligence, exceeds that of its designer.

(From now on we shall have to distinguish carefully between two problems: our problem, which is to design and build the solving-machine, and the solving-machine's problem – the one we want it to solve.)

4. Basic Design

Let us suppose for definiteness that the social and economic problem of the previous article is to be the solver's problem. How can we design a solver for it? The construction would in practice be a formidable task, but here we are concerned only with the principles. First, how is the selection to be achieved automatically?

SELECTION BY EQUILIBRIUM. We can take advantage of the fact that if any two determinate dynamic systems (X and S in Figure 2) are coupled through channels G and U so that each affects the other, then any resting state of the whole, (that is, any state at which it can stay permanently,) must be a resting state in each of the two parts individually, each being in the conditions provided by the other. To put it more picturesquely, each part has a power of veto over resting states proposed by the other. (The formulation can be made perfectly precise in the terms used in Article 6.)

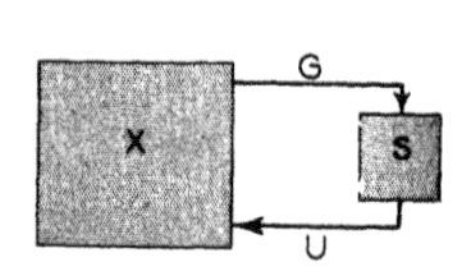

FIGURE 2

It is only a change of words to say that each part acts selectively towards the resting states of the other. So if S has been specially built to have resting states only on the occurrence of some condition ζ in S, then S's power of veto ensures that a resting state of the whole will always imply ζ in S. Suppose next that the linkage G is such that G will allow ζ to occur in S if and only if the condition η occurs in X. S's power of veto now ensures that any resting state of the whole must have condition η in X. So the selection of S and G to have these properties ensures that the only states in X that can be permanent are those that have the condition η.

It must be noticed that the selection of η, in the sense of its retention in X, has been done in two stages. The first occurred when the designer specified S and G and ζ. The second occurred when S, acting without further reference to the designer, rejected state after state of X, accepting finally one that gave the condition η in X. The designer has, in a sense, selected η, as an ultimate consequence of his actions, but his actions have worked through two stages, so the selectivity achieved in the second stage may be larger, perhaps much larger, than that used in the first.

The application of this method to the solving of the economic problem is, in principle, simple. We identify the real economic world with X and the conditions that we want to achieve in it with η. The selection of η in X is beyond our power, so we build, and couple to it, a system S, so built that it has a resting state if and only if its information through G is that η has occurred in a resting state in X. As time progresses, the limit of the whole system, X and S, is the permanent retention of η in X. The designer has to design and build S and G, and to couple it to X; after that the process occurs, so far as he is concerned, automatically.

5. The Homeostat

To see the process actually at work, we can turn briefly to the Homeostat. Though it has been described fully elsewhere [5], a description of how its action appears in the terms used here may be helpful in illustration. (Figure 3 is intended to show its principle, not its actual appearance.)

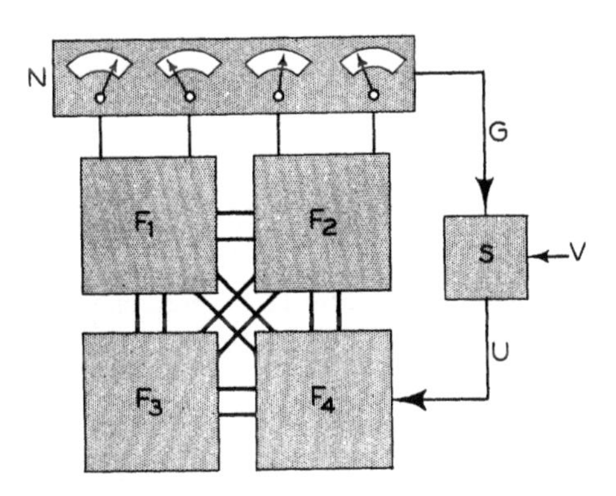

FIGURE 3

It consists of four boxes (F) of components, freely supplied with energy, that act on one another in a complex pattern of feedbacks, providing initially a somewhat chaotic system, showing properties not unlike those sometimes seen in our own society. In this machine, S has been built, and G arranged, so that S has a resting state when and only when four needles N are stable at the central positions. These are the conditions η. N and the F's correspond to X in Figure 2. S affects the F's through the channel U, whose activity causes changes in the conditions within the boxes.

Suppose now that the conditions within the boxes, or in the connexions between them, are set in some random way, say as a by-stander pleases; then if the conditions so set do not satisfy η, S goes into activity and enforces changes till η is restored. Since η may refer to certain properties of the stability within F, the system has been called "ultrastable", for it can regulate the conditions of its own stability within F. What is important in principle is that the combinations in F that restore η were not foreseen by the designer and programmed in detail; he provided only a random collection of about 300,000 combinations (from a table of random numbers), leaving it to S to make the detailed selection.

One possible objection on a matter of principle is that all the variation going to the trials in X seems, in Figure 2, to be coming from S and therefore from the designer, who has provided S. The objection can easily be met, however, and the alteration introduces an important technical improvement. To make the objection invalid, all the designer has to do is to couple S, as shown in Figure 3, to some convenient source of random variation V — a noisy vacuum tube say — so that the SV combination

(i) sends disturbance of inexhaustible variety along U if ζ is not occurring in S, and

(ii) keeps U constant, i.e., blocks the way from V to U, if ζ <u>is</u> occurring.

In the Homeostat, V is represented by the table of random numbers which determined what entered F along U. In this way the whole system, X and S, has available the inexhaustible random variation that was suggested in Article 2 as a suitable source for the solutions.

6. Abstract Formulation

It is now instructive to view the whole process from another point of view, so as to bring out more clearly the deep analogy that exists between the amplification of power and that of intelligence.

Consider the engineer who has, say, some ore at the foot of a mine-shaft and who wants it brought to the surface. The power required is more than he can supply personally. What he does is to take some system that is going to change, by the laws of nature, from low entropy to high, and he couples this system to his ore, perhaps through pistons and ropes, so that "low entropy" is coupled to "ore down" and "high entropy" to "ore up". He then lets the whole system go, confident that as the entropy goes from low to high so will it change the ore's position from down to up.

```
H1 ---- H2
 :       :
 :       :
C1      C2
```

FIGURE 4

Abstractly (Figure 4) he has a process that is going, by the laws of nature, to pass from state H_1 to state H_2. He wants C_1 to change to C_2. So he couples H_1 to C_1 and H_2 to C_2. Then the system, in changing from H_1 to H_2, will change C_1 to C_2, which is what he wants. The arrangement is clearly both necessary and sufficient.

The method of getting the problem-solver to solve the set problem can now be seen to be of essentially the same form. The job to be done is the bringing of X, in Figure 2, to a certain condition or "solution" η. What the intelligence engineer does first is build a system, X and S, that has the tendency, by the laws of nature, to go to a state of equilibrium. He arranges the coupling between them so that "not at equilibrium" is coupled to not-η, and "at equilibrium" to η. He then lets the system go, confident that as the passage of time takes the whole to an equilibrium, so will the conditions in X have to change from not-η to η. He does not make the conditions in X change by his own efforts, but allows the basic drive of nature to do the work.

This is the fundamental principle of our intelligence-amplifier. Its driving power is the tendency for entropy to increase, where "entropy" is used, not as understood in heat-engines but as understood in stochastic processes.

AXIOMATIC STATEMENT. Since we are considering systems of extreme generality, the best representation of them is given in terms of the theory of sets. I use the concepts and terminology of Bourbaki [6].

From this point of view a machine, or any system that behaves in a determinate way, can be at any one of a set of states at a given moment. Let M be the set of states and μ some one of them. Time is assumed to be discrete, changing by unit intervals. The internal nature of the machine, whose details are irrelevant in this context, causes a transformation to occur in each interval of time, the state μ passing over determinately to some state μ' (not necessarily different from μ), thereby defining a mapping t of M in M:

$$t: \quad \mu \longrightarrow \mu' = t(\mu).$$

If the machine has an input, there will be a set I of input states ι, to each of which will correspond a mapping t_ι. The states ι may, of course, be those of some other machine, or may be determined by it; in this way machine may be coupled to machine. Thus if machine N with states ν has a set K of inputs κ and transformations u_κ, then machines M and N can be coupled by defining a mapping ℓ of M in K, $\kappa = \ell(\mu)$, and a mapping m of N in I, $\iota = m(\nu)$, giving a system whose states are the couples (μ, ν) and whose changes with time are defined by the mapping, of $M \times N$ in $M \times N$:

$$(\mu, \nu) \longrightarrow \left(t_{m(\nu)}(\mu), u_{\ell(\mu)}(\nu)\right).$$

The abstract specification of the principle of ultrastability is as follows, Figure 5 corresponding to Figure 2:

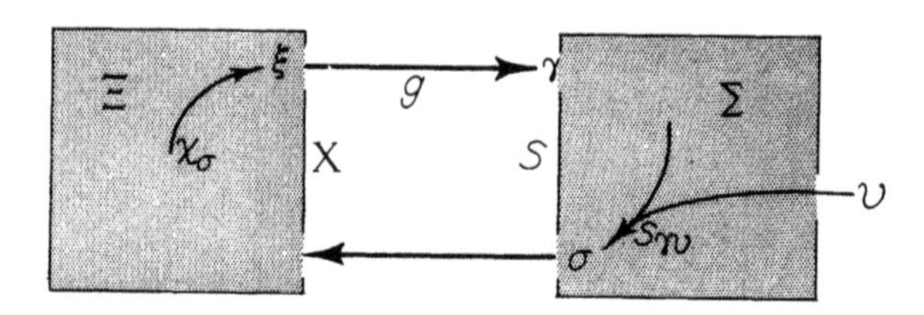

FIGURE 5

GIVEN:

(1) A set Γ consisting of two elements γ_1 and γ_2;

(2) A set Ξ of elements ξ;

(3) A mapping g of Ξ in Γ;

(4) A set Σ of elements σ;

(5) A family of mappings χ_σ of Ξ in Ξ;

(6) A random variable υ, inexhaustible in variety;

(7) A double family of mappings $s_{\gamma\upsilon}$ of Σ in Σ, with the property that, for all $\sigma \in \Sigma$ and all values of υ,

$$s_{\gamma_1\upsilon}(\sigma) \neq \sigma \qquad \text{and} \qquad s_{\gamma_2\upsilon}(\sigma) = \sigma;$$

(8) Time, advancing by discrete intervals, induces the operations χ_σ and $s_{\gamma\upsilon}$ and the successive values of υ simultaneously, once in each interval.

THEOREM: If the series of states of Ξ, induced by time, has a limit ξ^*, then $\xi^* \in g^{-1}(\gamma_2)$.

PROOF. The state of the whole system, apart from υ, is given by the couple (ξ, σ), an element in $\Xi \times \Sigma$. The passage of one interval of time induces the mapping $(\xi, \sigma) \longrightarrow \left(\chi_\sigma(\xi), s_{g(\xi),\upsilon}(\sigma)\right)$. If the series is at a limit-state (ξ^*, σ^*), then $s_{g(\xi^*),\upsilon}(\sigma^*) = \sigma^*$ for all values of υ. Therefore $g(\xi^*) = \gamma_2$, and $\xi^* \in g^{-1}(\gamma_2)$.

SECTION II

With this theorem our problem is solved, at least in principle. Systems so built can find solutions, and are not bounded by the powers of their designers. Nevertheless, there is still a long way to go before a man-made intelligence-amplifier will be actually in operation. Space prohibits any discussion of the many subsidiary problems that arise, but there is one objection that will probably be raised, suggesting that the method can never be of use, that can be dealt with here. It is of the greatest importance in the subject, so the remainder of the paper will be devoted to its consideration.

7. Duration of Trials

What we have discussed so far has related essentially to a process that, it is claimed, has "solution" as its limit. The question we are now coming to is, how fast is the convergence? how much time will elapse before the limit is reached?

A first estimate is readily made. Many of the problems require that the answer is an n-tuple, the solver being required to specify the value of each of n components. Thus the answer to an economic problem might require answers to each of the questions:

(1) What is the optimal production-ratio of coal to oil?
(2) What amount should be invested annually in heavy industry?
(3) What should be the differential between the wages of skilled and unskilled workers?

except that in a real system n would be far larger than three.

A first estimate of the number of states to be searched can be made by finding the number of states in each component, assuming independence, and then finding the product so as to give the number of combinations. The time, with a state by state search, will then be proportional to this product. Thus, suppose that there are on a chessboard ten White and ten Black men; each can move to one of six squares; how many possibilities will have to be searched if I am to find the best next two moves, taking White's two moves and Black's two into account? — With each man having about six moves (when captures are allowed for), the possible moves at each step are approximately 6^{10}; and the total possibilities are about 6^{40}. To find the best of them, even if some machine ran through them at a million a second, would take nearly a billion billion years — a prohibitively long time. The reason is that the time taken increases, in this estimate, exponentially with the number of components; and the exponential rate of increase is extremely fast. The calculation is not encouraging, but it is very crude; may it be seriously in error?

It is certainly in error to some extent, for it is not strictly an estimate but an upper bound. It will therefore always err by over-estimation. This however is not the chief reason for thinking that our method may yet prove practical. The reasons for thinking this will be given in the next three articles.

8. The Method of Models

The first factor that can be used to reduce the time of search is of interest because it is almost synonymous with the method of science itself. It is, to conduct the search, not in the real physical thing itself but in a model of it, the model being chosen so that the search can proceed in it very much more rapidly. Thus Leverrier and Adams, searching for a planet to explain the aberrations of Uranus, used pencil, paper and mathematics rather than the more obvious telescope; in that way they found Neptune in a few months where a telescopic search might have taken a lifetime.

The essence of the method is worth noticing explicitly. There is a set R, containing the solutions r, a subset of R; the task is to find a member of r. The method of models can be used if we can find some other set R' whose elements can be put into correspondence with those of R in such a way that the elements (a set r') in R' that correspond to those in r can be recognised. The search is then conducted in R' for one of r'; when successful, the correspondence, used inversely, identifies a solution in R. For the method to be worth using, the search in R' must be so much faster than that in R that the time taken in the three operations

(i) change from R to R',

(ii) search in R',

(iii) change back from r' to r

is less than that taken in the single operation of searching in R.

Such models are common and are used widely. Pilot plants are used for trials rather than a complete workshop. Trials are conducted in the drawing-office rather than at the bench. The analogue computer is, of course, a model in this sense, and so, in a more subtle way, is the digital computer. Mathematics itself provides a vast range of models which can be handled on paper, or made to "behave", far faster than the systems to which they refer.

The use of models with the word extended to mean any structure isomorphic with that of the primary system, can thus often reduce the time taken to a fraction of what might at first seem necessary.

9. Constraints

A second reason why the time tends to fall below that of the exponential upper bound is that often the components are not independent, and the effect of this is always to reduce the range of possibilities. It will therefore, other things being equal, reduce the time of search. The lack of independence may occur for several reasons.

CONSTRAINT BY RELATION. Suppose we are looking for a solution in the n-tuple $(a_1, \ldots, a_n)$, where a_1 is an element in a set A_1, etc. The solution is then one of a subset of the product set $A_1 \times A_2 \times \ldots \times A_n$. A relation between the a's, $\phi(a_1, \ldots, a_n)$, always defines a subset of the product space [6], so if the relation holds over the a's, the solution will be found in the subset of the product-space defined by ϕ. An obvious example occurs when there exist invariants over the a's. k invariants can be used to eliminate k of the a's, which shows that the original range of variation was over $n - k$, not over n, dimensions. More generally, every law, whether parliamentary, or natural, or algebraic [7] is a constraint, and acts to narrow the range of variation and, with it, the time of search.

Similarly, every "entity" that can be recognised in the range of variation holds its individuality only if its parts do not vary over the full range conceivable. Thus, a "chair" is recognisable as a thing partly because its four legs do not move in space with all the degrees of freedom possible to four independent objects. The fact that their actual degrees of freedom are 6 instead of 24 is a measure of their cohesion. Conversely their cohesion implies that any reckoning of possibilities must count 6 dimensions in the product or phase space, not 24.

It will be seen therefore that every relation that holds between the components of an n-tuple lessens the region of search.

CONSTRAINT BY CONTINUITY. Another common constraint on the possibilities occurs where there are functional relations within the system such that the function is continuous. Continuity is a restriction, for if $y = f(z)$ and f is continuous and a series of arguments z, z', z", ... has the limit z*, then the corresponding series y, y', y", ... must have the limit f(z*). Thus f is not free to relate y and z, y' and z', y" and z", ... arbitrarily, as it could do if it were unrestricted. This fact can also be expressed by saying that as adjacent values of z make the values of f(z) adjacent, these values of f(z) tend to be highly correlated, so that the values of f(z) can be adequately explored by a mere sampling of the possibilities in z: the values of y do not have to be tested individually.

A rigorous discussion of the subject would lead into the technicalities of topology; here it is sufficient to notice that the continuity of f puts restrictions on the range of possibilities that will have to be searched for a solution.

Continuity helps particularly to make the search easy when the problem is to find what values of $a, b, c, \ldots$ will make some function $\lambda(a,b,c, \ldots)$ a maximum. Wherever λ is discontinuous and arbitrary there is no way of finding the maximum, or optimum, except by trying every combination of the arguments individually; but where λ is continuous a maximum can often be proceded to directly. The thesis is well illustrated in the art of aircraft design, which involves knowing the conditions that will make the strength, lightness, etc., an optimum. In those aspects of design in which the behaviour of the aircraft is a continuous function of the variables of design, the finding of an optimum is comparatively direct and rapid; where however the relations are discontinuous, as happens at the critical values of the Reynolds' and Mach numbers, then the finding of an optimum is more laborious.

Chess, in fact, is a difficult game largely because of the degree to which it is discontinuous, in the sense that the "value" of a position, to White say, is by no means a continuous function of the positions of the component pieces. If a rook, for instance, is moved square by square up a column, the successive values of the positions vary, sometimes more or less continuously, but often with marked discontinuity. The high degree of discontinuity occurring throughout a game of chess makes it very unlike the more "natural" systems, in which continuity is common. With this goes the corollary that the test so often proposed for a mechanical brain — that it should play chess — may be misleading, in that it is by no means representative of the class of problem that a real mechanical brain will one day have to deal with.

The commonness of continuity in the world around us has undoubtedly played an important part in Darwinian evolution in that progress in evolution would have been far slower had not continuity been common. In this paper I have tended to stress the analogy of the solving process with that of the amplification of physical power; I could equally have stressed its deep analogy with the processes of evolution, for there is the closest formal similarity between the process by which adaptation is produced automatically by Darwinian selection and the process by which a solution is produced automatically by mechanical selection of the type considered in Article 4. Be that as it may, every stock-breeder knows that selection for good strains can procede much more rapidly when the relation between genotype and phenotype is continuous [8]. The time taken for a given degree of improvement to be achieved is, of course, correspondingly reduced.

To sum up: Continuity being common in the natural world, the time taken in the solution of problems coming from it may be substantially below that given by the exponential bound.

CONSTRAINT BY PRIOR KNOWLEDGE. It is also only realistic to consider, in this connexion, the effect on solving of knowledge accumulated in the past. Few problems are wholly new, and it is merely common sense that we, and the problem-solver, should make use of whatever knowledge has already been won.

The effect of such knowledge is again to put a constraint on the possibilities, lessening the regions that have to be searched, for past experience will act essentially by warning us that a solution is unlikely to lie in certain regions. The constraint is most marked when the problem is one of a class, of which several have already been solved. Having such knowledge about other members of the same class is equivalent to starting the solving process at some point that is already partly advanced towards the goal. Such knowledge can naturally be used to shorten the search.

"Solving a problem" can in fact be given a perfectly general representation. The phase- or sample-space of possibilities contains many points, most of them corresponding to "no solution" but a few of them corresponding to an acceptable "solution". Finding a solution then becomes a matter of starting somewhere in this unknown distribution of states and trying to find one of the acceptable states.

If the acceptable states are distributed wholly at random, i.e., with no recognisable pattern, then we are considering the case of the problem about which nothing, absolutely nothing, is known. After two hundred years of scientific activity, such problems today are rare; usually some knowledge is available from past experience, and this knowledge can be used to constrain the region of search, making it smaller and the search more rapid. Thus suppose, as a simple example, that there are 144 states to be examined and that 40 of them are acceptable, i.e., correspond to solutions. If they are really scattered at random, as are the black squares in I of Figure 6, then the searcher has no better resource than to start somewhere and to wander at random. With this method, in this particular example, he will usually require about 3.6 trials to find a black square. If, however, past experience has shown that the black squares are distributed as in II, advantage can be taken to shorten the search; for wherever the search may start, the occurrence of two white squares in succession shows the searcher that he is in an all-white quadrant. A move of

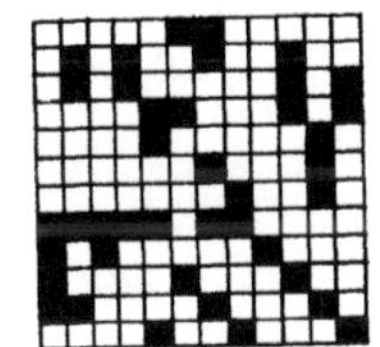

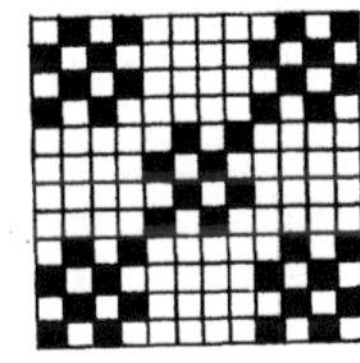

I FIGURE 6 II

four squares will take him to a checkered quadrant where he will find a dark square either at once or on the next move. In this case his average number of trials need not exceed about 2.4 (if we ignore the complications at the boundaries).

Information theory and the strategy of games are both clearly involved here, and an advanced knowledge of such theories will undoubtedly be part of the technical equipment of those who will have to handle the problem-solvers of the future. Meanwhile we can sum up this Article by saying that it has shown several factors that tend to make the time of search less than that given by the exponential bound.

10. Selection by Components

We now come to what is perhaps the most potent factor of all for reducing the time taken. Let us assume, as before, that the solver is searching for an element in a set and that the elements are n-tuples. The solver searches, then, for the n values of the n components.

In the most "awkward" problems there is no means of selection available other than the testing of every element. Thus, if each of the n components can take any one of k values, the number of states to be searched is k^n, and the time taken will be proportional, as we saw in Article 7, to the exponential bound. This case, however, is the worst possible.

Next let us consider the other extreme. In this case the selection can be conducted component by component, each component being identified independently of the others. When this occurs, a very great reduction occurs in the time taken, for now the number of states to be searched is kn.

This second number may be very much smaller than the first. To get some idea of the degree of change implied, let us take a simple example. Suppose there are a thousand components, each of which has a hundred possibilities. If the search has to be over every combination separately, the number of states to be searched is the exponential bound, 100^{1000}. If the states could be examined at one a second it would take about 10^{1993} years, a duration almost beyond thinking. If we tried to get through the same selection by the method of models, using some device that could test, say, a million in each microsecond, the time would drop to 10^{1981} years, which is practically no change at all. If however the components could be selected individually and independently, then the number of selections would drop to 100,000; and at one per second the problem could be solved in a little over a day.

This illustration will suffice to illustrate the general rule that selection by components is so much faster than selection by elements that a search that is utterly impractical by the one may be quite easy by the other.

The next question is to what extent problems are susceptible of this faster method. About all problems one can say nothing, for the class is not defined; but about the problems of social and economic type, to which this paper is specially directed, it is possible to form some estimate of what is to be expected. The social and economic system is highly dynamic, active by its own energies, and controllable only through a number of parameters. The solver thus looks for suitable values in the n-tuple of parameters. (Its number of components will not usually be the same as those of the variables mentioned in the next paragraph.)

Now a dynamic system may be "reducible"; this happens when the system that seems to be single really consists of two or more adjacent, but functionally independent, parts. If the parts, P and Q say, are quite independent, the whole is "completely" reducible; if part P affects part Q, but Q has no effect on P, then the whole is simply "reducible". In the equations of such a system, which can be given in the canonical form [5]

$$\frac{dx_1}{dt} = f_1(x_1, \ldots, x_n)$$
$$\cdots\cdots\cdots\cdots$$
$$\frac{dx_n}{dt} = f_n(x_1, \ldots, x_n)$$

reducibility is shown by the Jacobian of the n functions f_i with respect to the n variables x_j having, when partitioned into the variables that belong to P and those that belong to Q, one or more all-zero quadrants, being like I when reducible and like II when completely reducible:

$$\text{I}: \begin{bmatrix} J_1 & 0 \\ K & J_2 \end{bmatrix} \qquad \text{II}: \begin{bmatrix} J_1 & 0 \\ 0 & J_2 \end{bmatrix}$$

Such reducibility corresponds, in a real dynamic system, with the possibility of searching to some extent by components, the components being the projections [6] of the abstract space on those variables, or "coordinates", that occur together in one part (P, or Q) of the reducible system. The more a system is reducible, the more does it offer the possibility of the search being made by the quick method of finding the components separately.

Though the question has not yet been adequately explored, there is reason to believe that part-functions, i.e., variables whose f_i (above) are zero for many values of their arguments, are common in social and economic systems. They are ubiquitous in the physical world, as I have described elsewhere [5]. Whenever they occur, they introduce zeros into the Jacobian of the system (for if x_j is constant, over some interval of time, f_i must be zero and therefore so will be $\partial f_i/\partial x_j$); they therefore tend to introduce temporary reducibilities. This means that the problem may be broken up into a series of conditional and transient sub-problems, each of which has a solution that can be found more or less easily. The whole search thus has something of the method of search by components.

In this connexion it is instructive to consider an observation of Shannon's [9] on the practicability of using relays as devices for switching, for it involves a property closely related to that of reducibility. One of the problems he solved was to find how many elements each relay would have to operate if the network was to be capable of realising all possible functions on n variables. The calculation gave a number that was, apparently, ridiculously high for, did one not know, it would suggest that relays were unsuitable for practical use in switching. The apparent discrepancy proved to be due to the fact that the functions commonly required in switching are not as complicated as the class that can be considered in theory. The more they *look* complicated the more they tend to have hidden simplicities. These simplicities are of the form in which the function is "separable", that is to say, the variables go in sets, much as the variables in a reducible system go in sets. Separability thus makes what seems to be an impractical number of elements become practical. It is not unlikely that reducibility will act similarly towards the time of search.

11. The Lower Bound

It can now be seen that problem-solving, as a process of selection, is related to the process of message-receiving as treated in information theory [3, 10]. The connexion can be seen most readily by imagining that the selection of one object from N is to be made by an agent A who acts according to instructions received from B. As the successive elements appear, B will issue signals: "..., reject, reject, ..., accept" thereby giving information in calculable quantity to A. Usually the number of binary signals so given is likely to be greater than the number necessary, for the probabilities of the two signals, "reject" and "accept", are by no means equal. The most efficient method of selection is that which makes the two signals equally likely. This will happen if the whole set of N can be dichotomised at each act of selection. In this way we can find a lower bound to the time taken by the process, which will be proportional to log N. This time is none other than the least time in which, using binary notation, the solution can be written down, that is, identified from its alternatives.

We see therefore that, though the upper (exponential) bound is forbiddingly high, the lower bound is reassuringly low. What time will actually be taken in some particular problem can only be estimated after a direct study of the particular problem and of the resources available. I hope, however, that I have said enough to show that the mere mention of the exponential bound is not enough to discredit the method proposed here. The possibility that the method will work is still open.

SUMMARY

The question is considered whether it is possible for human constructors to build a machine that can solve problems of more than human difficulty. If physical power can be amplified, why not intellectual?

Consideration shows that:-

Getting an answer to a problem is essentially a matter of selection.

Selection can be amplified.

A system with a selection-amplifier can be more selective than the man who built it.

Such a system is, in principle, capable of solving problems, perhaps in the social and economic world, beyond the intellectual powers of its designer.

A first estimate of the time it will take to solve a difficult problem suggests that the time will be excessively long: closer examination shows that the estimate is biassed, being an upper bound.

The lower bound, achievable in some cases, is the time necessary for the answer to be written down in binary notation.

It is not impossible that the method may be successful in those social and economic problems to which the paper is specially addressed.

BIBLIOGRAPHY

[1] WECHSLER, D., Measurement of Adult Intelligence. Baltimore, Williams & Wilkins, 3rd. Ed., 1944.

[2] TERMAN, L. M. and MERRILL, M. A., Measuring Intelligence. London, Harrap & Co., 1937.

[3] SHANNON, C. E. and WEAVER, W., The Mathematical Theory of Communication. Urbana, University of Illinois Press, 1949.

[4] ASHBY, W. Ross, "Can a Mechanical Chess-player Outplay its Designer?" Brit. J. Phil. Sci., 3, 44: 1952.

[5] ASHBY, W. Ross, Design for a Brain. London, Chapman & Hall; New York, John Wiley & Sons, 1952.

[6] BOURBAKI, N., "Théorie des Ensembles." A. S. E. I. 1141. Paris, Hermann & Cie., 2nd. ed., 1951.

[7] BOURBAKI, N., "Structures Algébriques." A. S. E. I. 1144. Paris, Hermann & Cie., 2nd. ed., 1951.

[8] LERNER, I. M., Population Genetics and Animal Improvement. Cambridge University Press, 1950.

[9] SHANNON, C. E., "Synthesis of Two-terminal Switching Circuits." Bell System tech. J., 28, 59 - 98; 1949.

[10] WIENER, N., Cybernetics. New York, John Wiley & Sons, 1948.

THE EPISTEMOLOGICAL PROBLEM FOR AUTOMATA

D. M. MacKay

1. Introductory

An intelligent automaton is required to react within its field of activity as if it knew the current state of that field. This requires (at least in the present state of the art) that the relevant information shall have some physical representation within the mechanism of the artefact.

The present paper is concerned with the form which such a representation of the field of activity may take and with the corresponding limitations on the "universe of discourse" of the automaton. Is it true, for example, that the conceptual categories of an automaton must be limited to those of its designer? Or could an automaton develop and symbolize for itself new concepts as occasion might arise? Is it possible for an automaton to generate new (and non-trivial) hypotheses?

Such are some of the questions subsumed under our slightly whimsical title.

Our aim will be first to distinguish between two broad lines along which a solution can be sought (without necessarily implying that these are exhaustive). Thereafter, we shall discuss in a little greater detail what seems to be the more promising of these lines, and see what it offers by way of possible answers to our questions.

2. Two Approaches

It may be assumed that the automaton is equipped with receptive organs able to pick up all the necessary raw information from its field of activity. Our problem concerns the transition from the gross particularity of the information as received, to a physical representation of "that which is the case" in terms of universal categories. It is here that two fundamentally different possibilities arise.

We can on the one hand use as basic symbols a more or less arbitrary set of physical elements within the mechanism, their states being altered by the flux of incoming information so as to form a representation of the current state of the field of action in terms of concepts chosen by the designer. Some sort of coding-process must be interposed between received signals and symbolic representation, to ensure that the symbol for a concept shall not change under any transformation of the input with respect to which the concept is invariant.

The symbol for "circle", for example, cannot be simply the configuration of signals received when a circle is within the reception field, for this configuration would change with the size and position of the circle. The received signals must pass through a filter (in the generalized sense) which computes some sort of average over the group of transformations of the signal with respect to which the concept is invariant, and emits a standard symbol for the concept so identified.

But whatever the detailed principle adopted,[1] the final symbolic representation emerges as a kind of filtrate of the <u>input</u> signal. The chain of cause and effect moves uniformly inward from the receptors to the final representation. Between the latter and the activity of response (outwardly directed) there is a chain of calculation in terms of the categories represented by the basic symbols, aided by information stored in the same symbolic form (Figure 1).

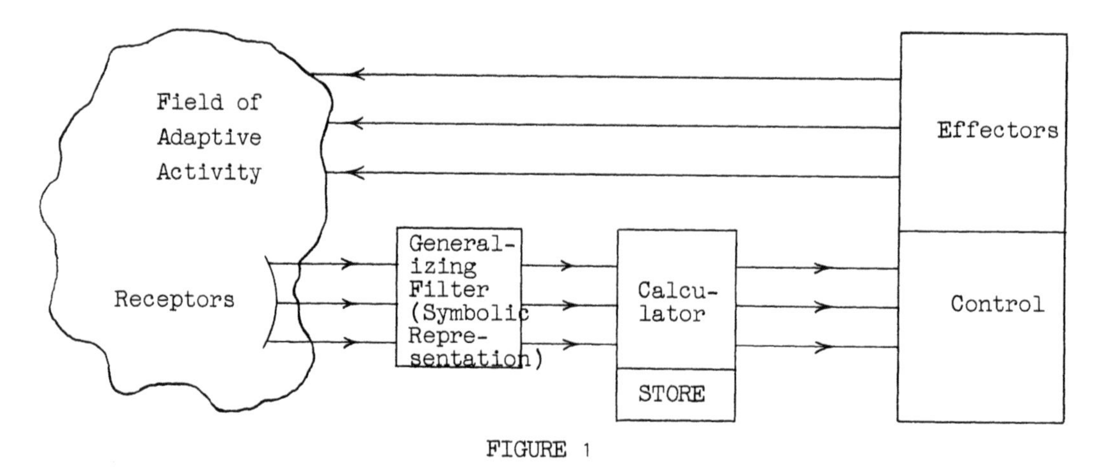

FIGURE 1

[1] A classic example being the McCulloch-Pitts scheme described in Reference 1.

In contrast with this is the second possibility, of allowing the incoming signals to stimulate an "imitative" internal response-mechanism, designed to adapt its activity to match or counterbalance internally what is received [2]. By continually modifying its activity to match the incoming signals, the mechanism may be thought of as symbolizing those features of the received information that have necessitated the modifications.

The symbol generated in this way for a given concept must again remain the same under all transformations of the input with respect to which the concept is invariant.

The pattern of control-signals by which a matching-response to a circle is organized, for example, must be the same in at least one diagnostic respect, whatever the size or position of the circle in the visual field.

The difference from the earlier case is most easily seen in the simplified flow diagram of Figure 2.

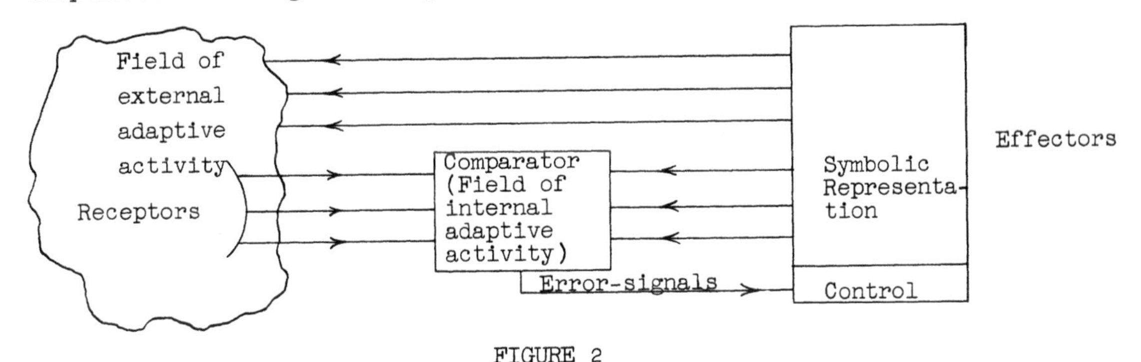

FIGURE 2

Here the internal activity evoked in matching-response to the incoming signals is <u>outward-directed</u>: the lines of cause-and-effect run from right to left to <u>meet</u> the incoming lines in the comparator; and it is the outward-directed activity of the elements organizing the internal matching-response, that constitutes the basic symbols for the conceptual vocabulary.

The details of this principle of representation will become clearer in the following sections. Meantime, we may sum up by saying that whereas on the first principle the world is described by symbols evoked as the end-product of a simple one-way causal chain, on the second, the descriptive symbols arise in a second causal chain adapted automatically (under evaluatory correction or "negative feedback") to match the activity in the first.

3. A Simple Example

It will best serve our purpose to consider first and in most detail the second and perhaps less familiar principle. By way of introduction to a simple example, let us first think of a man driving a car along a road at constant speed. The sequence of movements of his arms controlling the wheel may be said to form, in our sophisticated sense, a representation of the road, at least in its geometrical aspect. From the sequence of movements, if he is competent, we can infer the location and shape of all bends.

Let us now simplify still further and picture a car steered along a road made up only of straight stretches and corners of a standard angle, by means of a pair of buttons causing the car to turn through the required angle to left or right (L or R). Then the sequence of L's and R's would again form a running representation of the shape of the road. A given sequence, say LLRRRL, would represent the same shape (if the time-intervals were in the same proportions) whenever it occurred and whatever its total duration.

It is not difficult to imagine a servo-system to replace the driver here and produce these control signals as required, "feeling its way" under the correction of some sensory device that will signal the deviations from correct position on the road.

If the shape of the road is completely irregular, then no sequence of L's and R;s is more likely to recur than any other, and it is not possible for the control-process to benefit from past experience. It must be equally ready to try either L or R at all times.

Suppose, however, that the shape was regular, and required, for example, only a steady sequence LRLRLR ... . Then it could improve the accuracy of following if the control-system were more ready to try R after L and L after R. Indeed, in this simple case it would be best to arrange automatically that L and R should alternate.[2]

One way of doing this would be to have a secondary active system generating an alternating sequence of output signals, say L' and R', which could then be used to operate the L and R switches in anticipation of the corrective signals from the road. The activity of this system then provides a suitable symbol for the regularity that it matches.

Here we come to an important step. We have already taken it for granted that a servo-system under evaluatory correction can generate for itself the pattern of L and R control signals found necessary to match the sequence of corners in the road.

---

[2]For simplicity we make here the rather unrealistic assumption that no corrective signals are necessary while the road is straight.

It seems equally open to us to devise our secondary active system on the same lines, taking as its data the sequence of L and R control signals evoked in the primary system under corrective feedback from the road. Just as the goal of the primary system is to produce a sequence of movements of the car to match the corners in the road, so the goal of the secondary system can be to produce a sequence of L' and R' outputs to match the sequence of L and R control signals.

The details of such a system will concern us in the following sections. The present point is that the symbol for the abstraction (alternation of L and R corners) can be generated automatically as the activity of a second-order matching-response mechanism.

As one further step in this illustration, let us suppose that the shape of the road changes so that a sequence LLRRLLRR ... is now required to match it. We may suppose that the second-order mechanism would in due course be guided automatically to produce a new matching-pattern of outputs L'L'R'R'L'L'R'R' ... . The new rhythm has again the status of a symbol, for the new regular pattern. Let us call this pattern B, and the earlier one A.

Suppose now that over a long stretch of road patterns A and B were found to alternate. Once again it is evident that we could have a tertiary response-mechanism whose goal was the production of a rhythm of activity in imitation of the sequence ABAB ... . The activity of this tertiary system in effect symbolizes the hypothesis that the abstraction B alternates with the abstraction A.

The essential features of this hierarchy of responsive mechanisms are summed up in Figure 3.

(a) System 1 feels its way into adaptive response to the input. Regularities in the input show themselves as regularities in the response required and hence in the control-activity required in system 1.
(b) System 2 feels its way into a routine that matches the regular component of control-activity of system 1, and contributes its routine to that control-activity, so taking over control insofar as its routine is a successful match for the regular activity required. Regular changes in control routines show themselves as regularities in the control-activity required in system 2.
(c) System 3 feels its way into a routine that matches the regular component of control activity of system 2, ... and so forth.

This preliminary example may perhaps have made clear the distinction between

(a) symbols generated as an incoming transform of incoming signals, and

(b) symbols generated as the outward-directed (internal) response of an imitative or adaptive system to the incoming signals.

The distinction is not of course a merely verbal one. It determines in practice the part of the mechanism to which dependent logical networks are to be coupled.

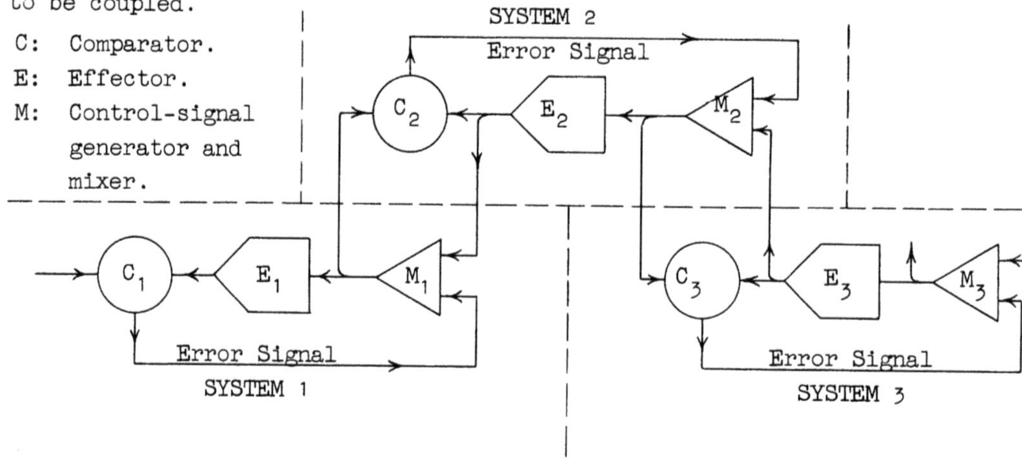

FIGURE 3

In type (a) connections have to be established between incoming signal circuits and outgoing response-circuits. In type (b) connections have to be established between one set of outgoing response-circuits and another.

In metaphorical terms we might picture the distinction by saying that for type (a) the act of knowing is an act of reception; for type (b) it is an act of response.

Our example illustrates also one principle on which an automaton could symbolize a hierarchy of abstract concepts, by the modes of response of a hierarchically organized adaptive system.

To all these points we shall return in due course. We must first digress slightly to clarify the principles on which error-sensitive systems in general may be designed.

Having distinguished between systems which are "fully-informed" and systems using trial and error, we shall see how the hierarchically-organized adaptive system we have illustrated above can be generalized, on a basis of statistically-controlled trial and error, so that abstract concepts of any order can be symbolized as required in the course of experience, with the minimum of restriction (or instruction) by the designer.

4. Error-Sensitive Automata

The most familiar error-sensitive automata are so designed that the nature and degree of error, as indicated by the error-signal, automatically prescribes the optimal corrective response. This requires that all independent degrees of freedom of the response system be determined by the error-signal, which must therefore have a corresponding information-content. In the simplest type of servo, in fact, the error-signal must have one logical degree of freedom for each that it determines in the response.

A thermostat, for example, requires an error-signal of only one dimension; a self-directed missile, in general, three; and so forth.

There are however two classes of error-sensitive automata in which the logical dimensionality or "logon-content"[3] [3] of the error-signal can be less than the required number of degrees of freedom. In the first class, the original error-indication is adequate (has the required dimensionality) to prescribe the required response uniquely, but it is transformed by a coding-process into an error-signal of lower dimensionality (logon-content) for transmission to the response-mechanism. At the receiving end it is then decoded into a form having the requisite number of degrees of freedom.

This type of automaton differs only trivially from the first. Although the logon-content of the error-signal may be less than that of the response, the selective information-content must be at least equal to that represented by the selection of the required response. In fact, if an immediate and a correct response is required from one given mechanism (leaving no time to take advantage of any redundancy, since this inherently entails delay [4])[4], then the error-signal must always comprise at least m bits of selective-information, where m is the number of binary decisions necessary in the response-mechanism to select the required response. If there are n possible responses of equal complexity, then $m > \log_2 n$.

---

[3]The term "logon" was first used by D. Gabor (J. Inst. Elec. Engrs., 93, III, 429-456, 1946) in a more restricted sense.

[4]In some cases delay can be reduced by having several response-mechanisms operating in parallel and arranging that at least one produces the required response (see below); but this does not get round the basic theorems for one given mechanism.

We can thus describe both these automata as fully-informed. Their corrective response can be immediate and can be deduced uniquely from the error-signal.

There is however a second class of automaton in which the error-signal is actually deficient in the selective information-content necessary to specify the required response uniquely.

This deficiency may arise through various physical limitations even when the logon-content is adequate. The noise-level, for example, or the resolving-power of the sensory system generating the error-signal, may be such that the optimal response cannot be uniquely determined from it. Indeed, in the strictest sense this is always the case, unless we "quantize" the set of required responses.

It may however arise simply because it is desired to use an error-signal of lower than adequate logon-content without the complexity of a coding and decoding-process. In the extreme case we may wish to use a simple one-dimensional two-valued ("go" - "no-go") error-signal to control a response-mechanism of many degrees of freedom.

But whatever the reason for the inadequacy of selective information-content, it is plain that the deficiency must be made good, if adaptive success is to be ensured, by a process of trial and error on the part of the response-mechanism. If only m bits are supplied by the error-signal, and one response must be selected out of n alternatives (supposed equally likely meanwhile for simplicity), then after receiving the error-signal the mechanism is still left with a choice from $(n/2^m)$ possible responses.

If these can be tried in the most economical way possible and full use is made of the information-capacity of the error signal at each trial, the minimum average number of further trials necessary will be $[\log_2 (n/2^m)]/m$, or $(\log_2 n)/m - 1$.
If, however, only a binary evaluation is possible at each later trial, then on the average at least $(\log_2 n - m)$ trials will be required in order to find the required response (if all are equally likely).

We may of course interpret this as just another way of saying that the number of bits of information by which the error-signal is deficient must be gained from the environment in one way or another if the final choice is to be adaptive to the required extent. Trial and error (optimally designed) can enable the missing bits to be gained from successive error signals, provided that the environment is not changing too rapidly.

This leads us to a further point. If the environment is changing, the adaptive response of the system must change to match it. The succession of changes of response ideally required may be thought of as defining a certain information-rate: so many bits (on the average) per second.

It follows that if the error-signal carries m bits per trial, and each trial requires Δt seconds, the system cannot respond completely adaptively to an environment requiring an information-rate greater than $(m/\Delta t)$ bits/sec. from the response-mechanism.

In terms of discrete changes, the environment must alter at a rate $((\log_2 n)/m)$ times slower than the adaptive response mechanism if the latter is to keep up with it.

Our chief point however is a positive one: it is possible for a response-mechanism to be guided adaptively by an error-signal of any selective-information-content, however small, to any required degree of accuracy, _given time_. But in a fluctuating environment the average selective information-content of successfully adaptive response cannot exceed the product of the rate of trial and error with the selective information-content of the error signal, or in general (and more obviously) the error-signal channel-capacity.

As a final general point, leading to our next section, we may note that where the probabilities of the adaptive responses required are not equal--where there is redundancy (in the sense of communication theory) in the response-pattern owing to the statistical structure of the pattern of events to which response is required--then the foregoing statements apply in the limit to the actual information-rates after allowing for redundancy.

It follows that it should be possible in principle to devise an automaton in which the adaptive response governed by an error-signal of given selective-information-content may be much more complex than that information-content might suggest, if advantage is taken of the redundancy in the response-pattern required.

If the statistical structure were known to the designer, this could be done by pre-setting the conditional probabilities of possible trial-responses to match that structure, so that when the automaton is confronted with a situation to which a certain response will be optimally adaptive, the average time-lag before that particular response is tried shall be the minimum made possible by fully exploiting redundancy.

From the discussion of Section 3, however, there is evidently another possibility. By generalizing the adaptive mechanism of Figure 3 to function on a basis of statistically-controlled trial and error, it should be possible, even in the absence of prior knowledge of conditional probabilities, to devise an automaton in which the optimal statistical structure for its trial-procedure could be evolved automatically as the result of experience.

Engineering details do not here concern us, but in the next section we shall consider briefly how such an automaton could function.

5. A Statistical Abstractive Mechanism

Automata have before been described, for example by Ashby [5], in which the elements are interconnected by deterministic links, but the error-signal functions by stimulating trial-and-error activity among relay-switches altering the interconnections.

The mechanism we shall here consider is of a more general type. We could describe it loosely as one in which the interconnections are not strictly deterministic, but have a variable probability of functioning. More precisely, it is a mechanism in which the probability of excitation of each element (in a given time interval) can be made to depend on continuously-variable physical factors as well as on the current states of any number of other elements linked to it [2, 6, 7, and 8].

In the limiting case an element may be spontaneously active, with a frequency depending in like manner on the current state of its (topological) environment.

In an automaton of the general type of Figure 3 constructed of such elements, adaptive activity can be guided by modifying the relative probabilities of excitation. Trial-and-error normally takes place spontaneously, and the error-signal functions by controlling the statistical structure of the trial process.

The term "threshold" is used to denote, roughly speaking, the resistance of an element to stimulation. In general each connecting link to one element from another can have its own particular threshold, determining the extent to which a signal in that link affects the probability of excitation. The higher the threshold the lower the probability, or the longer the interval of time that must elapse (with a given configuration of stimuli) before excitation.

Error-stimulated adjustment of thresholds, then, or evaluatory threshold-control for short, is the key notion on which we shall base our statistical abstractive mechanism.
Assuming that we have elements in which the thresholds may be at least semi-permanently modified according to the success or failure of current trials, we can envisage a system initially devoid of instruction evolving for itself a satisfactory pattern of adaptive activity.

To see how such statistically negative feedback can function, the simple illustration provided by the model of Figure 4 may be helpful. It was originally constructed to illustrate some points made in Reference [6], and has of course no pretensions to animal status.

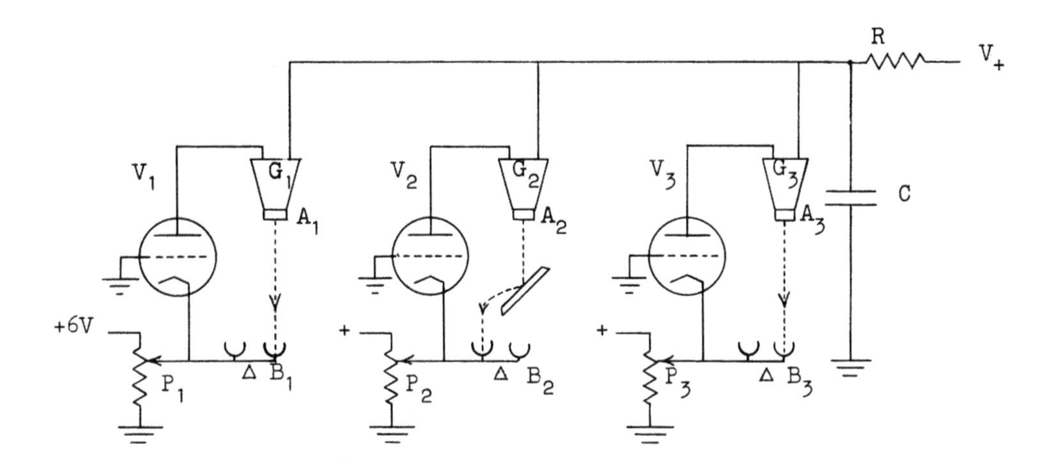

FIGURE 4

The model, illustrated in the figure, is made up of three identical units (the number is unrestricted in principle) each comprising a thyratron V which can actuate an electromagnetic gate G so as to drop a ball-bearing from an aperture A over a two-pan balance B. B is mechanically linked to a potentiometer P, which governs the bias on thyratron V. The anodes of all thyratrons are supplied from a common capacitor C charged through a resistor R.

If now C is allowed to charge slowly, the thyratron with the lowest effective bias, say V_n, will tend to fire first, discharging C and releasing a ball from A_n. Normally this will fall into the right-hand pan of B_n causing P_n to rotate and increase the bias on V_n. Thus the normal effect of a thyratron's firing is to reduce the probability of its firing next time.

If, however, a target T is placed under A_n so as to deflect any balls from A_n into the left-hand pan of B_n, the opposite effect will occur. The success of the model in striking the target when V_n fires will increase the probability that it will use V_n next time.

Evidently, no matter how the biases on different V's are set initially, the model will tend to an equilibrium state in which only the thyratron V_n fires and T is always hit. If, moreover, we move T to a new position, m say, the model will gradually "unlearn" its original "set" and eventually will be found firing only V_m.

In this example randomness is introduced only by the fluctuations of the thresholds of the various thyratrons, and the "evaluatory feedback" is quantized and rather inflexible. It illustrates even so the basic principle, whereby an automaton alters the transition probabilities governing its own activities according to its experience of their relative success (judged by some general criterion) and diminishes by a kind of "natural selection" the probabilities of all except those which have been successful often enough.

This is not the place to elaborate on engineering aspects of the application of this principle to the hierarchy of Figure 3. We must, however, satisfy ourselves that a trial-and-error process could converge quickly enough to be useful in the more general case.

It is first clear that if n independent thresholds must each be set to one of a levels, then up to $n \log_2 a$ binary choices may be required to change from one configuration of thresholds to another. Even an error-signal of m bits could guide such a change only after at least $(n/m) \log_2 a$ trials.

Clearly, until the higher-order statistical structure of events begins to be abstracted, the automaton must begin by searching for adaptive reactions from a very limited repertoire, if the search-process is to converge with any speed. It would seem sensible to begin with most thresholds low, so that in effect the number of degrees of freedom of response is small, the elements being coupled together in large groups. Spontaneous activity would then be initially fast and furious, and would of course, in a richly-interconnected system, include much random suppression of coupling elements. This dissociative activity would amount to random augmentation and alteration of the degrees of freedom of the response-system, and any changes found useful could be statistically favoured. (See page 244.)

As soon as simple adaptive activity began to show regularities, the second-order mechanism of Figure 3, on the same principle, could begin to develop simple organizing routines to improve the average overall success. At this stage matching-responses would be crude and perhaps not highly differentiated in their success-value from others; but as experience of success and failure accumulated, and useful second-and higher-order organizing routines began to develop, the speed of convergence on a rough matching-response would increase and leave progressively more time for exploratory development of finer response-structure before the environment changed sufficiently to demand a new response.

It will be noted that this trial process is not organized for maximum returns on simple betting principles. The response having maximum likelihood of success, on the information contained in the error-signal, is certainly most likely to be tried first, if the feedback from the error-signal on the threshold-configuration is effective. But it will not always take precedence; and any of the other responses with finite likelihoods of success may sometimes be tried first instead, in the spontaneous sequence of activity guided by the configuration of thresholds.

The reason for preferring this to the apparently more efficient policy is that in simple betting theory the categories of description (and usually the number) of competing alternatives are fixed; the optimal order of trial is then always the order of decreasing prior probability of success. But the case of our concept-forming automaton is different. Here at any time a hitherto unsuspected sequence of trials may prove to function as a group more effectively than any alone, and so itself acquire the status of a new competing alternative. The assumption of a closed set made in simple betting theory is thus inapplicable.

The point becomes obvious in a simple example. Suppose that among the pre-conceived possible responses A, B, C, ... the prior probability of success was always greatest for A, though perhaps small for all; and suppose that in the actual circumstances the response that would most often succeed would not be any of those alone, but the sequence of B followed by A. Clearly, the automaton could never discover this if it insisted on always trying A first. It would seem a reasonable compromise to make the probability of trial of each pre-conceived alternative proportional to its probability of success on the evidence of the error signal, but to set no other restriction on the order of trials.

At the same time we must remember that all organizing routines already symbolized can count among the alternatives open to trial; so that the chances of success of certain _sequences_ of simple responses will also be automatically evaluated, and their probabilities of trial adjusted, on the evidence of the error-signal. The same applies to any more complex statistical features of the response-pattern already found to be adaptive. Our point is simply that until a particular sequence has been found adaptive and becomes symbolized by a corresponding organizing-routine, it has not entered the conceptual vocabulary of the automaton, and cannot therefore be considered in its calculations as one of the responses competing for trial under the statistical guidance of the error-signal (operating on the corresponding thresholds).

It should also be realized that although for simplicity we have spoken mainly of a single time-series of responses, the principle of all that has been said applies _mutatis mutandis_ to combinations of simultaneous responses, for which conditional transition-probabilities are even more simply controlled.

Our policy then is to ensure speed of convergence of the trial-process by limiting the initial complexity of the response-pattern, and allowing it to grow only in proportion as the additional complexity can be controlled by internal organizing routines.

The limitations set by the information-content of the error-signals (discussed in Section 4) now govern, not the speed of convergence, but the rate of growth of complexity of response, and hence of conceptual vocabulary. It is evident that even a multidimensional error-signal of m bits can in the limit organize a vocabulary equivalent to not more than nm binary symbols after n independent trials. This would require the automaton to be presented with an ideally designed sample-flow of experience, and is of course many times greater than what might be expected if it were merely exposed to a typical human environment (including, of course, normal dialogue).

But even these superficial considerations suggest that a conceptual vocabulary of the same order as the human one could in principle be evolved in a time at least comparable with that required by a human being if a suitable flux of experience were provided, even without external evaluatory feedback from a human instructor.

Without such feedback there would of course be no guarantee that the conceptual analysis of the field of activity developed by the automaton would be in the same terms as the human one. Some of the concepts symbolized by the internal organizing routines might well be quite new, and barely capable of translation unless the relevant modes of adaptive response, and the criteria of "mismatch", were sufficiently analogous to the human.

6. Comparison of the Two Approaches

We are now in a position to make a few general remarks on the two principles of representation, by way of summary.

We may first note that even in the second type of automaton, the error-signal can itself be regarded as a representation working on the first principle: i.e., as a transform of the input, flowing from left to right. True, it represents, not the world in isolation, but (within the limits of its predetermined information-space) the nature and extent of the mismatch between the automaton and the world. But it is linked to the incoming signals by a simple one-way causal chain.

We have seen that the second type of automaton can function adaptively, given time, even if the error-signal is one-dimensional. The greater the statistical redundancy in the environmental activity to which continuous adaptation is required, the more complex the adaptive reactions attainable in given time with a given error-signal, and the less the complexity of the error-signal required for a given complexity and speed of response.

The more complex the error-signal, the less complex need be the statistical abstractive mechanism required to attain a given adaptive efficiency, and the more information can be taken in in a single "glimpse".

All this suggests that we might regard the first "fully-informed" type of automaton (Figure 1) as simply the other extreme case of the second type (Figures 2 and 3), in which the error-signal has been enlarged in information-content sufficiently to represent all that is necessary of the field of activity or "world of discourse".

The only "comparator" is in this case the field of activity itself. All the organizing routines with their operational significance as dynamic symbols of statistical expectations are missing.

Evidently the fully-informed automaton will give best returns for the complexity of its corrective information when dealing with situations of minimum redundancy.[5] But if redundancy is high enough, an automaton of the second type with a relatively simple error-indicating system can in principle respond as efficiently as a fully-informed automaton in the same situation. The second type is at its best in situations where redundancy is at a maximum.

Most significant, however, is the distinction between the conceptual range of the two types.

In the first the conceptual framework is predesigned, and "thinking", if such we call it, is confined to its categories. In the second the conceptual framework evolves to match the stationary statistical features of the world, and new hypotheses - in the form of tentative organizing routines - are always being framed.

If, moreover, the abstractive system is allowed to include in its data the activity (at various levels) of its own organizing routines, we have the possiblity of symbolic activity representing in effect <u>metalinguistic</u> concepts. It is difficult to set any limits to the types of conceptual activity that are thus within the scope of an automaton on this principle.

It is of some interest that the operational equivalent of an event of perception in the second case is the framing of a satisfactory internal matching-response, which is thereby distinguished from the events of reception.

---

[5]This is to assume that the representation "filtered" from the input signal can appear with negligible delay: a potentially costly requirement. In general the most economical filter would also benefit from redundancy in the input.

Operationally it is as if the automaton is conscious only of the features of the world symbolized by its own internal matching-responses, and only indirectly of the error-signals evoking them. It is the quantal pattern of the response-process that determines the way in which its "experience" is broken down into a succession of "known events". We can thus give operational meaning to metaphorical talk of the "changing contents of consciousness" of such an automaton.[6]

If, however, the first type is regarded as an extreme case of the second, in which the error-signal has been enlarged and the internal response-system eliminated, the operational correlate of such concepts is missing. The distinction may be worth pondering by those to whom the question of consciousness in automata is a live issue.

7. Conclusion

Our conclusion is that an automaton designed on statistical principles, which can evolve an internal organizing routine to respond adaptively to regularities of its sensory input, is capable in principle of developing its own symbols for concepts of any order of abstraction, including metalinguistic concepts, without prior instruction.

Any resemblance between such an automaton as described and the human brain is scarcely coincidental, but is logically inadmissible as evidence.

BIBLIOGRAPHY

[1] PITTS, W. and McCULLOCH, W. S., "How we know universals", Bull. Math. Biophys. 9, 124-147, 1947.

[2] MACKAY, D. M., "Mindlike behaviour in artefacts", Brit. J. for Phil. of Sci. 2, 105-121, 1951.

[3] MACKAY, D. M., "Quantal aspects of scientific information", Phil. Mag. 41, 289-311, 1950.

[4] SHANNON, C. E., "Mathematical Theory of Communication", Bell Syst. Tech. J. 27, 379-423; 623-658, 1948.

[5] ASHBY, W. R., "Design for a Brain", Electronic Engr. 20, 379-383, 1948.

[6] It is perhaps unnecessary to discuss here the various types of non-veridical "perception" and "imagination" possible in such an automaton. Reference 7 contains some indications of the possibilities.

BIBLIOGRAPHY

[6] MACKAY, D. M., "On the Combination of digital and analogical techniques in the design of analytical engines",(May 20, 1949), mimeographed.

[7] MACKAY, D. M., "Mentality in Machines", Proc. Arist. Soc. Suppt. 1952, 61-86.

[8] MACKAY, D. M., "Generators of Information", Communication Theory (ed. W. Jackson), Butterworths, 1953, 475-485.

CONDITIONAL PROBABILITY MACHINES AND CONDITIONED REFLEXES

Albert M. Uttley

ABSTRACT

An important characteristic of animal behaviour is that the same motor response can be evoked by a variety of different configurations of the world external to the animal. For the animal these configurations resemble one another in some respect. Similarly there can be variation in the motor response, and different responses resemble one another. In this paper, it is suggested that this "resemblance" is based on two known mathematical relations: the first is the _inclusive_ relation of Set Theory; the second relation is that of _conditional probability_. From these two relations are deduced the principles of design of a machine whose reactions to stimuli are similar in a number of ways to those of an animal; some similarities are suggested between the structure of the machine and that of a Nervous System.

The machine can assess resemblances between sets of input data, not between other phenomena from which that data was derived; the nervous system is limited similarly to assessing resemblances between signals in sets of fibres, not between sets of physical quantities external to this system, and from which those fibre signals were derived — between internal _representations_, not between external configurations.

In order to make a start on this problem, it has been found necessary to idealise the form of input, abstracting some important factors. The input signals must be all-or-none, active or inactive; the duration of activity is not considered. In a representation, inputs must be active simultaneously, so temporal pattern cannot be discussed. The neighbourhood of input fibres is not considered, so spatial pattern cannot be discussed; temporal and spatial patterns are considered in a separate paper. In consequence, the present paper is limited to considering representations such as taste, smell, and colour; these can be classified without reference to time, or direction.

The inclusive relation forms the basis of classification; a machine based on this principle will be called a classification machine. The relation of conditional probability arises out of the inclusive relation; a machine based on the two principles will be called a conditional probability machine; it must have all the properties and structure of a machine based only on the first principle. In consequence classification is discussed first.

The simplest form of classification machine can assess input signals which possess only two states, active and inactive, which may be said to indicate the presence and absence of properties; but it must make no use of the inactive state, to distinguish classes defined by the absence of properties; such a machine has been called "unitary".

Because a class of objects is defined by a set of properties, a classification machine must possess one unit for every possible combination of inputs; the unit must operate if they are all active, i.e., if a representation possesses the corresponding set of properties.

All units are of the same design, being two-state in nature, and they must be connected to inputs in all possible ways; such a machine can be constructed with random connections between units and inputs. For a classification machine, class recognition is instantaneous and correct, it does not grow or decay. Resemblance is limited to the determination that representations are of the same class, i.e., that for the two, there is a common set of active inputs; based only on Set Theory there can be no relation between representations with no such common set, other than that they are different, disjunct.

But a further relation can arise between disjunct representations, which is based on their relative frequency of joint occurrence; this is the relation of conditional probability and it measures a variable degree of resemblance. If the machine is extended so as to embody this principle, it must have two new design features. Each unit must possess a variable state; and there must be interconnections between units. The function of the variable state of a unit is to store the unconditional probability of the corresponding set of properties; this quantity may also be called the mean frequency of joint occurrence, and it can be time weighted in various ways. A conditional probability is the ratio of two unconditional probabilities; but in computing machines, division is a more difficult operation than subtraction, so machine design is simpler if unconditional probabilities are computed on a logarithmic scale. It can be shown that the unit must then possess two new properties; the stored quantity must grow in the absence of events; and a certain amount must be destroyed instantaneously if the set of properties occurs. From a growth equation describing both these functions it is possible to calculate the rate of growth and decay of the conditional probability relation between sets of properties.

For two sets of inputs J and K, whether disjunct or not, there must be connections in a conditional probability machine, from the J unit and the K unit to the $(J \cup K)$ unit, which stores the unconditional probability of the union of the two sets; these connections mediate a function of <u>supercontrol</u>, from a unit to a superunit. Rules have been deduced which determine which of the J and K units controls the $(J \cup K)$ unit. There must also be a separate system of interconnections from each $(J \cup K)$ unit to all subunits such as J and K, mediating <u>subcontrol</u> by the $(J \cup K)$ unit. Whichever unit is not effecting supercontrol is subcontrolled; there are other necessary rules.

Such a system of interconnections provides a physical path between sets of input channels whether disjunct or not; but this path is formed <u>to a certain degree</u>, and depends entirely on the nature of past representations. If a nervous system embodies the principles of a conditional probability machine, disjunct representations, such as "smell of food," "sight of food," or "sound of bell," can evoke a common response, e.g., "salivation". For such a system, the situtation must be described this way: if the conditional probability of "salivation," given "smell of food," is unity (deterministic, reflex behaviour), and if "sound of bell" and "smell of food" are presented jointly, then the conditional probability of "smell of food" given "bell" will rise. The unit storing this last conditional probability might be connected to an effector mechanism in different ways. The simplest possible effector mechanism would be no more than a threshold which caused an all-or-none reaction if the probability exceeded a certain value. The conditional probability machine could then be much simpler — a <u>conditional certainty machine</u>.

I. INTRODUCTION

An important characteristic of animal behaviour is that the same motor response can be evoked by a variety of different configurations of the world external to the animal. For the animal these configurations resemble one another in some respect. Similarly there can be variation in the motor response, and different responses resemble one another. In this paper, it is suggested that this "resemblance" consists of two known mathematical relations: the first is the <u>inclusive</u> relation of Set <u>Theory</u>; the second relation is that of <u>conditional probability</u>. From these two relations are deduced the principles of design of a machine, whose reactions to stimuli are similar in a number of ways to those of an animal; also, the structure

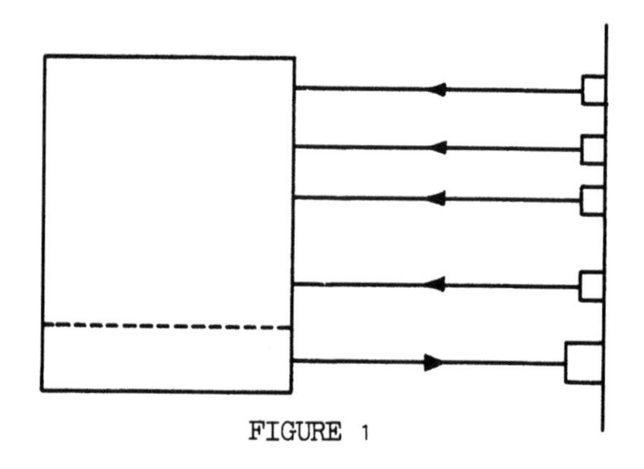

FIGURE 1

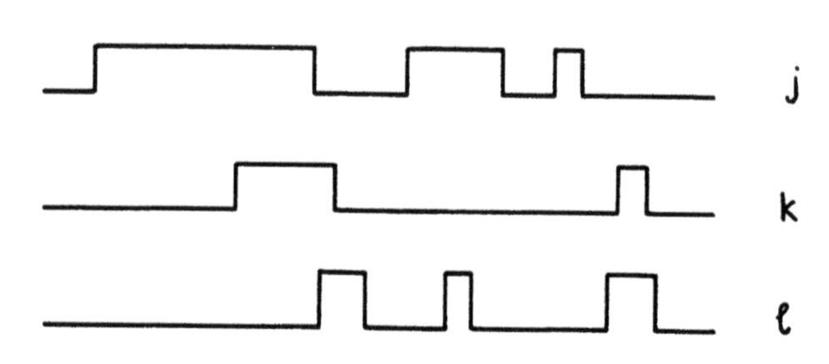

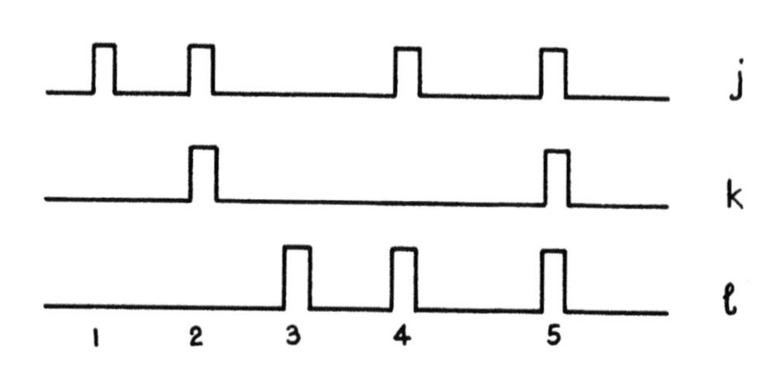

FIGURE 2

of the machine has some similarity to that of a Nervous System. [Sholl and Uttley, 1953]

The features of a Nervous System which are relevant to this discussion are shown in FIGURE 1. Signals enter the nervous system along discrete afferent channels, and after central delay, signals are sent out to motor units; the afferent pattern of signals will be considered first. At any instant, the external world will be said to be in a particular <u>configuration</u>. At the same instant there will be a set of signals in the afferent channels; each signal will be called an <u>input quantity</u> to the central mechanism; the complete set of input quantities will be called a "representation". The central mechanism can be concerned only with representations and their resemblance, not with the resemblance of configurations.

II. CLASSIFICATION

It has been suggested [Hayek, 1952] that when an animal gives the same reaction to two different representations, it is because they are of the same class or set. But the mathematics of Set Theory cannot be applied immediately to the input system described above. Firstly, the input quantities are known to contain a measure of intensity; and Set Theory is concerned with properties which are either possessed or not possessed. Secondly, the input signals are not independent; and in Set Theory no relation is considered between elements of a set other than that of inclusion in it.

II(a) <u>Some Necessary Abstractions</u>

Consider therefore the following simpler situation. Each input quantity has only a binary measure, i.e., it has one of two possible values, which will be called "active" and "inactive". If input j is active the representation will be said to possess property j. There is independence between all pairs of inputs, i.e., in a large ensemble of representations the fraction possessing property j is the same whether the ensemble possesses property k or not. A time record of the contents of three channels j, k, and l might take the form of FIGURE 2a.

Such an input is still too complex for simple classification. There must be three temporal abstractions. Firstly, there will be no consideration of the variation in duration of the active state of an input; secondly, the different inputs in a representation will be restricted to being active simultaneously; thirdly, the order of, and the interval between representations will not be considered — the representations will be treated as a timeless ensemble. The input to the machine will then take the form of FIGURE 2b. The third restriction will be removed later in this paper; but the first two restrictions, which still make the consideration of temporal pattern impossible, will be discussed in a further paper.

Lastly, the idea of neighbourhood between input channels will be abstracted; the transformation formed by the connections from receptors to the input points of the machine need not then be topological. With this restriction the consideration of spatial pattern is not possible, and it too, is the subject of a further paper.

Because of the adoption of binary measure, the machine is suited to inputs derived from absolute or differential threshold detectors; and without temporal or spatial aspects an important class of representations can still arise. For example, if there is a set of chemical detectors, each rendered active when a specific chemical exceeds a critical concentration, then from n such detectors 2^n different chemical representations can arise, the same number of different representations can arise if there are radiation detectors sensitive to n different wavebands. In neither example need the neighbourhood (adjacence) of receptors or input channels be considered. With the problem thus simplified, the classification of representations is straightforward.

II(b) Binary and Unitary Classification Machines

Any set of properties defines a class of representations; such a set of properties will be called a pattern. If there are a total of n properties, there are ${}_nC_r$ ways of forming a pattern of r properties. The total number of patterns is $\sum_{0}^{n} {}_nC_r$ which is 2^n. A machine, designed to classify representations on this principle, takes the form of FIGURE 3a.

Inputs must be combined in all possible ways; each combination of inputs must be connected to a unit; connections are either made or not made. A unit must operate if all the inputs connected to it are active; all units are of the same design.

There is an important limitation to the powers of classification of such a machine. Consider the class j_1 of representations possessing

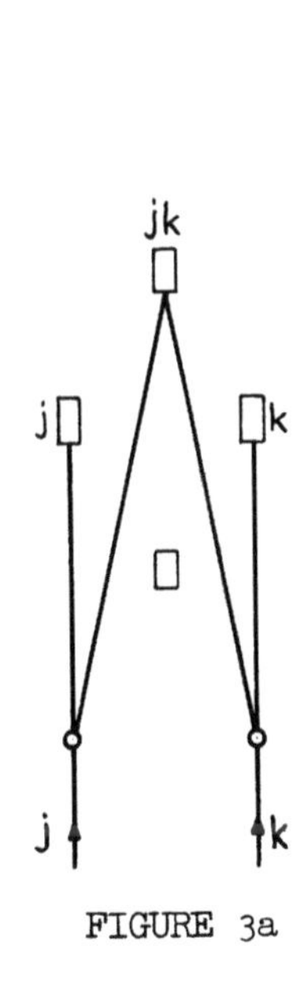

FIGURE 3a

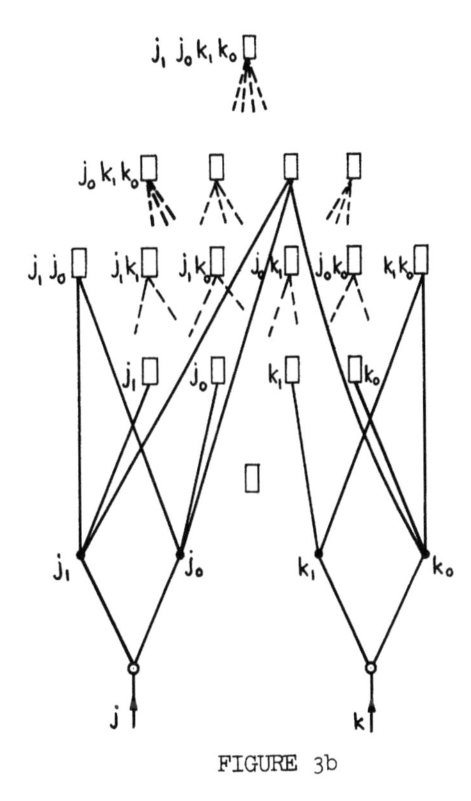

FIGURE 3b

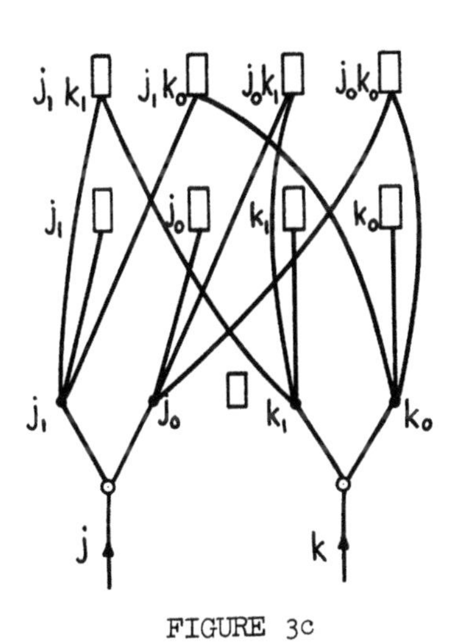

FIGURE 3c

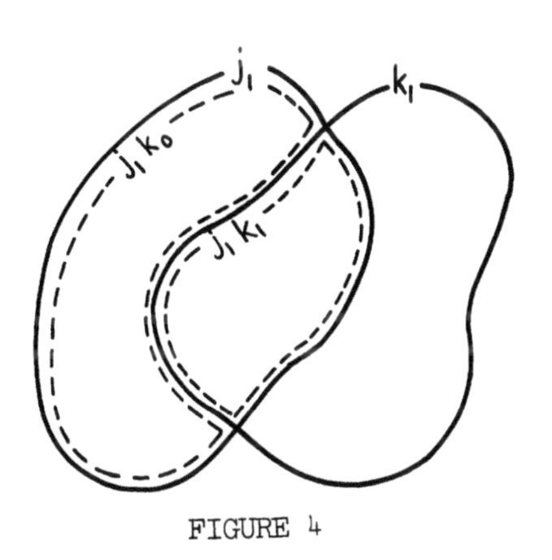

FIGURE 4

the property j; within this class there is a subclass $(j_1 k_1)$ possessing the set of properties j and k. These relations are shown in FIGURE 4. There is a class of representations $(j_1 k_0)$ possessing property j and not possessing property k. Class $(j_1 k_0)$ is the complement of class $(j_1 k_1)$, with respect to class (j_1). In the machine of FIGURE 3a there is no unit to indicate that a representation is of such a class. However, the necessary information does exist, within the machine, for such additional units to be included and controlled; but a principle of negation must be incorporated in the design of the extended machine. Each input quantity must then control two <u>points</u>. If the input is active, one point must be active and the other inactive; and conversely, if the input quantity is inactive. If units are connected to these points in all possible ways there will be 4^n units; a unit must operate if all the points connected to it are in the active state. Such a machine is shown in FIGURE 3b. But with this generality of connection, there will be many units connected to complementary points; such units can never operate. It would be possible to omit such units in the design of a machine; it can be shown that there would then be only 3^n units. The consequent constraint would cause a loss of the generality of connections. Such a machine is shown in FIGURE 3c.

The machine of FIGURE 3a will be called a <u>unitary</u> classification machine; those of FIGURES 3b and 3c will be called <u>binary</u> classification machines. There is now an arbitrary choice to be made in the form of the classification machine, which can be either

(a) unitary, with 2^n units distinguishing 2^n different patterns, and with generality of connection. Such a machine cannot indicate a class of representations which is distinguished by the absence of properties. Or

(b) binary, with 4^n units, and with generality of connection. This machine can distinguish 3^n patterns; only 3^n of the units will ever be used. Or

(c) binary, with 3^n units distinguishing 3^n patterns, and with many constraints in the connective system.

In this paper, only the simplest, unitary, machine will be considered further.

II(c) <u>Random Connections</u>

For a unitary machine of n inputs, there is one unit connected to input j which refers to a one-property pattern; there are n - 1 units connected to input j which refer to two-property patterns, and there are ${}_{n-1}C_{r-1}$ units connected to input j which refer to r-property patterns. The total number of units connected to input j is $\sum_{0}^{n} {}_{n-1}C_{r-1}$ which is 2^{n-1}; this is half the total number of units. The probability of a connection from any particular unit to any particular input is therefore $\frac{1}{2}$; this suggests that a statistical law might lead to a correct system of connections.

If the connections from units to inputs are entirely random, but so that P, the probability of connection from any unit to any input is $\frac{1}{2}$, then the probability of not being connected is also $\frac{1}{2}$; the probability of a unit being connected to a particular set of r inputs, and not being connected to the remaining n - r inputs is $(\frac{1}{2})^r(\frac{1}{2})^{n-r}$, which is independent of r. It follows, for a machine of n inputs thus randomly connected to 2^nK units, where K is a large number, that for each particular combination of connections there will be $K \pm \sqrt{K}$ units so connected. The apparently complicated connective system can be achieved in this simple manner, and all the different units will have approximately equal representation in the total population of randomly connected units. In part of the nervous system of many animals there are systems of connections which appear to be random, and it is suggested that their function is that of classification of inputs.

The simple law, $P = \frac{1}{2}$, contains no reference to the distance between unit and input; from this law it is not possible to determine how different units are arranged spatially. So that a spatial distribution of units can be found it is necessary to describe the random, root-like connections from the unit in the following way.

<u>For a unit at a given position relative to the input point, there is a probability of connection, P, which is a function of that position; this function will be said to define a probability field</u>. Such a function has been discovered by Sholl (1953) for the stellate and pyramidal cells of the visual area of the cat; it takes the form

$$P = a \exp(-r/r_0)$$

At any point, if P is the probability field for a single input, ρ is the total density of units, and ρc is the density of units connected to the input, then $\rho c = \rho . P$. If the spatial distribution of a set of inputs is known, then it is possible to calculate the density distribution of the units connected to this set [Uttley, 1955].

II(d) The Properties of Unitary Classification Machines

In a unitary classification machine, if a representation occurs, a unit will operate referring to the total set of properties possessed by the representation; in addition, all those units will operate which refer to subsets of the above set of properties; they indicate all the classes to which the representation belongs. If an effector mechanism were controlled by one such unit, it would be activated by all representations of the corresponding class. Much animal behaviour appears to involve the principle of classification; the behaviour of a bird may be the same towards all eggs of a certain form regardless of variations in markings, and even towards a stone of similar form. It is an important property of a classification machine that pattern recognition is instantaneous, depending only on the connections; there is no learning and no forgetting. But the same reaction to two different representations depends on there being a common set of input channels which are active for both representations; this cannot occur if the representations are derived from different sensory modes, or if the same light pattern falls on two different areas of a retina. More important still, the inclusive relation cannot arise between a set of interoceptor channels and an exteroceptor set, i.e., between a stimulus and a response. But such phenomena do occur in animals; the extended form of behaviour can be explained only by some relation between patterns, which is additional to that of possessing a common part.

III. THE RELATION OF CONDITIONAL PROBABILITY

Suppose that, in a past ensemble of representations, the input quantities j and k have been as follows:

$$\begin{array}{ll} j & 1111110000 \\ k & 1111110000 \end{array}$$

where 1 and 0 indicate the active and inactive state, respectively.

On this evidence alone it appears that the properties j and k are related. If now property k occurs alone, there is some justification for assuming that property j must have occurred as well, although not detected at the machine. The mathematical description of this situation is the $p(j/k)$, the conditional probability of j given k, is unity. A partial positive dependence is shown in the series

j 0001111000
k 1111110000

in this case, unconditionally, $p(j) = 0.4$ and $p(k) = 0.6$; but conditionally, $p(j/k) = 0.5$ and $p(k/j) = 0.75$. The occurrence of either property raises the probability of the other. Partial negative dependence is shown in the series

j 0000011110
k 1111110000

where the occurrence of either property lowers the probability of the other.

The relation of conditional probability is formed between properties, and sets of properties (patterns), on the basis of their joint occurrence in the same total representation, regardless of whether they possess a common part; it depends on their relative frequency of joint occurrence, so it is not formed instantaneously; it depends entirely on the nature of past representations, so it can be acquired and lost.

A machine can be designed which embodies this principle of relationship. In a classification machine, each unit has two possible states, to indicate the presence or absence of a pattern; the function of such a unit may be regarded as that of computing p the probability of the corresponding pattern, but with p restricted to the values 0 and 1. The proposed extension of principle in the machine is that p shall not be so restricted; each unit shall compute the probability of the corresponding pattern under all circumstances; three different circumstances can be distinguished. For a unit referring to the set of properties J.

(i) If the set J occurs, the unit must contain a quantity representing unit probability; this condition is met in the pure classification machine;

(ii) if no representations occur at all, the unit is to contain, not zero as in classification machines, but the unconditional probability of J as determined by past representations;

(iii) if the set of properties K occurs, the unit is to contain $p(J/K)$, the conditional probability of J, given that K occurs.

Such a machine will be called a conditional probability machine.

III(a) The Conditioned Reflex

Consider now a hypothetical experiment in conditioning. Suppose that the unconditioned and conditioned stimuli activate set of inputs J and K respectively, and that a series of representations is as in Table 1.

TABLE 1

Unconditioned stimulus J	0101	0011	0101	0101	0101
Conditioned stimulus K	0011	0011	0011	0011	0011
p(J/K)	$\frac{1}{2}$	$\frac{3}{4}$	4/6	5/8	6/10

In the first sequence of four representations there is independence, since $p(J) = \frac{1}{2}$ and $p(J/K) = \frac{1}{2}$. In the second sequence there is complete positive dependence; after it $p(J) = \frac{1}{2}$ as before, but p(J/K) has risen to $\frac{3}{4}$. In the next three sequences there is independence at first; the conditional probability falls, but more slowly than it rose. If, then on, there is always independence, p(J/K) tends to, but never equals, the limiting value of $\frac{1}{2}$. After the presentation of the dependent sequence the occurrence of the conditioned stimulus alone will always raise the probability of the unconditioned stimulus to some extent.

If, as in Table 2, there is complete negative dependence in sequence 4, i.e., the conditioned and the unconditioned stimuli never occur together, then the

TABLE 2

Unconditioned stimulus J	0101	0011	0101	1100	1101
Conditioned stimulus K	0011	0011	0011	0011	0011
p(J/K)	$\frac{1}{2}$	$\frac{3}{4}$	4/6	4/8	5/10

conditional probability falls rapidly to $\frac{1}{2}$; the effect of the negative dependent sequence has cancelled the effect of the positive dependent sequence; then on, the occurrence of the conditioned stimulus K does not increase the probability of the unconditioned stimulus J.

Now consider the unconditioned response. Suppose that it is aroused by signals in a set of efferent fibres L. With the limiting assumptions of this paper, these signals must be treated as two valued and simultaneous, as in FIGURE 2b. In consequence, temporal patterns of movement cannot arise, and the problem of serial order in behaviour [Lashley, 1951] cannot be considered.

The unconditioned reflex can be explained in terms of a classification machine, on the lines suggested in Section II(d). The system is shown in FIGURE 5. There is a set of input fibres J connected to a J unit; the output of the J unit is connected to the input of an L unit; when this input is active, the whole set of output fibres L is activated. Any representations which possess the set of properties J, and which may also possess additional properties, will evoke the response pattern L.

Suppose that the J unit forms part of a conditional probability machine; then not only will its output p(J) have unit value when the set J actually occurs, but it will have a variable value depending on the state of all the inputs. If after an ensemble of representations in which J and K are independent, J always occurs in the presence of K, then p(J/K) will tend to unity. The occurrence of K, the conditioned stimulus, will then produce an effect on the J unit which approaches that of the unconditioned stimulus J.

To explain the conditioned reflex, it is necessary to assume that the L unit is activated if the output of the J unit exceeds some threshold value less than unity. In this case, after a sufficient number of joint presentations of J and K, the unconditioned response L will occur in the presence of K; if J and K become independent the effect will disappear, but more slowly than it appeared.

The effect is produced because a physical path is formed in the machine from the J unit to the K unit; the way this path is formed is described in Section IV(b). The rate at which such an association will grow and decay is described in Section IV(c). In the conditioned reflex, there is a constant response to variable stimuli; the converse problem of variable response to a constant stimulus will be discussed elsewhere.

IV. THE DESIGN OF CONDITIONAL PROBABILITY MACHINES

For convenience, the design requirements of a conditional probability machine will be given again.

For the unit referring to the set of inputs J,

(i) if the set J occurs, the unit must contain a quantity representing unit probability; this condition is met in the pure classification machine;

(ii) if no representation occurs at all, the unit must contain, not zero as in the classification machine, but the unconditional probability of J as determined by past representations;

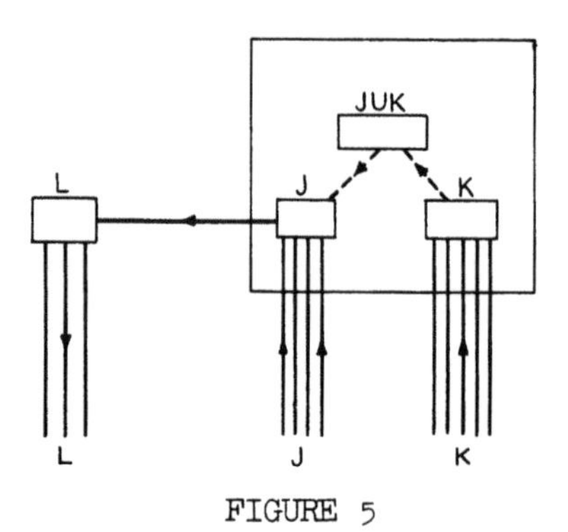

FIGURE 5

f
G
R
m
+
C
INPUT
(a)

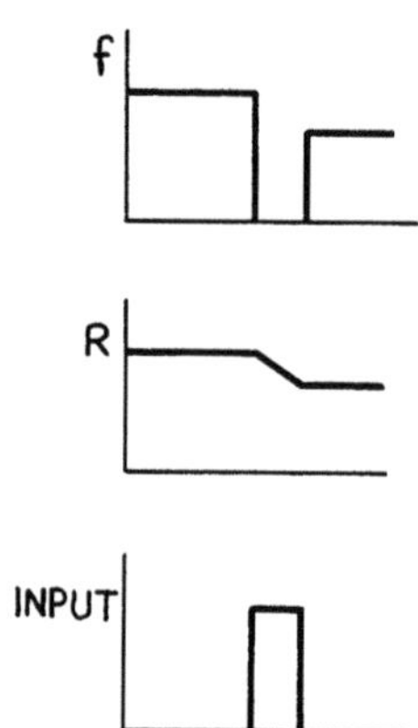

(b)

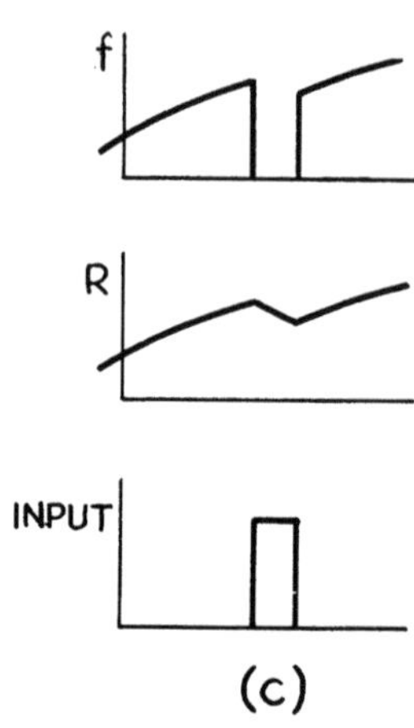

(c)

FIGURE 6

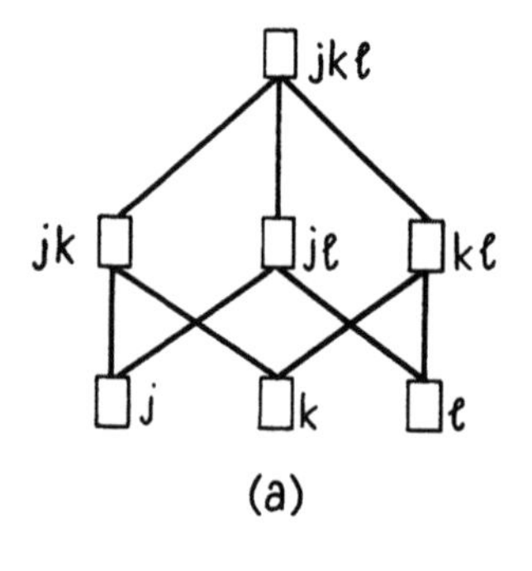

(a)

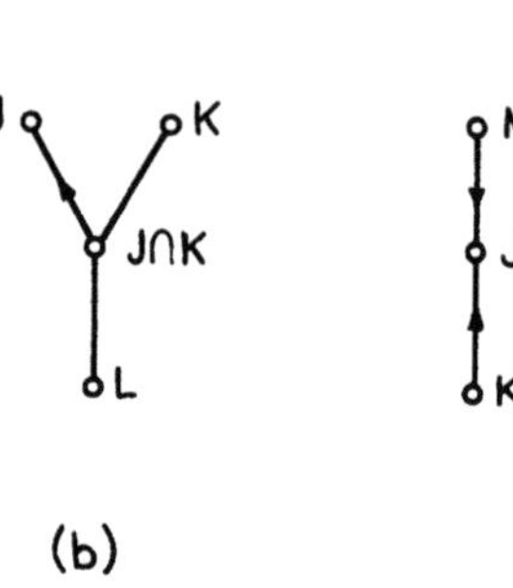

(b)

(a)

FIGURE 7

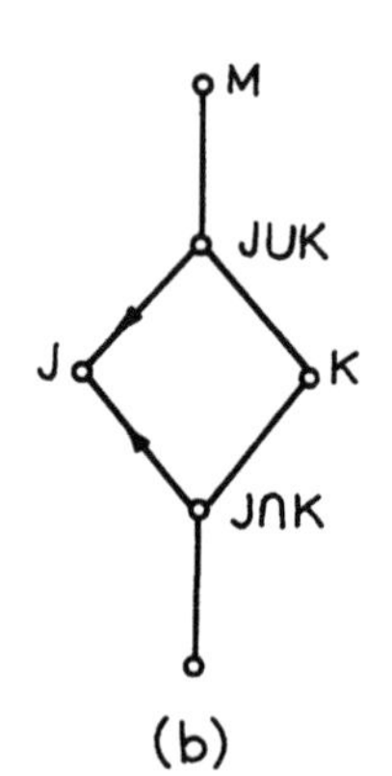

(b)

FIGURE 8

(iii) if the set of properties K occurs, the unit must contain $p(J/K)$.

It is probable that these requirements can be met in more than one way. A set of rules of machine design is proposed below, but there is no rigid proof of their necessity; the argument is concerned to show only that they are sufficient to meet all the requirements in all circumstances. Requirement (i) has been discussed. Requirement (ii) involves no change in the connective system, but a change in the design of the unit.

Regarding requirement (iii), the conditional probability $p(J/K)$ is the unconditional probability $p(J \cup K)$ divided by the unconditional probability $p(K)$: if condition (ii) has been met both these quantities will exist in the machine. No simple physical mechanism has been discovered, that will perform the operation of division, but numerous simple mechanisms can extract a difference between two like quantities. Accordingly, in meeting requirement (ii), unconditional probabilities will be computed on a logarithmic scale; and since probabilities lie between 0 and 1, the positive quantity $-\log p(J)$ will be computed. This function arises so frequently in the treatment, that it will be named the Rarity of J, and will be written $R(J)$. Zero rarity represents unit probability, i.e., certainty. If natural logarithms are used, a rarity of six units, for example, is a probability of 1 in e^6.

IV(a) The Computation of Unconditional Rarities

If the total number of past representations is m and the number possessing the pattern J is u, then

$$R(J) = \log m - \log u \qquad (1a)$$

so

$$\delta R(J) = \frac{\delta m}{m} - \frac{\delta u}{u} \qquad (1b)$$

Because of the first term on the right-hand side, the quantity stored in the J unit must grow logarithmically with the number of past representations; this is true for all units. Because of the second term, if pattern J occurs (i.e., $\delta u = 1$), then the quantity stored in the J unit must decrease by an amount $\frac{1}{u}$; this decrease in the stored quantity is the act of counting.

A functional diagram of a unit connected to a single input j is shown, with the associated waveforms, in FIGURES 6a and b. The unit computes $-\log u$; by the addition of $\log m$ the unconditional rarity of j is formed. But, to meet requirement (i) above, the unit output must be zero during the occurrence of j. Calling the unit output $f(j)$, when the input is inactive $f(j) = R(j)$, and when the input is active $f(j) = 0$. These requirements are met by a gate G between $R(j)$ and $f(j)$. The active input has a double function, to increase u by 1, and to close the gate so as to cause $f(j)$ to become zero; this function will be called counting control.

Time and the weighting of past events. So far, the representations have been treated as timeless ensemble. As a first step towards removing this abstraction, let the representations occur uniformly in time at intervals T; then n the total number of representations, since the time $t = 0$ is t/T. The probability p of a pattern then becomes uT/t, which is the probability that the pattern will occur in time T. Since u/t is the mean frequency of occurrence of the pattern, uT/t may also be regarded as the mean frequency occurrence of the pattern, relative to $\frac{1}{T}$, the frequency of occurrence of representations.

The design of the unit now becomes simpler.

$$R = \log \frac{t}{T} - \log u \tag{2a}$$

$$\frac{\partial R}{\partial t} = \frac{1}{t} - \frac{1}{u}\frac{du}{dt} \tag{2b}$$

It is no longer necessary to count m; instead, in each unit the computed quantity R must grow logarithmically with time; but a counter of u is still required. It is a property of such a unit that all events are weighted equally, regardless of when they occurred, or of the intervals between them. As in equation (1), $\frac{\partial R}{\partial u} = \frac{1}{u}$; so the extent to which R can be altered by an event, becomes less as u increases, i.e., as t increases.

If a weighting function is introduced, so that the effect of a past event is less than that of a present one, there are significant changes in the nature and properties of the unit. The quantity $\frac{\partial R}{\partial t}$ need no longer be a function of t, u and $\frac{du}{dt}$, but only of the present quantities R and $\frac{du}{dt}$. Since it is not necessary to count or store the quantities t and u, the design of the unit is simplified. The changed properties are as follows.

(a) There is no particular time defined by $t = 0$.

(b) There is no particular state defined by $u = 0$, so the indeterminate quantity $-\log 0$ does not arise.

(c) The extent to which R can be altered by an event is independent of time, and depends only on R.

Any quantity which grows in the absence of events, and decays during events, will provide some measure of the mean frequency of occurrence; the scale may not be logarithmic and the weighting function can take any form. An ionic concentration within a neuron might possess such properties, and so provide a measure of the mean frequency of impulses.

The principles of design, which have been proposed so far, are as follows:

(a) The machine consists of numbers of identical units connected in all possible ways to inputs.

(b) The J unit counts if the set of inputs J which are connected to it, are all in the active state.

(c) The J unit computes the weighted unconditional rarity $R(J)$ of the set of properties J. In the absence of events, $R(J)$ grows; if J occurs $R(J)$ decreases by a certain amount.

(d) If $f(J)$ is the output of the unit,
$f(J) = R(J)$ if J does not occur.
$f(J) = 0$, if J occurs.

A machine which embodies these principles will be called an Unconditional Probability Machine.

IV(b) The Computation of Conditional Rarities

The third requirement is that, when the set K occurs, the unit J must contain $R(J/K)$ instead of $R(J)$. Since $R(J/K) = R(J \cup K) - R(K)$, there must be a comparison between the contents of the $(J \cup K)$ and the (K) units; consequently a connection will be required between them. The difference having been calculated, it must be transferred to the (J) unit; so a connection is required between the $(J \cup K)$ unit and the (J) unit. These interconnections between units form a new feature of the machine design. The present section is concerned to deduce their nature from their necessary function. By means of the interconnections a physical path is formed between sets of inputs, whether they are disjunct or not, via the unit which corresponds to their union. Such a path, from K to J via the $(J \cup K)$ unit is indicated by dotted lines in FIGURE 5.

In what follows, it will be convenient to call a J unit a superunit or a subunit of a K unit if $J \supset K$, or $J \subset K$, respectively. The computation of $R(J/K)$ will be considered for three possible relations between J and K:

J is a subset of K,

J is a superset of K;

J is neither a superset nor a subset of K.

If J is a subset of K, then if the set K occurs the set J occurs, so $R(J) = 0$; this is effected by counting control, and is discussed in Section IV(a).

(i) SUPERCONTROL

J is a superset of K. Since $J \cup K = J$, $R(J/K) = R(J) - R(K)$; it follows that if K occurs, the content of the K unit must be subtracted from that of its superunit J.

The subtraction of the content of a unit, when its associated set of properties occurs, from the content of one of its superunits will be called "supercontrol". So that there may be minimal constraints in the system of interconnections, it will be laid down, as a principle of design, that there shall be supercontrol by a unit of <u>all</u> its superunits.

The supercontrol connections for a 3-input machine are shown in FIGURE 7.

If no other conditions are laid down, there will be two incorrect consequences. Firstly, if the set K occurs, then L, any proper subset of K occurs also. Consequently, if there were general control of supersets, $R(L)$ would be subtracted from the content of the J unit; this is in conflict with the requirement that $R(K)$ be subtracted. But

$$L \not\subset K, \qquad \therefore\ p(L) \geq P(K), \qquad \therefore\ R(L) \leq R(K).$$

The conflict can be resolved correctly, if it is laid down that <u>when there is multiple supercontrol of a unit, that which would cause the maximum decrease in rarity shall be effective</u>.

Next consider any set J which is not a superset of K i.e., such that $J \cap K \not\subset K$, as in FIGURE 8a. Then there will be supercontrol of J from any set L for which $J \cap K \supseteq L$. But $R(J \cap K) \geq R(L)$; so, by the above rule regarding multiple supercontrol, effective supercontrol will be from $(J \cap K)$ causing the content of the J unit to become $R(J) - R(J \cap K)$; but this is quite incorrect, since $R(J/K) = R(J \cup K) - R(K)$. There is no general inequality relating $R(J) - R(J \cap K)$ to $R(J \cap K) - R(K)$; but <u>this false supercontrol</u> can be detected because it arises from a unit whose rarity has decreased by an amount $R(J \cap K)$ for which $R(J \cap K) \leq R(K)$.

To summarise, for a machine to have the function of effective supercontrol, the design rules are:

(1) Counting control, which demands zero rarity, overrules supercontrol.
(2) There must be a connection from every unit to each of its superunits.
(3) If a set of properties occurs, the content of the unit associated with that set must be subtracted from the content of each of its superunits.

(4) If there is multiple supercontrol at a unit, that which demands the maximum decrease in rarity shall be effective.

The way in which false supercontrol is suppressed will be discussed in the next section.

(ii) SUBCONTROL

If J is not a superset of K, it is necessary to consider the set $J \cup K$ which is a superset of K. If K occurs, supercontrol by the K unit of the $J \cup K$ unit will cause its content to become $R(J \cup K) - R(K)$; this quantity is $R(J/K)$. It is therefore necessary to transfer the content of the $(J \cup K)$ unit to the (J) unit, if K occurs.

The transfer of the content of a unit to one of its subunits will be called "subcontrol". So that there shall be minimal constraints in the system of interconnections for subcontrol, it will be laid down, as a principle of design, that there shall be subcontrol by a unit of <u>all</u> its subunits. The system of subcontrol connections is a duplicate of the supercontrol system, but it has a different function, operating in the reverse direction. As in the case of supercontrol, if there is subcontrol of <u>all</u> subunits, incorrect effects will occur. There will be subcontrol of J by any unit M for which $M \supset J$: so there must be additional conditions to ensure that there is subcontrol only if $M = J \cup K$.

It has been shown in the previous section that if $J \not\supset K$ there can be false supercontrol of the (J) unit by the $(J \cap K)$ unit. The following rules are therefore necessary.

(a) Subcontrol overrules false supercontrol.

(b) Correct supercontrol, since it is correct, overrules subcontrol.

These rules impose a further condition upon M. If $M \supset K$ it can experience correct supercontrol, which will overrule subcontrol by rule (b); so M will contain $R(M) - R(K)$. But $M \supset J$, so subcontrol by M will then cause the content of J to be $R(M) - R(K)$.

If $M \not\supset K$, it can experience false supercontrol, therefore, by rule (a), it will experience subcontrol from some other unit N; for which $N \supset M$; in consequence it is not possible to describe the content of the M unit in terms of $R(M)$ and $R(K)$, but only to say that its content equals that of the N unit. This situation will be repeated in considering the N unit, unless $N \supseteq K$, in which case the content of the N unit will be $R(N) - R(K)$. There is now subcontrol by the N unit of the M unit, and by the M unit of the J unit, causing the content of the J unit to become $R(N) - R(K)$. The same result occurs because of direct subcontrol of the J unit by the N unit. The effect of the two rules, therefore, is to ensure that the J unit can be subcontrolled only by a unit M for which $M \supseteq K$. But subcontrol connections ensure that $M \supset J$, $\therefore M \supseteq J \cup K$.

There is now multiple subcontrol of J by every unit M for which $M \supseteq J \cup K$. But $R(M) \geq R(J \cup K)$, *therefore correct subcontrol is from the unit which would cause the greatest decrease in the rarity of J*.

(iii) THE SUPPRESSION OF FALSE SUPERCONTROL

Consider units J, K, and M, for which $J \supset K$, and $M \supset J$, as in FIGURE 8b. There can be correct supercontrol and subcontrol of J, by K and M respectively, which demand that the content of J change to $R(J) - R(K)$ and $R(M) - R(K)$, respectively. The two forms of control are from units containing rarities which decrease by the same amount, $R(K)$, when K occurs. Supercontrol is correct; it will therefore be laid down, as a principle of design, that

if supercontrol and subcontrol are from units suffering equal decreases in rarity supercontrol overrules subcontrol.

Now suppose that J does not contain K, as in FIGURE 8b. False supercontrol is from the $(J \cap K)$ unit, whose decrease is $R(J \cap K)$. Correct subcontrol is from the $(J \cup K)$ unit, whose decrease is $R(K)$; and $R(J \cap K) \leq R(K)$. It will be laid down that

if supercontrol is from a unit suffering a smaller decrease in rarity than does the subcontrolling unit, then it is overruled by subcontrol.

This rule covers the case of $R(J \cap K) < R(K)$, but not that of $R(J \cap K) = R(K)$, which must now be considered.

False supercontrol of J by the $(J \cap K)$ unit would cause the content of the J unit to become $R(J) - R(J \cap K)$. Subcontrol of J would cause its content to become $R(J \cup K) - R(K)$. But if

$$R(J \cap K) = R(K),$$

then $$R\left\{(K - J \cap K)/(J \cap K)\right\} = 0;$$

$$\therefore \quad R\left\{(K - J \cap K)/J\right\} = 0;$$

$$\therefore \quad R(J \cup K) = R(J);$$

$$\therefore \quad R(J) - R(J \cap K) = R(J \cup K) - R(K);$$

so false supercontrol and subcontrol compute the same correct quantity for $R(J/K)$. In this case, there is objection to rule (a), which causes supercontrol to override subcontrol.

To summarise, for a machine to possess the function of correct subcontrol, there are the following design rules:

(1) Counting control overrides subcontrol.

(2) There are subcontrol connections from every unit to all its subunits.

(3) By subcontrol, the content of a unit is transferred to its subunit.

(4) Where there is multiple subcontrol, that which demands the maximum decrease in rarity shall be effective.

(5) Subcontrol overrides supercontrol if it is from a unit experiencing a greater decrease in rarity than does the unit demanding supercontrol; otherwise supercontrol overrides subcontrol.

IV(c) The Properties of Conditional Probability Machines

A conditional probability machine differs in structure from a classification machine in two significant ways. Firstly, the basic unit is changed from a two-state mechanism to one possessing a variable state, which records an unconditional probability on a logarithmic scale. Secondly, there are interconnections between units, which make possible the computation of the differences between the states of units, i.e., the logarithms of conditional probabilities. For any two sets of inputs J and K, whether disjunct or not, there is a physical path between the J and K units via the $(J \cup K)$ unit.

If these interconnections were always fully made, all conditional probabilities would be unity, i.e., if one input became active it would be as if all were active. But because the units possess variable states, the activity of one input, or set of inputs, only modifies the probability of activity of all other inputs; some may become more probable, others less; many will not be modified at all because, based on the past, they are statistically independent of the active set of inputs. For a conditional probability machine, the statement that two representations resemble one another can be replaced by the statement that the occurrence of one causes a high probability of the other.

If a nervous system operates on this principle it can be seen how a conditioned stimulus can cause a growing probability of the unconditioned stimulus, hence of the unconditioned response. If the unit storing this probability is connected via a threshold device to an effector mechanism, then the probability can be converted into an occurrence.

A conditional probability machine has been built on the basis of the rules devised in this section, and it has all the properties described; its description will appear elsewhere.

V. CONDITIONAL CERTAINTY MACHINES

A conditional probability machine computes the probability of every set of properties under all circumstances. Such a machine could control effector mechanisms in many ways; but if it is to control them in all-or-none manner when their probabilities exceed a threshold value, the machine can be of a simpler design.

For each set of properties J, it is necessary that the machine compute only a binary digit which is 0, say, if $R(J/K) < \epsilon$, and is 1 if $R(J/K) \geq \epsilon$. Such a machine will be called a conditional certainty machine.

Supercontrol and Subcontrol

The computation of the difference between $R(J)$ and $R(K)$ is not necessary. The reduced function of supercontrol is to form a digit in the superunit, as determined by the sign of $R(J) - R(K) - \epsilon$. For this purpose, only an approximately logarithmic function of probability need be computed in the basic unit; so there is greater freedom in the choice of growth equations. It is still necessary that the connective system communicate a variable quantity, rarity, from a unit to a superunit.

Subcontrol is concerned with the communication of binary digits only; so the connective system is much simpler in nature than that of supercontrol.

For the above reasons it is much easier to conceive a conditional certainty machine, in neurological terms than a conditional probability machine; but evidence from conditioning experiments might point the other way. If, from the data of such experiments, it is possible to compute the conditional probability of "unconditioned response" given "conditioned stimulus"; and if some measure of the response is found to be a function of that probability, then there must be, in the system, some kind of conditional probability machine.

Acknowledgement

Acknowledgement is made to the Chief Scientist, the British Ministry of Supply, and to the Controller of H.B.M. Stationery Office for permission to publish this paper.

REFERENCES

[1] HAYEK, F. A., "The Sensory Order," London 1952, Routledge and Kegan Paul.

[2] LASHLEY, K. S., "Cerebral Mechanisms in Behaviour," Hixon Symposium, New York, John Wiley, pp. 112-146.

[3] SHOLL, D. A., "Dendritic Organisation in the Neurons of the Visual and Motor Cortices of the Cat," J. Anat. 1953, 87, pp. 387-406.

[4] SHOLL, D. A. and UTTLEY, A. M., "Pattern Discrimination and the Visual Cortex," Nature 1953, 171, pp. 387-388.

[5] UTTLEY, A. M., "The Probability of Neural Connections," Proc. Roy. Soc. B., 1955, 144, No. 916.

TEMPORAL AND SPATIAL PATTERNS IN A CONDITIONAL PROBABILITY MACHINE

Albert M. Uttley

ABSTRACT

If there are direct paths and delayed paths between input signal channels and a conditional probability machine, then all aspects of temporal patterns can be distinguished by the machine, except that of duration.

If the property of pattern completion, which is possessed by a conditional probability machine, is considered for temporal patterns, it is found to be very similar to that of prediction.

If the property of temporal pattern completion is considered for proprioceptor inputs, properties of the machine emerge which appear to resemble those of animal motor patterns.

Consider a patch of light falling on a surface composed of receptors which are inputs to the machine; if this light patch is consistently changed from one shape into another, for example, by translation, rotation or magnification, then the machine will compute a high probability of the second shape, given only the first. It is suggested that this property forms the basis of spatial pattern generalisation; in consequence it is a phenomenon which must be learnt and can be forgotten.

INTRODUCTION

In the preceding paper the phenomenon was considered of an animal giving the same response to two different stimuli; it was suggested that this could arise out of two different mathematical principles; upon these principles the design of a machine was based. Firstly, two such stimuli might evoke the same response because they possess common properties, i.e., were of the same class or set; a machine based on this principle was called a classification machine, its recognition was instantaneous. Secondly, the conditional probability of one stimulus, given the other,

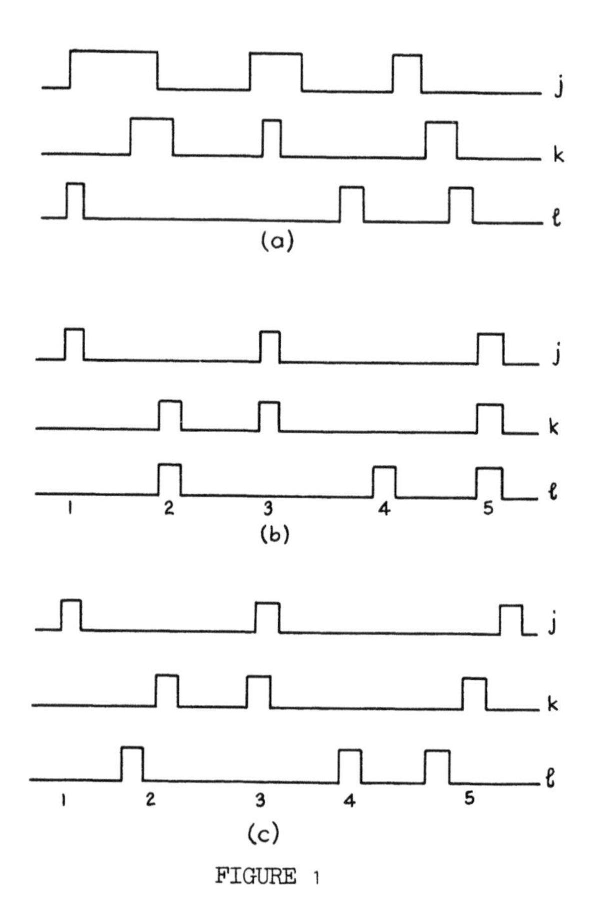

FIGURE 1

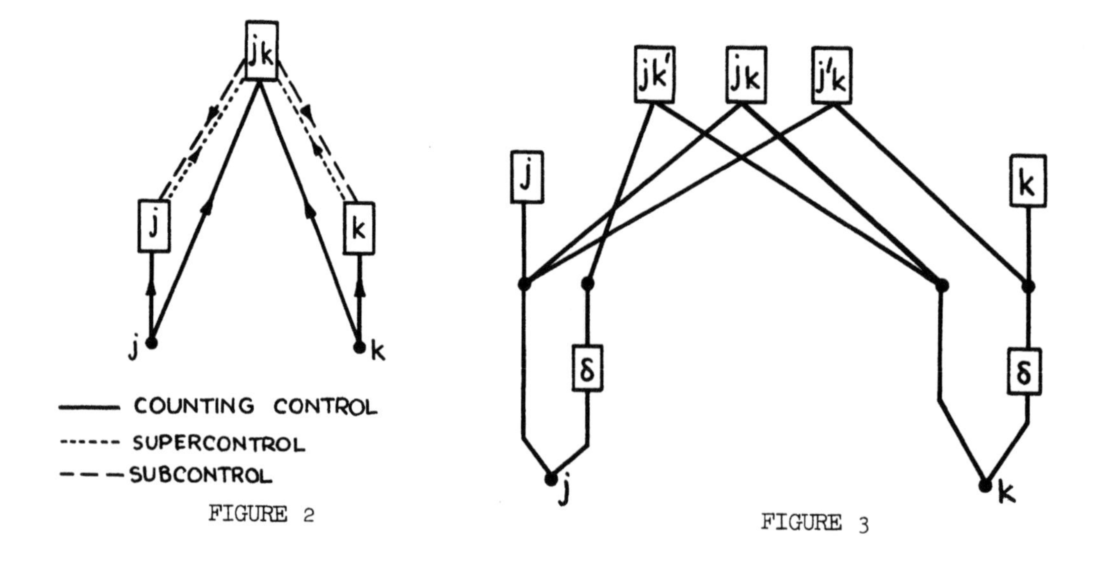

FIGURE 2

FIGURE 3

might be high; this relation would depend entirely on the nature of past representations, so it could be acquired or lost; a machine embodying this principle was called a conditional probability machine.

Regarding the stimulus, it was necessary to distinguish the configuration of the world external to the animal and the corresponding internal representation; this representation consists of signals in sets of fibres. In order to discuss the problem mathematically, it was necessary, initially to consider only very simple signals. Firstly, each signal, called an input, was restricted to be in either an active or an inactive state, and no form of time coding was considered, e.g., pulse frequency to represent intensity. If the input j was active the representation was said to possess property j, i.e., to belong to class j. A set of such properties was said to form a pattern. A time record of the state of these inputs might take the form of FIGURE 1a. Secondly, there were three temporal abstractions—the duration of the active state of an input was not considered; the different inputs in a representation were active simultaneously; the order of, and intervals between, representations was not considered—in consequence temporal patterns could not be discussed. Such a restricted input is shown in FIGURE 1b. Thirdly, the neighbourhood of inputs was not considered so spatial patterns could not be discussed. In consequence, the theory was applicable only to patterns such as smell or colour.

It is the aim of this paper to extend the theory to cover some of the temporal and spatial aspects of pattern; the binary input is retained. It has not been found possible to treat spatial pattern without first considering temporal pattern; the aspect of duration of inputs is not discussed. The material of the two papers is discussed less formally elsewhere [Uttley, 1954b].

TEMPORAL PATTERNS

For only two inputs, j and k, the conditional probability machine described in the earlier paper takes the form of FIGURE 2; there are (j), (k) and (jk) units corresponding to these three patterns; the connections for counting control, supercontrol and subcontrol are shown in the figure. Units count only if all associated inputs are active. If the waveforms of FIGURE 1b were presented to such a machine, the pattern (jk) would be counted twice, in representations 3 and 5.

Consider a time sequence of inputs which are not simultaneously active, which vary in temporal separation; but whose duration of activity

(a)

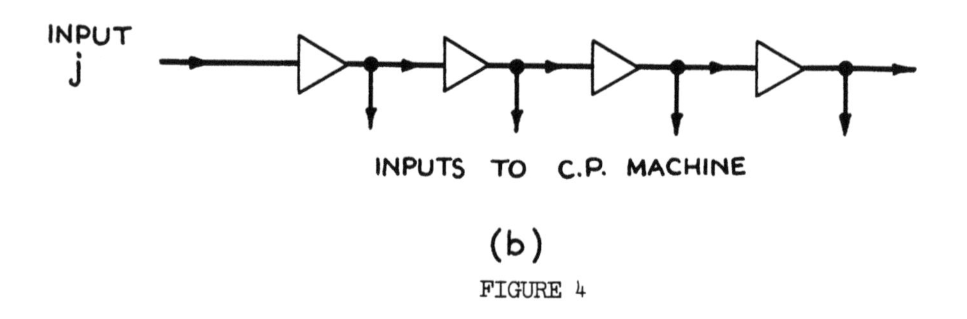

(b)

FIGURE 4

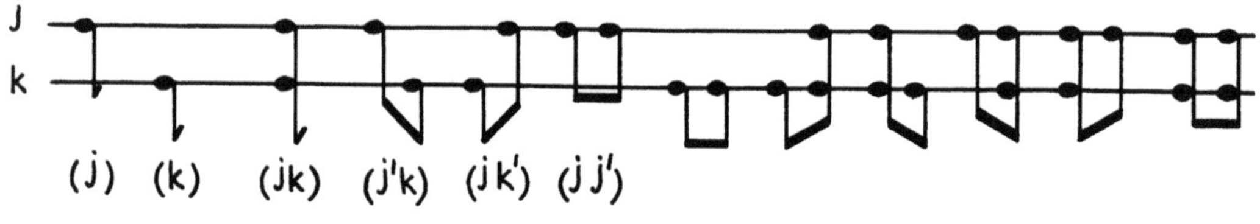

FIGURE 5

is constant; an example is shown in FIGURE 1c. If these waveforms were presented, to the above machine, the pattern (jk) would be counted only once, in representation 3; the pattern (j after k), which occurs in representation 4, would not be counted at all.

It is necessary to consider what are the simplest additions which must be made to the machine, so that it can treat (jk) and (j after k) as different patterns. To this end, the measure of temporal separation will take what appears to be its simplest form, distinguishing only <u>before</u>, <u>simultaneous</u> and <u>after</u>.

Consider two inputs j and k, and the patterns

(j);
(k);
(j before k), written as (j'k)
(j and k simultaneous), written as (jk); and
(k before j), written as (jk')

For the last three patterns there must be three different machine units connected to the j and k inputs; between these units and the inputs there must be elements which are different in nature. Suppose that there is an element, producing a time delay δ between the j input and the (j'k) unit, and between the k input and the (jk') unit. Let all other connections be direct; then the machine takes the form of FIGURE 3, where only counting connections are shown.

Let all inputs have a duration of activity τ; since units count only if all inputs connected to them are active, the (jk) unit will count if k commences at any time t after j, such that

$$\tau > t > -\tau \qquad (1a)$$

i.e., for all such intervals, pairs of inputs will be treated simultaneous. Similarly the (j'k) unit will count if

$$\delta + \tau > t > \delta - \tau \qquad (1b)$$

and the (jk') unit will count if

$$-\delta + \tau > t > -\delta - \tau \qquad (1c)$$

For intervals greater than $\delta + \tau$ patterns relations cannot arise. These time relations are shown in FIGURE 4a. If there is to be no interval t between the values $\pm$ $(\delta + \tau)$ for which no count can occur, it is necessary that $\delta < 2\tau$; if $\delta > 2\tau$, there will be intervals for which the (jk) unit and either the (jk') unit or the (j'k) unit will count.

The system of FIGURE 3 can be regarded as part of a conditional probability machine with inputs j, j' k and k'; such a machine can handle all patterns which are subsets of (jj'kk') whereas the timeless system can handle only subsets of (jk). The additional patterns are not only (j'k) and (jk'), but also (j'), (k'), (jj'), (kk'), (j'k'), (j'kk'), (jkk'), (jj'k), (jj'kk'). But these patterns are not all independent; (j'), (k') and (j'k) are identical to (j), (k) and (jk). But the fact that p(j/j'), for example, is unity raises no problem in a conditional probability machine. The above temporal patterns can be represented simply by means of musical notation as in FIGURE 5.

The principle of classifying temporal patterns in a machine by means of delayed inputs, can be extended to distinguish degrees of temporal separation; each input j must suffer a series of delays, to form the series of inputs j', j'', j"', etc.; this is shown in FIGURE 4b. This method distinguishes only a finite number of classes, not a continuum of temporal separation.

PATTERN COMPLETION AND PREDICTION

It is a most important property of a conditional probability machine that if one pattern occurs, the conditional probability of all other patterns is computed; this computation is based on all past occurrences. Suppose that, in the past, pattern (j) was always accompanied by pattern (k); if now, (j) occurs alone the probabilities of (jk) and (k) become unity; <u>for these units</u>, it is as if (k) has occurred. In a conditional probability machine this property of pattern completion can be partial, i.e., p(k/j) can have all values from 0 to 1. This property also applies to temporal pattern; for example, in FIGURE 5 the patterns [musical notation] [musical notation] [musical notation] [musical notation] all contain the subpattern [musical notation] . Suppose that, in the past, the four patterns occurred with equal frequency; if [musical notation] now occurs, the conditional probability of [musical notation] is $\frac{1}{4}$. But if in the past [musical notation] had always been followed by [musical notation] then, for the machine, if [musical notation] occurs, it is as if [musical notation] occurs.

So long as all inputs are connected to a conditional probability machine both directly and via delays, the property of temporal pattern completion will be possessed for all possible patterns, whatever their source, and however many elements they possess. This property appears to be closely allied to that of <u>prediction</u> [Craik, Lashley].

THE TEMPORAL PATTERN OF MOTOR RESPONSES

In an animal, suppose that the signals from all receptors are converted into the binary propositional form, in a set of fibres which form inputs to a conditional probability machine; suppose that each input is connected to the machine directly and via delays δ', δ'', δ''', etc. The total set of inputs which are active at any instant form a total pattern which may be divided into two components, P' the pattern arising from all proprioceptors, and K' the union of all other sensory patterns. Then, on the basis of all past representations there is computed and stored in the machine $p(P_2 \cup P' \cup K')$, the joint probability of every possible future motor pattern P_2 in every possible circumstance, i.e., for every possible present motor pattern P' and every possible present environment K'; this is done for all delays δ', δ'', δ''', etc. This statement refers to the internal state of the system, quite apart from what is now occurring.

If a particular motor pattern P_1', occurs, there is afferent feedback, from the corresponding set of proprioceptors to the machine, whereby it computes the conditional probability $p(P_2/P_1' \cup K_1')$ of every possible future motor pattern P_2; there is also feedback of K_1', the union of all other sensory patterns which are now occurring. This feedback determines the probability of the pattern of motor behaviour for a time ahead equal to the greatest delay δ in the system; but the computation is pattern by pattern, not element by element.

The pattern P' was defined as the total pattern arising, at any instant, from all proprioceptors. Consider Q' a subset of P': then Q' defines a class of motor patterns of which P' is a member. Since $P' \supseteq Q'$, then $p(P') < p(Q')$; i.e., the probability of a class of patterns cannot be less than that of one of its members, and in general it will be greater. In consequence, <u>the probability of a class of motor patterns can rise to a particular value upon less data than can any member of that class</u>; given further data, the probability of one member of the class may rise to this particular value and so be distinguished from other members of the class.

In the system here discussed there are afferent signals and central computation, but no efferent signals, in consequence, there has been no mention of the control of motor units to cause motor patterns, but only a discussion of their probabilities. To consider the design of a complete control system, it is necessary to make a further hypothesis relating the probability of a motor pattern, as computed from afferent signals, to a pattern of efferent signals emitted by the machine; such a hypothesis will not be discussed in this paper.

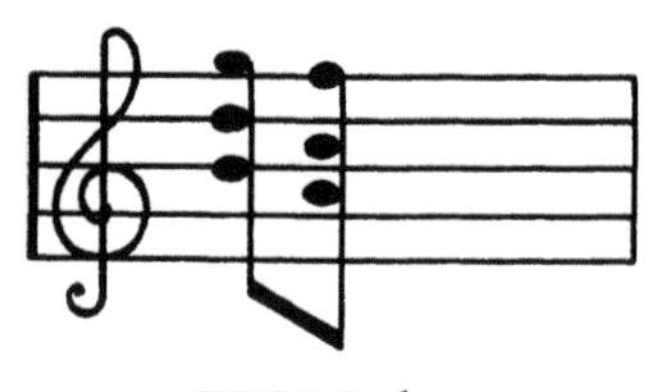

FIGURE 6

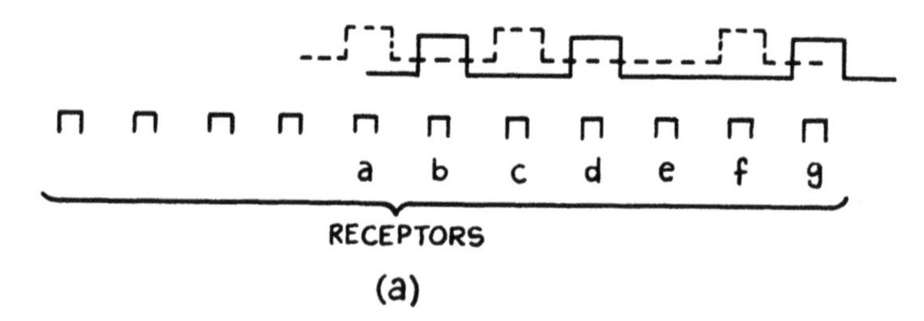

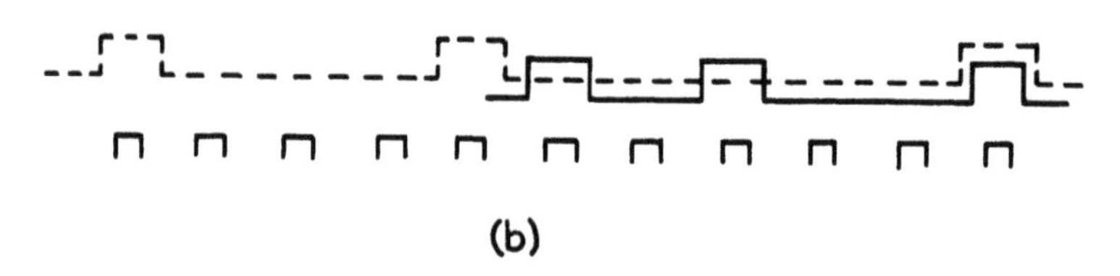

FIGURE 7

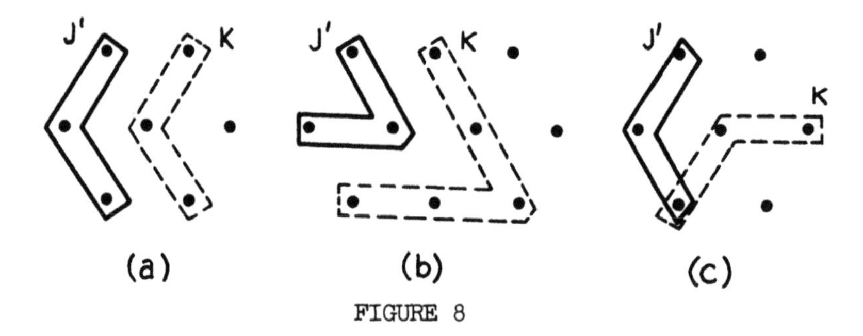

FIGURE 8

SPATIAL PATTERN

Consider a machine with a large number of inputs a,b,c,d, e,f,g, each connected to it directly, and via a single delay δ; suppose that, among all temporal patterns that the machine has received in the past, the first chord of FIGURE 6, i.e., (b'd'g') has always been followed by the second, (a c f), to form the total pattern (b'd'g'a c f). Then the probability of the second chord, given only the first, is unity; the two chords or stimuli become equivalent. This situation can arise if the inputs of FIGURE 6 are considered as light receptor, and are placed in a line as in FIGURE 7a. If a patch of light of suitable form moves repeatedly over the receptors from right to left the patterns of FIGURE 6 will occur; such a situation can arise if an eye moves in relation to external objects.

Suppose that the pattern (a c f) had become a conditioned stimulus to some reflex response, then if p(a c f /b'd'g') becomes unity, the pattern (b'd'g') will arouse the same response as (a c f). It is suggested that this effect forms the basis of spatial pattern generalisation in animals; if this is so, the effect must be learned all over a retina. But approximate fixation will introduce an economy, whereby this learning process need be acquired over only a small area. By this principle, spatial pattern generalisation is achieved, not by present scanning, but by the stored effects of past scanning. The same principle can be applied to other forms of movement of a light patch over an array of receptors. In FIGURE 7b a light patch suffers, not translation, but linear magnification. The principle can be extended to two dimensions; the motion of a light patch in translation, magnification and rotation is shown in FIGURE 8a, b and c respectively. In each case, due to the repetition of similar pattern charges, the occurrence of a first pattern will give rise to a high probability of a second.

Acknowledgement

Acknowledgement is made to the Chief Scientist, the British Ministry of Supply, and to the Controller of H.B.M. Stationery Office for permission to publish this paper.

BIBLIOGRAPHY

1. CRAIK, K. J. W., Theory of the Human Operator in Control Systems.
2. LASHLEY, K. S., Cerebral Mechanisms in Behaviour, Hixon Symposium, John Wiley, 1951, 112-146.
3. UTTLEY, A. M., Unpublished Memorandum.
4. UTTLEY, A. M., The Classification of Signals in the Nervous System, E.E.G. Clin. Neurophysiol., 1954(b).

GPSR Authorized Representative: Easy Access System Europe - Mustamäe tee 50, 10621 Tallinn, Estonia, gpsr.requests@easproject.com

www.ingramcontent.com/pod-product-compliance
Ingram Content Group UK Ltd.
Pitfield, Milton Keynes, MK11 3LW, UK
UKHW051927010425
456954UK00016B/259
* 9 7 8 0 6 9 1 0 7 9 1 6 5 *